Soybean - Recent Advances in Research and Applications

Edited by Takuji Ohyama, Yoshihiko Takahashi, Norikuni Ohtake, Takashi Sato and Sayuri Tanabata

Published in London, United Kingdom

Soybean - Recent Advances in Research and Applications
http://dx.doi.org/10.5772/intechopen.98162
Edited by Takuji Ohyama, Yoshihiko Takahashi, Norikuni Ohtake, Takashi Sato and Sayuri Tanabata

Contributors
Sinan Meriç, Alp Ayan, Tamer Gümüş, Çimen Atak, Bhagwan Asewar, Megha Jagtap, Gopal Shinde, Shivaji Mehtre, Madan Pendke, Takuji Ohyama, Sayuri Tanabata, Norikuni Ohtake, Takashi Sato, Kuni Sueyoshi, Yoshihiko Takahashi, Shinji Ishikawa, Yuki Ono, Natsumi Yamashita, Akinori Saito, Sokichi Shiro, Yuichi Saeki, Summra Siddique, Azhar Mat Easa, Moses Ojukwu, Mohammad Sohidul Islam, Imam Muhyidiyn, Md. Rafiqul Islam, Md. Kamrul Hasan, Golam Hafeez, Moaz Hosen, Hirofumi Saneoka, Akihiro Ueda, Liyun Liu, Misbah Naz, Celaleddin Barutçular, Javeed Lone, Muhammad Ammar Raza, Murat Erman, M. Kaium Chowdhury, Ayman EL Sabagh, Canon E.N. Engoke Norris Savala, Stephen Kyei-Boahen, David Chikoye, Lilian Hasegawa Florentino, Rayane Nunes Lima, Mayla D.C. Molinari, Carlos Gilberto Raetano, Matheus Mereb Negrisoli, Flávio Nunes da Silva, Danilo Morilha Rodrigues, Evandro Pereira Prado, Md. A. Mannan, Ismot Rima, Abdul Karim, António André Chivanga Barros, Ana Kiesse Zeleme, Fernando Luiz Pellegrini Pessoa, Ewerton Emmanuel da Silva Calixto, Bruna Aparecida Souza Machado, Ana Lucia Barbosa de Souza, Hugo Gomes D'amatto Villardi, Fumiaki Takakai, Takemi Kikuchi, Tomomi Sato, Masato Takeda, Saki Kanamaru, Yasuhiro Aono, Shinpei Nakagawa, Kentaro Yasuda, Yoshihiro Kaneta

First published in London, United Kingdom, 2022 by IntechOpen
IntechOpen is the global imprint of INTECHOPEN LIMITED, registered in England and Wales, registration number: 11086078, 5 Princes Gate Court, London, SW7 2QJ, United Kingdom

British Library Cataloguing-in-Publication Data
A catalogue record for this book is available from the British Library

Additional hard and PDF copies can be obtained from orders@intechopen.com

Soybean - Recent Advances in Research and Applications
Edited by Takuji Ohyama, Yoshihiko Takahashi, Norikuni Ohtake, Takashi Sato and Sayuri Tanabata
p. cm.
Print ISBN 978-1-80355-699-4
Online ISBN 978-1-80355-700-7
eBook (PDF) ISBN 978-1-80355-701-4

Meet the editors

Takuji Ohyama is the chairman of the Fertilization Research Foundation. He obtained a Ph.D. in Agriculture from the University of Tokyo in 1980. His previous positions include professor in the Faculty of Agriculture, Dean of the Graduate School of Science and Technology, and dean of the Faculty of Agriculture, all at Niigata University, Japan. He was also a Professor in the Faculty of Applied Biosciences, Department of Agricultural Chemistry, Tokyo University of Agriculture (2017 -2022). He was president of the Japanese Society of Soil Science and Plant Nutrition from 2007–2009. His research interests are nitrogen fixation and metabolism in soybean plants, new technology of deep placement of slow-release nitrogen fertilizers for soybean cultivation, and nitrogen and carbon metabolism in rice, tulip, curcuma, and cucumber. He is also interested in the use of stable isotopes and positron-emitting radioisotopes.

Yoshihiko Takahashi obtained a Ph.D. degree in agriculture from the Tokyo University of Agriculture in 1995. He was a professor in the Faculty of Agriculture, Niigata University, Japan from 2006 to 2020. He was also a researcher at the Niigata Prefectural Agricultural Research Institute from 1998 to 2003. He studies the high-quality cultivation of paddy rice and soybean by improving fertilizer management on a field scale. He developed a new fertilization technology, a deep placement of slow-release nitrogen fertilizers for soybean cultivation, which did not repress nitrogen fixation activity by root nodules.

Norikuni Ohtake is a professor in the Faculty of Agriculture, Niigata University, Japan. He obtained a Ph.D. degree in agriculture from the Niigata University of graduate school the same university in 1998. He was a senior researcher at the Cabinet Office, Government of Japan in 2007–2008. He studies metabolism and accumulation regulation analysis in developing soybean seeds, with a focus on seed storage protein accumulation.

Takashi Sato is a professor in the Faculty of Bioresource Sciences, Department of Environmental Science, Akita Prefectural University, Japan. He obtained a Ph.D. degree in agriculture from Niigata University, Japan in 1999. His research interests are soybean yield increase technology in a paddy-upland rotation system and soil amendment by legume green manure.

Sayuri Tanabata is an associate professor at the Center for International Field Agriculture Research and Education, College of Agriculture, Ibaraki University, Japan. She obtained a Ph.D. degree in agriculture from Niigata University, Japan in 2008. Her research interests are autoregulation of nodulation in legume plants, nitrogen and carbon metabolism in soybean, and the development of the use of organic fertilizers. She is also interested in the use of stable isotopes and radioisotopes.

Contents

Preface

This book covers all the areas of scientific research and development of soybean, especially physiology and metabolism, seed production, breeding and genetics, food from soybean seeds, and new soy foods. The demand for soybean seeds has been increasing due to their unique nutritional values. Annual soybean production increased to 350 Mt in 2017, which was followed by that of corn (1100 Mt), wheat (800 Mt), and rice (800 Mt). Soybean seeds contain a large amount of protein (about 40%) and oil (about 20%) and the matured seeds do not contain starch, which is the major nutrient in corn, wheat, and rice seeds. This book includes fourteen chapters in five sections.

Section 1: “Role of Soybean for Food Security”

Section 2: “Physiology and Biochemistry of Soybean Plants”

Section 3: “Cultivation and Productivity of Soybean”

Section 4: “Breeding and Biotechnology of Soybean”

Section 5: “Food and Biodiesel Industry”

Section 1 includes Chapter 1, in which Dr. Mohammad Sohidul Islam et al. summarize the important role of soybean for sustainable agriculture and global food security. Human beings and animals need an adequate amount of nutrients, such as carbohydrates, protein, lipids, minerals, vitamins, and so on. The increasing population and the expected decrease in crop production due to climate changes and land damage may be mitigated by changing from animal meat to plant protein, especially soybean protein.

Section 2 includes Chapter 2, in which Takuji Ohyama et al. review the effects of the application of various forms of nitrogen on the growth of nodules and roots related to carbon and nitrogen metabolism. Soybean plants obtain nitrogen from the fixed N2 by the root nodules and the absorbed inorganic nitrogen by the roots from soil or fertilizers. To obtain a high yield of soybean, good modulation and high and long-lasting nitrogen fixation activity are important because the availability of soil N is limited in many regions. However, it is well known that a high concentration of mineral N represses nodule formation and nitrogen fixation activity, especially nitrate, the most abundant inorganic nitrogen in upland fields, which severely inhibits nodulation and nitrogen fixation activity of soybean plants. The authors find that nitrate represses nodule growth rapidly and reversibly by the addition of nitrate in the culture solution. Also, ammonium, urea, and glutamine repress the nodule growth as well as nitrate. In Chapter 3, Dr. Md. Mannan reviews the physiological and biochemical basis of stress tolerance in soybean, especially responses to drought stress. Plants can adapt in response to water scarcity situations by altering cell metabolism and activating various defense mechanisms. Higher salt tolerance in resistant soybean genotypes was associated with better water relation, salt dilution by juiciness, and better osmotic adaptation with an accumulation of more amino acids, sugars, and proline. In addition, less damaging chlorophylls, higher photosynthetic efficiency and cell membrane stability, and higher calcium content contributed to the higher salt tolerance of soybean genotypes.

Section 3 begins with Chapter 4, in which Dr. Fumiaki Takakai et al. report the nitrogen budget in a paddy-upland rotation field with soybean cultivation by lysimeter experiments and reveal that considerable loss of N occurred in both the upland soybean and paddy rice cultivation periods. They concluded that to mitigate the N loss in the rotation system, N supply from organic matter applications such as leguminous green manure or livestock manure compost is required. In Chapter 5, Dr. Savala Canon reports that the inoculation formulation and applications determined nitrogen availability and water use efficiency in soybean production in Mozambique. The author concludes that the inoculants applied to soil or seed increase the amount of biologically fixed nitrogen and have the potential of improving soybean productivity in Mozambique. In Chapter 6, Dr. Raetano Carlos reviews the optimization of application technology for plant protection from diseases and insects in Brazil. The chapter describes the main application techniques of chemical or biological products in the control of agents by technological innovations involving remote sensing, unmanned aerial vehicles, and other techniques. In Chapter 7, Dr. Asewar Bhagwan et al. discusses climate-resilient technologies for enhancing the productivity of soybean in India. Recent research reveals that climate-resilient technologies such as selecting suitable varieties, sowing within the proper sowing period, dry spell management practices to reduce moisture stress, rainwater harvesting, soil conservation through broad bed furrow method of planting, and reuse of harvested and stored rainwater during moisture stress conditions can mitigate the effects of climate change and enhance the productivity of soybean in a sustainable manner.

Section 4 beings with Chapter 8, in which Dr. Sinan Meriç et al. discuss current strategies and the future of mutation breeding in soybean improvement. Currently, there are numerous alternative breeding approaches, including conventional breeding as well as biotechnological approaches. Among these, mutation breeding is still a major method to produce new alleles and desired traits within the crop genomes. The chapter highlights common bacterial, fungal, and viral pathogens and the role that biotechnology can play in controlling the disease. In Chapter 9, Dr. Sokichi Shiro and Dr. Yuichi Saeki examine the breeding of *Rj* gene-accumulated soybean genotypes and their availability for improving soybean productivity. Some soybean varieties harbor *Rj* genes, which regulate nodulation by preventing infection and nodulation by specific rhizobial strains. Soybean genotypes carrying several *Rj* genes may enhance the occupancy of useful bradyrhizobia, which exhibits potent nitrogen fixation ability and exhibit nodulation compatible with the *Rj* genotype of soybean. In Chapter 10, Dr. Summra Siddique reviews soybean quality improvement through CRISPR/Cas9. The fields of functional genomics and crop improvement have been transformed by advances in genome editing tools. CRISPR/Cas9 (clustered regularly interspaced short palindromic repeat)-Cas9 is a versatile genetic engineering tool.

Section 5 begins with Chapter 11, in which Dr. Lilian Hasegawa Florentino et al. review soybean functional proteins and synthetic biology. Despite soybean-producing high and elevated nutritive and functional proteins, it also produces allergenic proteins, harmful secondary metabolites, and carcinogenic elements. Therefore, synthetic biology is a powerful and promising science field for the development of new devices and systems. It represents a new disruptive frontier that will allow better use of soybean functional proteins, both for animal and human food and for the pharmaceutical and chemistry industry. In Chapter 12, Dr. Moses Ojukwu and Dr. Azhar Mat Easa discuss the innovative application of soy protein isolate and combined crosslinking technologies to enhance the structure of gluten-free rice noodles. Soy protein isolate may serve as a health-enhancing functional ingredient in many food products. The authors combined crosslinking and cold gelation of soy protein isolate to improve the structure, texture, and cooking properties of rice flour noodles. In Chapter 13, Dr. Ana Kiesse Zelema Geremias and Dr. António André Chivanga Barros discuss biodiesel production using a reactive distillation column based on intensification processes. For the preparation of biofuels that have the same characteristics as petroleum derivatives, a reactive distillation column was designed, constructed, installed, and operated using process intensification principles. It was applied in the production of biodiesel, using residual frying oil as the raw material, by the transesterification reaction in a continuous regime. In Chapter 14, Dr. Fernando Luiz Pellegrini Pessoa et al. discuss the routes to the aggregate value of soybean biofuel. The traditional production routes, such as pyrolysis, allow for the generation of bio-oils with high burning power by mixing oils and plastics. One example of an alternative route is single-step interesterification, where the methyl acetate reacts with the triglycerides in the oil, forming fatty acid methyl esters and triacetin as a byproduct.

This book is a useful resource for researchers, students, persons in the agriculture and food industries, and policymakers.

We greatly appreciate all the contributors and editors. We also thank Author Service Manager Ms. Jelena Vrdoljak at IntechOpen for her kind support.

Takuji Ohyama
The Fertilization Research Foundation,
Tokyo, Japan

Yoshihiko Takahashi and Norikuni Ohtake
Niigata University,
Niigata, Japan

Takashi Sato
Akita University,
Akita, Japan

Sayuri Tanabata
Ibaraki University,
Mito, Japan

Section 1

Role of Soybean for Food Security

Chapter 1

Soybean and Sustainable Agriculture for Food Security

Mohammad Sohidul Islam, Imam Muhyidiyn, Md. Rafiqul Islam, Md. Kamrul Hasan, ASM Golam Hafeez, Md. Moaz Hosen, Hirofumi Saneoka, Akihiro Ueda, Liyun Liu, Misbah Naz, Celaleddin Barutçular, Javeed Lone, Muhammad Ammar Raza, M. Kaium Chowdhury, Ayman El Sabagh and Murat Erman

Abstract

Global food security is under-challenged due to over increasing human population, limited cropland, and risk of climate change. Therefore, an appropriate agricultural policy framework needs to be developed for food security that should be sustainable economically and ecologically. Nitrogen (N) is a crucial element that controls the growth productivity of crop plants. N accounts for around 78 volume per cent of the atmosphere but all crop plants cannot use it directly. Agricultural land is mostly dominated by cereals (e.g. rice, wheat, maize) which have specifically high N demand as compared to food legumes. Soybean exemplifies the most significant and cultivated food legume, presently cultivated worldwide under varying climatic conditions. It plays a significant role in global food security as well as agricultural sustainability due to a high seed protein and oil concentration, and low reliance on N fertilization. Soybean enriches soil health by fixing atmospheric N through biological nitrogen fixation (BNF), the most productive and economical system for N fixation and crop production, associated with more intensive production systems. However, the efficiency of BNF depends on several factors. This study is focused to develop more reliable guidelines for managing BNF by using the potential of natural agro-ecosystems.

Keywords: soybean, food security, biological nitrogen fixation, climate change, agro-ecosystems

1. Introduction

The global population is predicted to reach 8.6 billion in 2030, 9.8 billion in 2050, and 11.2 billion in 2100 [1]. This expanding population and their subsequent consumption will lead to an increase in the global food demand, and it will be great challenge for food security under climate change and land-use scenarios. Exclusively, abiotic and biotic stresses caused by the global climate change progressively affected the cropping systems which will pose serious intimidations for global food production [2]. Most developed countries have had to embrace modern-day agricultural technologies to achieve food security for increasing populations, as well as to support agri-business and income generation. Currently, scientists have been propagating to explore crop diversification as an alternative strategy for developing countries. The ecological consequences of technologically focused agricultural systems that have been adopted and appreciated for many decades without the consideration of the environment, and the impact on the ecosystem is now coming into focus and scrutiny with the vivid and negative environmental impact of modern-day agriculture, and how it has greatly contributed to climate change. The current agricultural practices are not sustainable due to their misuse of valuable resources and environmental degradation. Hence, the philosophy of basic plant science research, and the direction of demand-based plant breeding should be changed to allow the plants for growing well in normal and limited resources in a sustainable way. In these reflections, it is suggested to grow soybean due to its higher adaptation and mitigation approaches in changing climates and multiplicity effects.

In history, soybean (*Glycine max* L. Merr.) was domesticated in China and afterward introduced into the USA and Brazil [3]. Currently, Argentina, Brazil, and the USA are the top soybean-producing countries at the global scale comprising 16, 32, and 33%, respectively [4]. Globally soybean production is projected to increase 311.1 and 371.3 million metric tons in 2020 and 2030, respectively. The annual growth rates of soybean from 2005 to 2007 to 2010, and 2010 to 2020 were 2.9 and 2.5%, respectively, and the rate is projected to increase 1.8% from 2020 to 2030 [5]. All the same, it is estimated that the demand in 2030 will be increased approximately 1.7 times greater than that of 2005–2007 [5]. Climate change has the potential to allow a significant increase in soybean production in Africa, irrespective of which production scenario becomes reality in the future [6]. Despite biotic and abiotic stresses, soybean production is continuing to increase over time [7, 8]. Soybean is known as the 'Africa's Cinderella crop' owing to the increasing demand for soybean production in recent years in Africa [9]. This chapter will focus on the potential role of soybean in agriculture for food security.

2. Soybean for sustainable development

2.1 Potential source of food and health benefits

Soybean is one of the most valuable crops in the world due to its multiple uses as a least expensive source of protein, healthy unsaturated fats and carbohydrate for the human diet, livestock and aquaculture feed, and biofuel. It is predominantly grown worldwide for high-quality, inexpensive proteins, and oil. It is highly nutritious food commodity as a source of vegetable protein and low cholesterol at an affordable price and is considered as a good substitute for animal protein due to containing essential amino acids required for human nutrition. The approximate composition of soybean is 36% protein, 19% oil, 35% carbohydrate including 17% dietary fiber, 5% minerals, and several other components including vitamins [10]. Soybean oil contains 16% saturated fatty acids, 23% monounsaturated fatty acids, and 58% polyunsaturated fatty acids [11]. In addition to edible oil, soybean is used as many processed foodstuffs such as soybean sprouts, toasted soy protein flours, soy milk, tofu, tempeh, miso, natto, soybean paste, and soy sauce [12], and also, bean curd, oncom, tauco, soybean cake, ice cream, soy flour, etc. [13]. Soybeans are also an important food commodity after rice and maize. Soybean is by far the cheapest source of protein for the poor smallholders as compared to other quality foods that are rich in proteins such as animal meat, fish, eggs, and milk. Based on the protein quality (protein digestibility corrected amino acid score), the value of soybean protein is equivalent (whole soybeans 96, soybean milk 91) to eggs (97) [14]. Several bioactive compounds like isoflavones, peptides, flavonoids, phytic acid, soy lipids, soy phytoalexins, soyasaponins, lectins, hemagglutinin, soy toxins, and vitamins are isolated from soybean and soy food products [15]. It has been reported earlier in many studies that consumption of soybean in different forms provides bioactive compounds as well which significantly lowered the risks for several cancers including breast [16], prostate [17], lung [18], colon [19], liver [20], and bladder [21], hypercholesterolemia and cardiovascular diseases [22], osteoporosis [23], hypertension [24], and blood pressure [25]. The consumption of protein from soybean sources by human beings is currently low worldwide, although there is increasing public and commercial interest since the crop could be a major source of dietary protein for the future. Malnutrition is a major global health problem, especially for developing countries, and food insecurity is the prime factor for malnutrition [26]. However, soybean-based foods are cheaper and readily available which can solve the problems.

2.2 Imperative source of animal feed

Soybean is not only a good source of high-quality edible oil and proteins for human beings but also a high-quality forage protein in animal feed worldwide. Feed is a key pillar in the journey of improving the productivity of livestock. Quality feed is the fundamental factor to increase the productivity of livestock. Soybean is also widely used as high quality and protein-rich animal feed [13] due to its auspicious attributes such as relatively high protein content, suitable amino acid profile except for methionine, and minimal variation in nutrient content. Soybean byproducts (raw materials and soybean meal) are used as a source of protein feedstuff for domestic animals including pig, chicken, cattle, horse, sheep, and fish feed and many prepackaged meals [10]. Soybean meal (SBM) contributes about 30% to poultry feeds [27]. It represents two-thirds of the total world output of protein feedstuffs [28]. Its feeding value is unparalleled by any other plant protein source [29]. SBM usually contains 47–49% crude protein (CP) and 3% crude fiber (CF) [30]. SBM is considered superior to other vegetable protein sources in terms of CP content and exceeds them in both total and digestible amino acid content [31]. The protein digestibility of SBM in poultry is approximately 85% [32]. Among the vegetable protein sources, SBM is used to meet the feed requirement of animals for limiting amino acids in cereal-based diets due to being the most cost-effective source of amino acids [33]. Therefore, the production of soybean, which is used extensively as animal feed, must be increased beyond the current production level due to meet the animal protein demand of overgrowing population in the world.

2.3 Contribution in biological nitrogen fixation

Nitrogen is a critical limiting element for growth and development by increasing chlorophyll as well as photosynthesis in crop plants. It is also the most abundant element in the atmosphere and exists in the diatomic form (N_2) but the plant cannot uptake and use N_2 directly. Only a group of plants known as legumes under the family of Fabaceae are well-known for being able to harvest N_2 from the atmosphere and incorporated it into the soil which is termed biological nitrogen fixation (BNF). BNF is firstly discovered by Beijerinck in 1901 [34]. The conversion of atmospheric dinitrogen (N_2) to ammonia (NH_3) under the combined action of biological and chemical activities is known as BNF [35]. It is a chemical process by which molecular N_2, with a strong triple covalent bond, in the air is converted into ammonia (NH_3) or related nitrogenous compounds, typically in soil or aquatic systems [36]. It is an important microbially mediated process that converts N_2 gas to NH_3 using the nitrogenase protein complex [37]. Some bacteria contain enzymes that can reduce N_2 and turn it into ammonia. Consequently, the NH_3 is used to produce essential elements, and it is a process known as BNF [38, 39]. The BNF can be symbiotic (mutualistic associations between plant species and fixing microorganisms, mainly rhizobia), or asymbiotic (when transmitted by free-living fixing microorganisms, like the species of the genera Azotobacter and Beijerinckia) [40].

Soybean also improves soil fertility, another benefit of soybean cultivation, by fixing atmospheric nitrogen through BNF [41, 42]. Soybean plants can freely assimilate NH_3 to produce nitrogenous biomolecules. These prokaryotes include aquatic organisms (cyanobacteria), free-living soil bacteria (*Azotobacter*), bacteria (*Azospirillum*) which make associative relationships with plants, and bacteria (*Rhizobium* and *Bradyrhizobium*) to build up highly significant symbiotic relationships with legumes and other plants [43]. The productivity of soybean largely depends on the BNF, the most important source to supply N in the soil. It has been reported earlier [44] that soybean seed yield is strongly linked to the N fixation process and N uptake of seed. It has been estimated that the contribution of N fixation to plant N demand ranges from 40 to 70% depending on the plant growth conditions (environments) and the association with the host-bacteria symbiosis [45–47]. In soybean, N derived from the atmosphere (NDFA) via BNF is recorded by 0–98% of the total N uptake, equivalent to 0–337 kg N ha^{-1} [48], and the total N uptake greatly depends on the activity of rhizobia. The yield of soybean increased over time in the last decades [49] by maintaining a high seed protein and N fixation process. Therefore, the N fixation process has become a growing concern on a global scale [44]. This BNF would be a major benefit to smallholder farming systems in developing countries where soil degradation and nutrient depletion have gradually increased because of high cropping intensity, and now pose serious threats to sustainable food production. Soybean farming is considered as one of the most cost-effective ways for sustaining soil fertility, especially for smallholder farmers which helps them to promote improved living standards and food security. Hence, soybean production and commercialization would be a milestone for improving food and nutritional security as well as to meet sustainable agriculture.

The fixation of atmospheric nitrogen is a complex process that requires a large input of energy to carry on [43]. For fixing nitrogen microorganisms require 16 moles of adenosine triphosphate (ATP) to reduce each mole of nitrogen [50]. Microorganisms obtain this energy by oxidizing organic molecules, such as non-photosynthetic free-living microorganisms obtain from other organisms, photosynthetic microorganisms (Cyanobacteria) obtain from sugars (photosynthetic product), and associative and symbiotic nitrogen-fixing microorganisms obtain from their host plants' rhizospheres [50, 51]. The BNF process is affected by several factors [52] like abiotic stresses water deficit or excess water, salinity, temperature, heavy metals, and biocides [53], mineral elements such as high soil nitrate concentration [54], phosphorous [55] and sulfur [56, 57], acidity [58] and alkalinity [59], and biotic factors like ineffective rhizobia [60], plant diseases [61], and weeds [62]. At pH 7.0, we observed that there was low nitrogenase activity.

2.4 Contribution as rotational crop

Crop rotation is an important agronomic management practice that is followed to sustain soil fertility and reduce pests and diseases. It also enhances to form some beneficial soil microbes with the following crops when the rotational crops are legumes specially soybean, which pointedly increased the growth and productivity of the crops. It has been well established that cultivation of soybean in 2- and 3-year rotations with corn and wheat in agriculture is highly profitable and advantageous for soil [63]. It has also been reported that soybean as a rotational crop is significantly cost-effective and beneficial to soil health [64, 65]. It has been established earlier that crop rotation recovers soil health and resilience by increasing soil organic carbon (SOC) [66–68], improving soil structure [69], enhancing nutrient availability [70], decreasing pests and pathogens in crops, increasing the population disease-defeating soil microbes [71, 72], and consequently increases yield of crops [73]. It is well documented that crop rotation as corn-soybean-wheat increased soybean yield in 1-year out of 3-year rotations as compared to growing every other year in corn-soybean rotations [74]. The high frequency of soybean in a crop rotation has decreased the SOC storage, and reduced macro aggregation owing to low residue inputs of soybean [75–77]. Soybean provided N through BNF as well as exploiting soil N from chemical sources [78, 79]. The soybean yield is meaningfully increased under rotations of corn-soybean in 2 years as compared to growing continuously [80–82]. Soybean in a 2-year rotations with corn increased grain yield by 9.2 and 12% over continuous soybean growing under no-tillage and conventional tillage conditions, respectively [80]. Rotation of soybean with traditional crops such as maize increases soil fertility by fixing nitrogen in the soil consequently increasing yield by 10–20% [83].

As compared to cereal crops, the residues of soybean contain a low C to N ratio, which promotes the decomposition of residues rapidly [76, 84]. However, accumulation and sequestration of C in a stable soil aggregate from soybean residues is lower over the corn and cereal residues, indicating a lower C to N ratio, and lower phenolic acid content of soybean residues [85, 86].

Corn-soybean rotation including winter wheat increased soybean yield over mono-cropping soybean due to higher infestation of pest predation and/or soil-borne plant pathogens as well as reduced SOC levels owing to lower aboveground and belowground biomass from continuous soybean cropping [64, 87]. The soybean yield is significantly increased with rotation as compared to continuous soybean due to increasing soil organic matter plus improving soil properties [88, 89], increasing the resource available for heterotrophic soil microbial communities, and increasing C and N cycling [89, 90]. Moreover, the strong rotational benefits were observed by Giller *et al*. [91] for maize-soybean rotation as the crop broke the cycle of continuous maize cultivation and fixed nitrogen to the soil, and support to build up sustainable soil fertility systems and profitability. As well, soybean-maize rotation deducted N fertilizer which helps to reduce carbon emissions that ensure sustainable agricultural production. It has been reported earlier [92] that soybean contains the climatic resilience and native *Bradyrhizobium* strains which are well apposite to the current crop rotation system.

2.5 Impact on soil fertility

Soil is a nonrenewable resource that may be degraded due to inappropriate management practices. Intercropping systems allow to enhance resource-use efficiency and crop productivity which promote multiple ecosystem services [93]. Integration of legume crops is fundamental in many intercropping systems [94], and legume-based cropping systems improve soil fertility in many ways, such as increasing SOC and humus content, N and P availability, etc. [95]. It has been documented earlier that grain legumes are weak suppressors of weeds, but the mixing of crop species in the same cropping system improves the ability of the crop to suppress weeds [96, 97]. Soybean is characterized as a major economic crop in smallholder farming systems due to sustaining soil fertility [42], providing feed for livestock, and improving rural household nutrition and income. Inoculation of *Bradyrhizobium japonicum* strain 61-A-101 and mycorrhizal fungi with soybean potentially augmented the N and P uptake by the host plant through efficient colonization of *Glomus mosseae* [98]. It has significant agronomic benefits to refresh the soils such as the crop canopies protecting the soil from recurrent erosion, decaying root residues improving soil fertility, and fixing atmospheric nitrogen into the soil which leads to higher levels of sustainable agriculture with minimal input requirements. Soybean is primarily grown as an intercrop with maize, sorghum, finger millet, sugarcane, which may be a suitable approach for sustainable agriculture.

2.6 Impact on greenhouse gas emission

There are a number of the impact that grain legumes have on the environment and the soil in regards to quality. Meanwhile, the role of legumes like soybean to alleviate the negative effects caused by climate change has been rarely addressed. The emission of greenhouse gases (GHG) such as carbon dioxide (CO_2) and nitrous oxide (N_2O), methane (CH_4), etc. are the causes of global warming. Legumes reduce the emission of GHG in agricultural systems by reducing mineral N fertilization, sequestration of carbon in soils, and the overall fossil energy inputs in the system [99].

N_2O is much more active than CO_2 which represents nearly 5–6% of the total atmospheric gases [100]. Around 60% N_2O emission is occurred by agricultural practices which exemplify as the main source of emission [101], and the production of crops and animals are the main source of emission [102]. In crop production, the application of nitrogenous fertilizers is the birthplace of the majority of these emissions [101]. It has been estimated that about 1.0 kg of N is emitted as N_2O from every 100 kg of N fertilizer [95]. The amount of N_2O emission largely depends on several factors including N application rate, soil organic C content, soil pH, and texture [103, 104]. In most of cropping and pasture systems, de-nitrification is the leading source of N_2O emission [104–106]. Several studies in recent years have been signified the role of legumes in the reduction of GHG emissions. For example, it has been reported that legumes discharge around 5–7 times less GHG per unit area compared with other crops [107]. Generally, the losses N_2O from soils under legume crops are undoubtedly lower than those from both N_2O fertilized in grasslands and non-legume crops [95]. Among legumes, soybean most efficiently produced and provided the maximum protein (g) per GHG emission out of 22 plant and animal protein sources [108]. Adoption of sustainable agricultural systems mitigate the emission of GHG such as conservation agriculture systems, which is suitable for the cultivation of both grain, and green-manure legumes lessen the emission of GHG.

2.7 Socio-economic aspects

Food and water security will be a major global issue focus in the coming decades due to climate change and population pressure. Malnutrition, predominantly protein deficiency, is prevalent in many parts of the world. Therefore, appropriate technology should be addressed by lawmakers and scientists for food security, and the cultivation of legumes majorly soybean is a first step to address the food security issues worldwide. Soybeans produce the highest amount of protein per hectare [109] and are well positioned to meet the need of future global protein. Conventional protein sources are highly expensive as well as a vulnerable population is unable to purchase from these sources. Hence, soybean-based protein foods are an important strategy to relieve malnutrition and hunger problems. Since it has been evidenced that smallholder farmers have limited capability to overcome crop production challenges due to changing climate [110]. They produce soybean for gaining higher yields, family demand, and net profits with minimum N fertilizer input which eventually improved their living standards as well as food security [111].

3. Conclusion

A sustainable agricultural system is the only way to sustainably intensify food crop production without causing damage to human and environmental health. Soybean and other nitrogen-fixing legumes should be a viable crop included in all forms of cropping systems as they can efficiently utilize atmospheric nitrogen through the process of BNF. The most important thing is the integration of soybean and another legume across different cropping systems which would effectively reduce the usage of chemical nitrogenous fertilizers, and conserve soil fertility. It is important to focus on the cultivation of crops that provides higher yield, economic return by maintaining soil health as well as environmental balances. Some priority areas seem to emerge, and these areas require deeper investigation to fully understand how the BNF dynamics, and how to utilize BNF in best way for sustainable agriculture. Thus, soybean crops should be grown to reduce hunger, malnutrition, and poverty as well as to bring food security by sustaining agriculture in light of climate and population challenges.

Author details

Mohammad Sohidul Islam[1*], Imam Muhyidiyn[1], Md. Rafiqul Islam[2], Md. Kamrul Hasan[1], ASM Golam Hafeez[3], Md. Moaz Hosen[4], Hirofumi Saneoka[5], Akihiro Ueda[5], Liyun Liu[6], Misbah Naz[7], Celaleddin Barutçular[8], Javeed Lone[9], Muhammad Ammar Raza[9], M. Kaium Chowdhury[10], Ayman El Sabagh[9,11*] and Murat Erman [9]

1 Department of Agronomy, Hajee Mohammad Danesh Science and Technology University, Dinajpur, Bangladesh

2 Agronomy Division, Regional Agricultural Research Station, BARI, Pabna, Bangladesh

3 Department of Agricultural Finance, Bangladesh Agricultural University, Mymensingh, Bangladesh

4 Department of Agricultural Chemistry, Hajee Mohammad Danesh Science and Technology University, Dinajpur, Bangladesh

5 Graduate School of Integrated Sciences for Life, Hiroshima University, Hiroshima, Japan

6 Department of Biotechnology, Osaka University, Suita, Osaka, Japan

7 Institute of Environment and Ecology, School of Environment and Safety Engineering, Jiangsu University, Zhenjiang, China

8 Faculty of Agriculture, Department of Field Crops, Çukurova University, Adana, Turkey

9 Faculty of Agriculture, Department of Field Crops, Siirt University, Siirt, Turkey

10 Agricultural Training Institute, Department of Agricultural Extension, Gaibandha, Bangladesh

11 Faculty of Agriculture, Department of Agronomy, Kafrelsheikh University, Kafrelsheikh, Egypt

*Address all correspondence to: shahid_sohana@yahoo.com and ayman.elsabagh@agr.kfs.edu.eg

References

[1] Anonymous. The World Population Prospects: The 2017 Revision. United Nations, New York, USA: UN Department of Economic and Social Affairs Population Division; 2018

[2] Yadav SS, Hunter D, Redden B, Nang M, Yadava DK, Habibi AB. Impact of climate change on agriculture production, food, and nutritional security. In: Redden R, Yadav SS, Maxted N, Dulloo MS, Guarino L, Smith P, editors. Crop Wild Relatives and Climate Change. New Jersey, USA: Wiley; 2015. pp. 1-23

[3] López-López A, Rosenblueth M, Martínez J, Martínez RE. Rhizobial symbioses in tropical legumes and non-legumes. In: Dion P, editor. Soil Biology and Agriculture in the Tropics. Soil Biology. Vol. 21. Berlin, Heidelberg, Germany: Springer; 2010. pp. 163-184. DOI: 10.1007/978-3-642-05076-3_8

[4] USDA NASS. United States and all state data–crops. 2017. Avaiable on: http://www.nass.usda.gov/QuickStats/ [Accessed: June, 2017]

[5] Siamabele B. The significance of soybean production in the face of changing climates in Africa. Cogent Food and Agriculture. 2021;7(1):1933745. DOI: 10.1080/23311932.2021. 1933745

[6] Fodor N, Challinor A, Droutsas I, Ramirez-Villegas J, Zabel F, Koehler AK, et al. Integrating plant science and crop modeling: Assessment of the impact of climate change on ssybean and maize production. Plant and Cell Physiology. 2017;**58**(11):1833-1847. DOI: 10.1093/pcp/pcx141

[7] Grassini P, Torrion JA, Cassman KG, Yang HS, Specht JE. Drivers of spatial and temporal variation in soybean yield and irrigation requirements in the western US corn belt. Field Crops Research. 2014;**163**:32-46. DOI: 10.1016/j.fcr.2014.04.005

[8] Specht JE, Diers BW, Nelson RL, Toledo JF, Torrion JA, Grassini P. Soybean [*Glycine max* (L.) merr.]. In: Smith JSC, Carver B, Diers BW, Specht JE, editors. Yield Grains in Major US Field Crops: Contributing Factors and Future Prospects. USA: ASA-CSSA-SSSA; 2014. pp. 311-355

[9] Kolapo AL. Soybean: Africa's potential cinderella food crop. In: Tzi-Bun NG, editor. Soybean: Biochemistry, Chemistry and Physiology. UK: InTech Open; 2011

[10] Liu KS. Chemistry and nutritional value of soybean components. In: Soybean: Chemistry, Technology, and Utilization. New York: Chapman & Hall, USA; 1997. pp. 25-113

[11] Wolke RL. Where There's Smoke, There's a Fryer. Washington D.C, USA: The Washington Post; 2007

[12] Fournier DB, Erdman JW, Gordon GB. Soy, its components, and cancer prevention: A review of the in vitro, animal, and human data. Cancer Epidemiology, Biomarkers and Prevention. 1998;7(11):1055-1065

[13] Pagano MC, Miransari M. The importance of soybean production worldwide. In: Miransari M, editor. Abiotic and Biotic Stresses in Soybean Production: Volume 1. Cambridge, Massachusetts, USA: soybean production. Academic Press; 2016. pp. 1-26. DOI: 10.1016/B978-0-12-801536-0.00001-3

[14] FAO/WHO. Protein Quality Evaluation: Report of the Joint FAO/WHO Expert Consultation. Bethesda, MD (USA): Food and Agriculture Organization of the United Nations (Food and Nutrition Paper). 1989, 51

[15] Davis J, Iqbal MJ, Steinle J, Oitker J, Higginbotham DA, Peterson RG. Soy protein influences the development of the metabolic syndrome in male obese ZDFxSHHF rats. Hormone and Metabolic Research. 2007;**37**:316-325

[16] Boyapati SM, Shu XO, Ruan ZX, Dai Q, Cai Q, Gao YT, et al. Soyfood intake and breast cancer survival: A follow up of the Shanghai breast Cancer study. Breast Cancer Research and Treatment. 2005;**92**:11-17

[17] Jacobsen BK, Knutsen SF, Fraser GE. Does high soy milk intake reduce prostate can- cer incidence? The Adventist health study (United States). Cancer Causes & Control. 1998;**9**(6):553-557

[18] Swanson CA, Mao BL, Li JY, Lubin JH, Yao SX, Wang JZ, et al. Dietary determinants of lung-cancer risk results from a case-control study in Yunnan province, China. International Journal of Cancer. 1992;**50**(6):876-880

[19] Azuma N, Machida K, Saeki T, Kanamoto R, Iwami K. Preventive effect of soybean resistant proteins against experimental tumorigenesis in rat colon. Journal of Nutritional Science and Vitaminology. 2000;**46**(1):23-29

[20] Kanamoto R, Azuma N, Miyamoto T, Saeki T, Tsuchihashi Y, Iwami K. Soybean resistant proteins interrupt an enterohepatic circulation of bile acids and suppress liver tumorigenesis induced by azoxymethane and dietary deoxycholate in rats. Bioscience, Biotechnology, and Biochemistry. 2001;**65**(4):999-1002

[21] Sun CL, Yuan JM, Arakawa K, Low SH, Lee HP, Yu MC. Dietary soy and increased risk of bladder cancer: The Singapore Chinese health study. Cancer Epidemiological Biomarkers Preview. 2002;**11**(12):1674-1677

[22] Carroll KK. Hypercholesterolemia and atherosclerosis: Effects of dietary protein. Federation Proceedings. 1982;**41**:2792-2796

[23] Messina M, Messina V. Soyfoods, soybean isoflavones, and bone health: A brief over- view. Journal of Renal Nutrition. 2000;**10**:63-68

[24] Kim SJ, Jung KO, Park KY. Inhibitory effect of Kochujang extracts on chemically induced mutagenesis. Journal of Food Science and Nutrition. 1999;**4**:38-42

[25] Welty FK, Lee KS, Lew NS, Zhou JR. Effect of soy nuts on blood pressure and lipid levels in hypertensive, prehypertensive, and normotensive postmenopausal women. Archives of Internal Medicine. 2007;**167**:1060-1067

[26] Betebo B, Ejajo T, Alemseged F, Massa D. Household food insecurity and its association with nutritional status of children 6-59 months of age in east Badawacho District, South Ethiopia. Journal of Environmental and Public Health. 2017;**2017**:6373595. DOI: 10.1155/2017/6373595

[27] Leesons S, Summers J. Commercial Poultry Nutrition. Guelph, Ontario, Canada: Nottingham University Press, UK; 2005

[28] Oil World. Oil World Annual 2015. Hamburg, Germany: ISTA Mielke GmbH; 2015

[29] Cromwell GL. Soybean meal - the "gold standard". The Farmer's Pride, KPPA News. 10 Nov 1999;**11**(20)

[30] Cromwell GL. Soybean Meal - an Exceptional Protein Source. Ankeny, IA, USA: Soybean Meal InfoCenter; 2012

[31] Ajinomoto Heartland Lysine LLC Revision 7. True digestibility of essential amino acids in poultry. Available on: http://www.lysine.com/new/Technical%20Reports/Poultry/PoultryDigTableV7.pdf

[32] Woodworth JC, Tokach MD, Goodband RD, Nelssen JL, O'Quinn PR, Knabe DA, et al. Apparent ileal digestibility of amino acids and digestible and metabolisable energy content of dry extruded-expelled soybean meal and its effect on growth performance of pigs. Journal of Animal Science. 2001;**79**:1280-1287

[33] Kerley MS, Allee GL. Modifications in soybean seed composition to enhance animal feed use and value: Moving from dietary ingredient to a functional dietary component. AgBioforum. 2003;**6**(1&2):14-17

[34] Beijerinck MW. Über oligonitrophile Mikroben. Zentralblatt fur Bakteriologie. 1901;7:561-582

[35] Franche C, Lindström K, Elmerich C. Nitrogen-fixing bacteria associated with leguminous and non-leguminous plants. Plant and Soil. 2009;**321**:35-59. DOI: 10.1007/s11104-008-9833-8

[36] Postgate J. Nitrogen Fixation. 3rd ed. Cambridge: Cambridge University Press. UK; 1998

[37] Streicher SL, Gurney EG, Valentine RC. The nitrogen fixation genes. Nature. 1972;**239**(5374):495-499. DOI: 10.1038/239495a0

[38] Hungria M, Campo RJ, Mendes IC. A importância do processo de fi xação biológica do nitrogênio para a cultura da soja: componente essencial para a competitividade do produto brasileiro. Documentos 283, Embrapa Soja, Londrina, Brazil; 2007. p. 80

[39] Di Ciocco C, Coviella C, Penón E, Díaz-Zorita M, López S. Biological fixation of nitrogen and N balance in soybean crops in the pampas region. Spanish Journal of Agricultural Research. 2008;**6**(1):114-119

[40] Freitas SS. Rizobactérias promotoras de crescimento de plantas. In: Silveira APD, Freitas SS, editors. Microbiota do solo e qualidade ambiental. Campinas: Instituto Agronômico de Campinas; 2007. pp. 1-20

[41] Mpepereki S, Javaheri F, Davis P, Giller KE. Soybeans and sustainable agriculture; promiscuous soybeans in southern Africa. Field Crops Research. 2000;**65**:137-149

[42] Chianu JN, Ohiokpehai O, Vanlauwe B, Adesina A, De Groote H, Sanginga N. Promoting a versatile but yet minor crop: Soybean in the farming Systems of Kenya. Journal of Sustainable Development in Africa. 2009;**10**(4):324-344

[43] Postgate JR. The Fundamentals of Nitrogen Fixation. New York, USA: Cambridge University Press; 1982

[44] Ciampitti IA, Salvagiotti F. New insights into soybean biological nitrogen fixation. Agronomy Journal. 2018;**110**(4):1185-1196. DOI: 10.2134/agronj2017.06.0348

[45] Pauferro N, Guimarães AP, Jantalia CP, Urquiaga S. 15N natural abundance of biologically fixed N_2 in soybean is controlled more by the Bradyrhizobium strain than by the variety of the host plant. Soil Biology and Biochemistry. 2010;**42**(10):1694-1700. DOI: 10.1016/j.soilbio.2010.05.032

[46] Collino DJ, Salvagiotti F, Perticari A, Piccinetti C, Ovando G, Urquiaga S, et al. Biological nitrogen fixation in soybean in Argentina: Relationships with crop, soil, and meteorological factors. Plant and Soil. 2015;**392**:239-252. DOI: 10. 1007/ s11104-015-2459-8

[47] Santachiara G, Borrás L, Salvagiotti F, Gerde JA, Rotundo JL. Relative importance of biological nitrogen fixation and mineral uptake in high yielding soybean cultivars. Plant and Soil. 2017;**418**(1-2):191. DOI: 10.1007/s11104-017-3279-9

[48] Salvagiotti F, Cassman KG, Specht JE, Walters DT, Weiss A, Dobermann A. Nitrogen uptake, fixation and response to fertilizer N in soybeans: A review. Field Crops Research. 2008;**108**:1-13. DOI: 10.1016/j. fcr.2008.03.001

[49] Rincker K, Nelson R, Specht J, Sleper D, Cary T, Cianzio SR, et al. Genetic improvement of U.S. soybean in maturity groups II, III, and IV. Crop Science. 2014;**54**:1419-1432. DOI: 10.2135/cropsci2013.10.0665

[50] Hubbell DH, Kidder G. Biological Nitrogen Fixation. Florida, USA: University of Florida, Institute of Food and Agricultural Sciences (IFAS) Extension Publication SL16; 2009. pp. 1-4

[51] NRC (National Research Council). Biological Nitrogen Fixation: Research Challenges. Washington, DC, USA: National Academy Press; 1994

[52] Peoples MB, Ladha JK, Herridge DF. Enhancing legume N_2 fixation through plant and soil management. Plant and Soil. 1995;**174**:83-101

[53] Helemish FA, Abdel-Wahab SM, El-Mokadem MT, Abou-El-Nour MM. Effect of sodium chloride salinity on the growth, survival and tolerance response of some rhizobial strains. Ain Shams Scientific Bulletin. 1991;**28B**:423-440

[54] Saito A, Tanabata S, Tanabata T, Tajima S, Ueno M, Ishikawa S, et al. Effect of nitrate on nodule and root growth of soybean (*Glycine max* (L.) merr.). International journal of molecular. Science. 2014;**15**:4464-4480. DOI: 10.3390/ijms15034464

[55] Chalk P. Integrated effects of mineral nutrition on legume performance. Soil Biology and Biochemistry. 2000;**32**:577-579. DOI: 10.1016/ S0038-0717(99)00173-X

[56] Divito GA, Sadras VO. How do phosphorus, potassium and Sulphur affect plant growth and biological nitrogen fixation in crop and pasture legumes? A meta-analysis. Field Crops Research. 2014;**156**:161-171. DOI: 10.1016/j.fcr.2013.11.004

[57] Borja Reis AF, de Rosso LHM, Davidson D, Kovács P, Purcell LC, Below FE, et al. Sulfur fertilization in soybean: A meta-analysis on yield and seed composition. European Journal of Agronomy. 2021;**127**:126285. DOI: 10.1016/j.eja.2021.126285

[58] Lin M, Gresshoff PM, Ferguson BJ. Systemic regulation of soybean nodulation by acidic growth conditions. Plant Physiology. 2012;**160**:2028-2039. DOI: 10.1104/pp.112.204149

[59] Ferreira TC, Aguilar JV, Souza LA, Justino GC, Aguiar LF, Camargos LS. pH effects on nodulation and biological nitrogen fixation in *Calopogonium mucunoides*. Brazilian Journal of Botany. 2016;**39**(4):1015-1020. DOI: 10.1007/ s40415-016-03000-0

[60] Thies JE, Singleton PW, Bohlool B. Influence of the size of

indigenous rhizobial populations on establishment and symbiotic performance of introduced rhizobia on field-grown legumes. Applied and Environmental Microbiology. 1991;**57**:19-28. DOI: 10.1128/AEM.57.1.19-28.1991

[61] Brockwell J, Bottomley PJ, Thies JE. Manipulation of rhizobia microflora for improving legume productivity and soil fertility: A critical assessment. Plant and Soil. 1995;**174**:143-180

[62] Li L, Zhang L-Z, Zhang F-Z. Crop mixtures and the mechanisms of overyielding. In: Levin SA, editor. Encyclopedia of Biodiversity, Vol. 2. 2nd Ed. Cambridge, Massachusetts, USA: Waltham: Academic Press; 2013. pp. 382-295

[63] Gaudin ACM, Tolhurst T, Ker A, Janovicek K, Tortora C, Martin RC, et al. Increasing crop diversity mitigates weather variations and improves yield stability. PLoS One. 2015;**10**:1-20. DOI: 10.1371/journal.pone.0113261

[64] Gaudin ACM, Janovicek K, Deen B, Hooker DC. Wheat improves nitrogen use efficiency of maize and soybean-based cropping systems. Agriculture Ecosystems and Environment. 2015;**210**:1-10. DOI: 10.1016/j.agee.2015.04.034

[65] Hoss M, Behnke GD, Davis AS, Nafziger ED, Villamil MB. Short corn rotations do not improve soil quality, compared with corn monocultures. Agronomy Journal. 2018;**110**:1274-1288. DOI: 10.2134/agronj2017.11.0633

[66] Poeplau C, Don A. Carbon sequestration in agricultural soils via cultivation of cover crops-a meta-analysis. Agriculture Ecosystems and Environment. 2015;**200**:33-41. DOI: 10.1016/j.agee.2014.10.024

[67] Lal R. Soil health and carbon management. Food and Energy Security. 2016;**5**:212-222. DOI: 10.1002/fes3.96

[68] McDaniel MD, Grandy S. Soil microbial biomass and function are altered by 12 years of crop rotation. The Soil. 2016;**2**:583-599. DOI: 10.5194/soil-2-583-2016

[69] Campbell CA, Zentner RP. Soil organic matter as influenced by crop rotations and fertilization. Soil Science Society of America Journal. 1993;**57**:1034-1040. DOI: 10.2136/sssaj1993.03615995005700040026x

[70] Copeland PJ, Crookston RK. Crop sequence affects nutrient composition of corn and soybean grown under high fertility. Agronomy Journal. 1992;**84**:503-509. DOI: 10.2134/agronj1992.00021962008400030028x

[71] Krupinsky JM, Bailey KL, McMullen MP, Gossen BD, Turkington TK. Managing plant disease risk in diversified cropping systems. Agronomy Journal. 2002;**94**:198-209. DOI: 10.2134/agronj 2002.1980

[72] Peralta AL, Sun Y, McDaniel MD, Lennon JT. Crop rotational diversity increases disease suppressive capacity of soil microbiomes. Ecosphere. 2018;**9**:1-16. DOI: 10.1002/ecs2.2235

[73] Karlen DL, Varvel GE, Bullock DG, Cruse RM. Crop rotations for the 21st century. Advances in Agronomy. 1994;**53**:1-45. DOI: 10.1016/S0065-2113(08)60611-2

[74] Lund MG, Carter PR, Oplinger ES. Tillage and crop rotation affect corn, soybean, and winter wheat yields. Journal of Production Agriculture. 1993;**6**:207-213. DOI: 10.2134/jpa1993.0207

[75] Studdert GA, Echeverría HE. Crop rotations and nitrogen fertilization to manage soil organic carbon dynamic. Soil Science Society of America Journal.

2000;**64**:1496-1503. DOI: 10.2136/sssaj2000.6441496x

[76] Wright AL, Hons FM. Soil aggregation and carbon and nitrogen storage under soybean cropping sequences. Soil Science Society of America Journal. 2004;**68**:507-513. DOI: 10.2136/sssaj 2004.5070

[77] Wright AL, Hons FM. Soil carbon and nitrogen storage in aggregates from different tillage and crop regimes. Soil Science Society of America Journal. 2005;**69**:141-147. DOI: 10.2136/sssaj2005.0141

[78] Herridge D, Peoples M, Boddey R. Global inputs of biological nitrogen fixation in agricultural systems. Plant and Soil. 2008;**311**:1-18

[79] Córdova SC, Castellano MJ, Dietzel R, Licht MA, Togliatti K, Martinez-Feria R, et al. Soybean nitrogen fixation dynamics in Iowa, USA. Field Crops Research. 2019;**236**:165-176. DOI: 10.1016/j.fcr.2019.03.018

[80] Houx JH, Wiebold WJ, Fritschi FB. Rotation and tillage affect soybean grain composition, yield, and nutrient removal. Field Crops Research. 2014;**164**:12-21. DOI: 10.1016/j.fcr.2014.04.010

[81] Al-Kaisi MM, Archontoulis S, Kwaw-Mensah D. Soybean spatiotemporal yield and economic variability as affected by tillage and crop rotation. Agronomy Journal. 2016;**108**:1267-1280. DOI: 10.2134/agronj2015.0363

[82] Mazzilli SR, Ernst OR. Soybean yield increases when maize is included in the cropping system. Agrosystems, Geosciences & Environment. 2019;**2**:1-6. DOI: 10.2134/age2018.09.0033

[83] TechnoServe. Southern Africa Regional Soybean Roadmap: Final Presentation. Southern Africa Trade Hub. Bill and Melinda Gates Foundation. Agland Investment Services, Inc. Global Agriculture, Food and Resources Consultants; 2011

[84] Halvorson AD, Schlegel AJ. Crop rotation effect on soil carbon and nitrogen stocks under limited irrigation. Agronomy Journal. 2012;**104**:1265-1273. DOI: 10.2134/agronj2012.0113

[85] Martens DA. Plant residue biochemistry regulates soil carbon cycling and carbon sequestration. Soil Biology & Biochemistry. 2000;**32**:361-369. DOI: 10.1016/S0038-0717(99)00162-5

[86] Martens DA. Management and crop residue influence soil aggregate stability. Journal of Environmental Quality. 2000;**29**:723-727. DOI: 10.2134/jeq2000.00472425002900030006x

[87] Gagnon B, Pouleur S, Lafond J, Parent G, Pageau D. Agronomic and economic benefits of rotating corn with soybean and spring wheat under different tillage in eastern Canada. Agronomy Journal. 2019;**111**:3109-3118. DOI: 10.2134/agronj2018.10.0653

[88] Smith RG, Gross KL, Robertson GP. Effects of crop diversity on agroecosystem function: Crop yield response. Ecosystems. 2008;**11**:355-366. DOI: 10.1007/s10021-008-9124-5

[89] McDaniel MD, Tiemann LK, Grandy AS. Does agricultural crop diversity enhance soil microbial biomass and organic matter dynamics? A meta-analysis. Ecological Applications. 2014;**24**:560-570. DOI: 10.1890/13-0616.1

[90] Bullock DG. Crop rotation. Critical Reviews in Plant Sciences. 1992;**11**:30926. DOI: 10.1080/07352689209382349

[91] GillerKE,MurwiraMS,DhliwayoDKC, Mafongoya PL, Mpepereki S. Soyabeans and sustainable agriculture in southern Africa. International Journal for Sustainable Agriculture. 2011;**9**:50-58. DOI: 10.3763/ijas.2010.0548

[92] Mapfumo P, Mtambanengwe F, Giller KE, Mpepereki S. Tapping indigenous herbaceous legumes for soil fertility management by resource poor farmer in Zimbabwe. Agriculture, Ecosystems & Environment. 2005;**09**(3-4):221-233. DOI: 10.1016/j.agee.2005.03.015

[93] Li L, Tilman D, Lambers H, Zhang F-S. Biodiversity and over yielding: Insights from belowground facilitation of intercropping in agriculture. New Phytologist. 2014;**203**:63-69

[94] Hauggaard-Nielsen H, Jensen ES. Facilitative root interactions in intercrops. Plant and Soil. 2005;**274**:237-250

[95] Jensen ES, Peoples MB, Boddey RM, Gresshoff PM, Hauggaard-Nielsen H, Alves BJ, et al. Legumes for mitigation of climate change and the provision of feedstock for biofuels and biorefineries. A review. Agronomy for Sustainable Development. 2012;**32**:329-364

[96] Siddique KH, Johansen C, Turner NC, Jeuffroy MH, Hashem A, Sakar D, et al. Innovations in agronomy for food legumes: A review. Agronomy for Sustainable Development. 2012;**32**:45-64

[97] Šarūnaitė L, Deveikytė I, Kadžiulienė Ž. Intercropping spring wheat with grain legume for increased production in an organic crop rotation. Žemdirbystė-Agriculture. 2010;**97**:51-58

[98] Xie Z, Staehelin C, Vierheili H, Wiemken A, Jabbouri S, Broughton WJ, et al. Rhizobial nodulation factors stimulate mycorrhizal colonization of undulating and non-nodulating soybeans. Plant Physiology. 1995;**108**(4): 1519-1525. DOI: 10.1104/pp.108.4.1519

[99] Stagnari F, Maggio A, Galieni A, Pisante M. Multiple benefits of legumes for agriculture sustainability: An overview. Chemical and Biological Technologies in Agriculture. 2017;**4**:1-13. DOI: 10.1186/s40538-016-0085-1

[100] Crutzen PJ, Mosier AR, Smith KA, Winiwarter W. N_2O release from agrobiofuel production negates global warming reduction by replacing fossil fuels. Atmospheric Chemistry and Physics Discussions. 2007;**7**:11191-11205

[101] Reay DS, Davidson EA, Smith KA, Smith P, Melillo JM, et al. Global agriculture and nitrous oxide emissions. Nature Climate Change. 2012;**2**:410-416

[102] Voisin AS, Guéguen J, Huyghe C, Jeuffroy MH, Magrini MB, Meynard JM, et al. Legumes for feed, food, biomaterials and bioenergy in Europe: A review. Agronomy for Sustainable Development. 2014;**34**:361-380

[103] Peoples MB, Hauggaard-Nielsen H, Jensen ES. The potential environmental benefits and risks derived from legumes in rotations. In: Emerich DW, Krishnan HB, editors. Nitrogen Fixation in Crop Production. Madison: American Society of Agronomy, Crop Science Society of America, Soil Science Society of America; 2009. pp. 349-385

[104] Rochester IJ. Estimating nitrous oxide emissions from flood irrigated alkaline grey clays. Australian Journal of Soil Research. 2003;**41**:197-206

[105] Peoples MB, Boyer EW, Goulding KWT, Heffer P, Ochwoh VA, Vanlauwe B, et al. Pathways of nitrogen

loss and their impacts on human health and the environment. In: Mosier AR, Syers KJ, Freney JR, editors. Agriculture and the Nitrogen Cycle, the Scientific Committee on Problems of the Environment (SCOPE). Covelo, Washington, D.C, USA: Island Press; 2004. pp. 53-69

[106] Soussana JF, Tallec T, Blanfort V. Mitigating the greenhouse gas balance of ruminant production systems through carbon sequestration in grasslands. Animal. 2010;**4**:334-350

[107] Jeuffroy MH, Baranger E, Carrouée B, Chezelles ED, Gosme M, Hénault C. Nitrous oxide emissions from crop rotations including wheat, oilseed rape and dry peas. Biogeosciences. 2013;**10**:1787-1797

[108] González AD, Frostell B, Carlsson-Kanyama A. Protein efficiency per unit energy and per unit greenhouse gas emissions: Potential contribution of diet choices to climate change mitigation. Food Policy. 2011;**36**:562-570

[109] Hartman GL, West ED, Herman TK. Crops that feed the world 2. Soybean-worldwide production, use, and constraints caused by pathogens and pests. Food Security. 2011;**3**:5-17

[110] Idrisa YL, Ogunbameru BO, Amaza PS. Influence of farmers' socio-economic and technology characteristics on soybean seeds technology adoption in southern Borno state, Nigeria. African Journal of Agricultural Research. 2010;**5**(12):1394-1398. DOI: 10.5897/AJAR09.734

[111] Osmani MAG, Islam MK, Ghosh BC, Hossain ME. Commercialization of smallholder farmers and its welfare outcomes: Evidence from Durgapur Upazila of Rajshahi District, Bangladesh. Journal of World Economic Research. 2014;**3**(6):119-126

Section 2

Physiology and Biochemistry of Soybean Plants

Chapter 2

Effects of Application of Various forms of Nitrogen on the Growth of Soybean Nodules and Roots Related to the Carbon and Nitrogen Metabolism

Takuji Ohyama, Sayuri Tanabata, Norikuni Ohtake, Takashi Sato, Kuni Sueyoshi, Yoshihiko Takahashi, Shinji Ishikawa, Yuki Ono, Natsumi Yamashita and Akinori Saito

Abstract

Soybean plants require a large amount of nitrogen either from nitrogen fixation in nodules or nitrogen absorption from roots. It is known that nitrate, a major inorganic nitrogen compound in upland soils, represses nodule growth and nitrogen fixation. Rapid and reversible inhibition of nodule growth and nitrogen fixation activity was found in the hydroponically cultivated soybeans after changing the nutrient solution with or without nitrate. Isotope tracer analysis revealed that the major cause of this inhibition depended on the changes in the partitioning of photo-assimilate between nodules and roots and was not directly related to the transported N compounds. Transcriptome and metabolome analyses supported that nitrate strongly promotes nitrogen and carbon metabolism in the roots but represses them in the nodules. The application of ammonium, glutamine, or urea also inhibited the nodule growth and nitrogen fixation like nitrate, although the inhibition was lower than that of nitrate. The degree of inhibition was related to the decrease in carbon isotope partitioning into the nodules, rather than the import of nitrogen isotope to nodules. Urea was detected in xylem sap and all parts of soybean, and some urea might be originated from ureide degradation.

Keywords: soybean, nitrogen fixation, nodules, roots, nitrate, ammonium, urea, ureides

1. Introduction

1.1 Characteristics of nitrogen assimilation in soybean

Currently, soybean seed production is increasing worldwide, and the annual production in 2019 was 334 million tons [1]. The percentage composition of soybean seeds is proteins (35%), lipids (19%), carbohydrates (28%), minerals (5%), and water (13%) [2]. Soybean plants originate from East Asia, and Asian people including the Japanese eat various kinds of traditional soy foods such as Tofu, Miso, Shoyu, Natto, etc. [3]. Traditional Eastern soy foods are now gradually accepted by Western people for promoting their health, as well as meat substitutes made from soybean. In addition, soybean seeds are very important for feeding livestock.

Because soybean seeds contain a higher amount of protein than cereals and other legume seeds, soybean requires a large amount of nitrogen for high seed yield [4]. One of soybean seeds production requires 70–90 kgN assimilation, therefore, the world average yield of 2.77 t ha^{-1} in 2019 requires as high as 200–250 kgN ha^{-1}. Soybean plants depend on nitrogen fixation by root nodules (Ndfa), and the nitrogen is absorbed from soil (Ndfs) or fertilizer (Ndff) when applied. The N availability from soil mineralization during the soybean cultivation period varies widely depending on the soil fertility, but it is about 50 kgN ha^{-1} in Japan. Therefore, nitrogen fixation is the main source, and about 60–75% of N was derived from N derived from Ndfa in Niigata, Japan [5]. The root nodule is a symbiotic organ with soil bacteria, rhizobia. **Figure 1A** shows the photograph of nodulated roots of hydroponically cultivated soybean. Soybean nodules can be visible from 8 days after planting and grow and start to fix N_2 around 15–20 DAP, when rhizobia were inoculated to seeds [6]. The cross-section of a soybean nodule is shown in **Figure 1B**. Infected rhizobia live in the red central zone of the nodule due to a high concentration of leghemoglobin, which bind to O_2 supporting respiration and nitrogen fixation by rhizobia [6].

Another characteristic of N assimilation in soybean is that soybean needs about 80% of N after the beginning of flowering, which is quite different from the paddy rice which assimilates only 20% of N is assimilated after heading [5]. Therefore, a long-lasting high nitrogen fixation activity after flowering is essential for high soybean seed yield. However, nitrogen fixation activity tends to decrease during the pod filling stage, due to competition of nutrients mainly photo-assimilate between seeds and nodules. Therefore, supplemental application of N fertilizers may be beneficial to support vigorous shoot growth and photosynthetic activity during the pod filling stage. However, a basal application of chemical N fertilizers often inhibits the nodule growth and nitrogen fixation activity, and they will be lost by nitrate leaching and

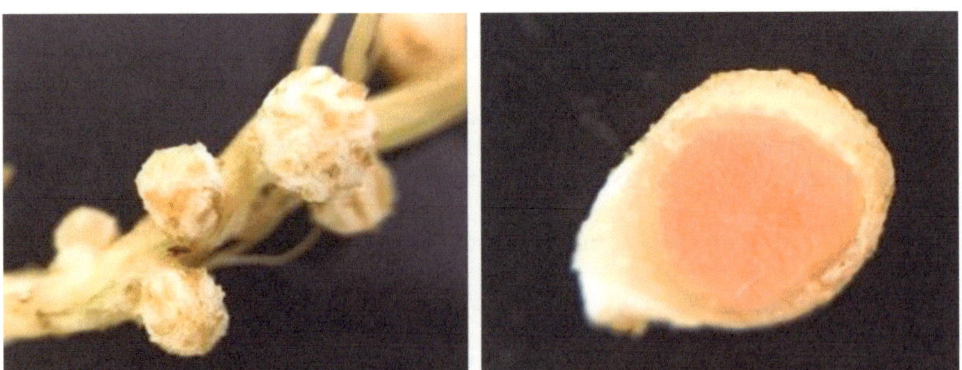

(A) (B)

Figure 1.
Photographs of root nodules of hydroponically cultivated soybean. (A) Root nodules attached to the roots of soybean. (B) Cross-section of a soybean nodule.

denitrification. So, it is necessary to harmonize the N fertilization and N fixation to obtain a constant and high seed yield of soybean [7, 8].

1.2 Nitrogen inhibition on the growth and nitrogen fixation activity of soybean nodules

The inhibitory effects of combined nitrogen, especially nitrate, on nodule formation and nitrogen fixation of legumes have been studied for over 100 years [9]. According to Streeter's review, the responses can be divided into three classes, the number of nodules per root, nitrogenase activity per unit mass of the nodules, and the nodule mass per plant [10]. The effects of nitrate concentration on the magnitude of the three responses are different. A relatively high nitrate concentration is required for the inhibition of nodule number per plant, followed by nitrogenase activity per nodule mass. The concentration effect on nodule mass is more sensitive than nodule number and nitrogen fixation activity, while low levels of nitrate stimulate nodule growth through the promotion of shoot growth.

Concerning the effects of nitrate on nodule growth, there are two different situations, the first is a direct effect or local effect in which nodules are in direct contact with the solution containing nitrate, and the second is an indirect or systemic effect in which nodules are not directly contacted with nitrate and nitrate is absorbed from the distant part of the roots [11]. In the direct effect, when soybean plants were hydroponically cultivated, the addition of 5 mM nitrate in culture solution rapidly stopped the individual nodule growth within one day as well as decreased N_2 fixation activity measured by C_2H_2 reduction activity [12, 13].

Figure 2 shows the effect of 5 mM nitrate supply to the culture solution on the nodule growth of soybean plants. Soybean seeds were inoculated with *Bradyrhizobium diazoefficiens* (strain USDA110), germinated in a vermiculite bed, and the seedlings were cultivated in an N-free culture solution. The inoculated plants which had been cultivated with N-free culture solution were grown with 0 mM or 5 mM nitrate from 11 DAP (days after planting). The nodule growth with 0 mM nitrate grew from 1.23 mm diameter on 11 DAP to 3.00 mm on 19 DAP. On the other hand, the nodule with 5 mM nitrate grew from 1.03 mm on 11 DAP to 1.33 mm on 13 DAP, but the growth was almost completely stopped from 13DAP to 19 DAP. When 5 mM nitrate was removed from the culture solution back to N-free condition, the nodule growth and nitrogen fixation activity were quickly recovered in a day (**Figure 3**). The rapid and reversible inhibition of nodule growth by nitrate was similarly observed in the large size nodules and small nodules [12]. The quick and reversible nitrate inhibition on soybean nodules by nitrate supply was due to the decrease in the photo-assimilate supply from the shoot to nodules and it was conversely increased to the roots by isotope tracer experiments [13]. Imsande earlier reported the short-term exposure of 4 mM NO_3^- in hydroponic solution reversibly inhibited nitrogenase activity and nodule dry weight [14].

As for the indirect or systemic effect of nitrate, Tanaka et al. [15] reported that nitrate supplied to one side of the split root system of soybean did not inhibit the nodule growth and nitrogen fixation activity of the other side of the roots supplied N-free medium. The upper and lower root systems were separated vertically by a two-layered pot system, the concentration and period of nitrate supply from lower roots gave a different effect on the nodule growth in the upper roots [16]. The long-term supply of a high concentration (5 mM) of nitrate from the lower roots inhibited the nodule growth (DW) of the upper roots, but the continuous supply of a low

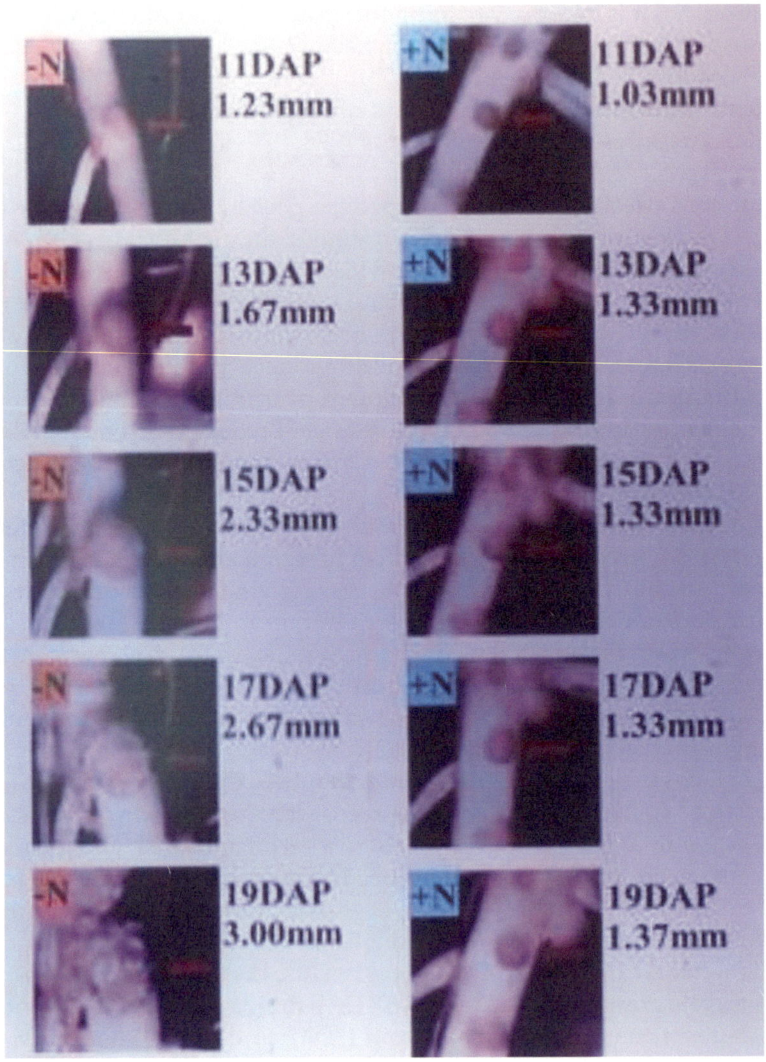

(A) (B)

Figure 2.
Effect of 0 mM (A) or 5 mM (B) nitrate on the growth of nodules of hydroponically cultivated soybean from 11 days to 19 days after planting. (A) Root nodules attached to the roots of soybean cultivated with 0 mM NO_3^-. (B) Root nodules attached to the roots of soybean cultivated with 5 mM NO_3^-. (A) 0 mM NO_3^-, (B) 0 mM NO_3^-.

concentration (1 mM) of nitrate in the lower roots promoted the nodule growth in the upper roots through the increased in shoot growth and photosynthetic activity.

Concerning the chemical forms of N, the inhibitory responses were more sensitive to nitrate than to ammonium, and urea was only slightly inhibitory [17, 18]. The physiological meaning of the different responses to nitrogen compounds is not well understood. In this review, we would like to introduce recent advances in nitrate inhibition to nodule growth and nitrogen fixation activity first, and the effect of nitrate, ammonium, urea, and glutamine on the inhibition of nodule growth was compared.

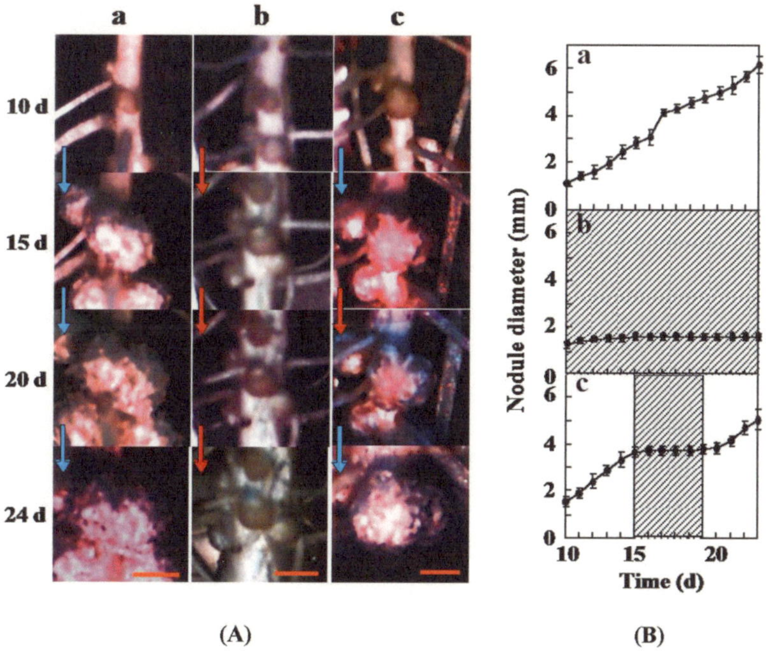

Figure 3.
Growth response of soybean nodules to 0 mM or 5 mM nitrate application in the culture solution. (A) Photographs of nodulated roots after nitrate treatments to 0 mM (blue arrows) or 5 mM nitrate (red arrows) from 10 DAP to 24 DAP. (B) Graphs of the changes in nodule diameter after nitrate treatments from 10 DAP to 24 DAP. (a) 0 mM NO_3^-, (b) 5 mM NO_3^-, and (c) 0,5,0 mM NO_3^-, white background: 0 mM NO_3^-, gray background: 5 mM NO_3^-. From Fujikake et al. [13].

2. Short-term effect of nitrate supply on nodule and root growth of soybean

Saito et al. [19] investigated the short-term responses of nodule growth to the 5 mM nitrate supply at one-hour intervals for 16 h under light and dark conditions (**Figure 4A**). Inoculated soybean plants were cultivated in a photo chamber under 16 h light at 28°C and 8 h dark at 18°C conditions. The nodule size is indicated by the nodule area (mm^2) measured by a time-lapse camera and the nodule area measuring software NODAME [20, 21]. Under light conditions, the nodule growth of 13 DAP plants was constant from the start of light period to the end of the 16 h light. When 5 mM NO_3^- was applied under light conditions, the increase in the nodule area was the same until the initial 2 h and became significantly repressed after 7 h. The increase in the nodule area with 5 mM NO_3^- during the 16 h was about 60% of that with 0 mM NO_3^- under light conditions. This result indicated that nitrate inhibition on nodule growth begins very rapid at a few h after the addition of nitrate to the culture solution. Under dark conditions with 0 mM NO_3^- the nodule growth was more severely depressed (42% of 0 mM NO_3^- under light conditions at 16 h) than that treated with 5 mM NO_3^- under light conditions. The nodule growth was most severely repressed under dark conditions with 5 mM NO_3^- (25% of 0 mM NO_3^- under light conditions at 16 h) among treatments. These results support the former hypothesis that the addition of NO_3^- on the nodulated roots is repressed

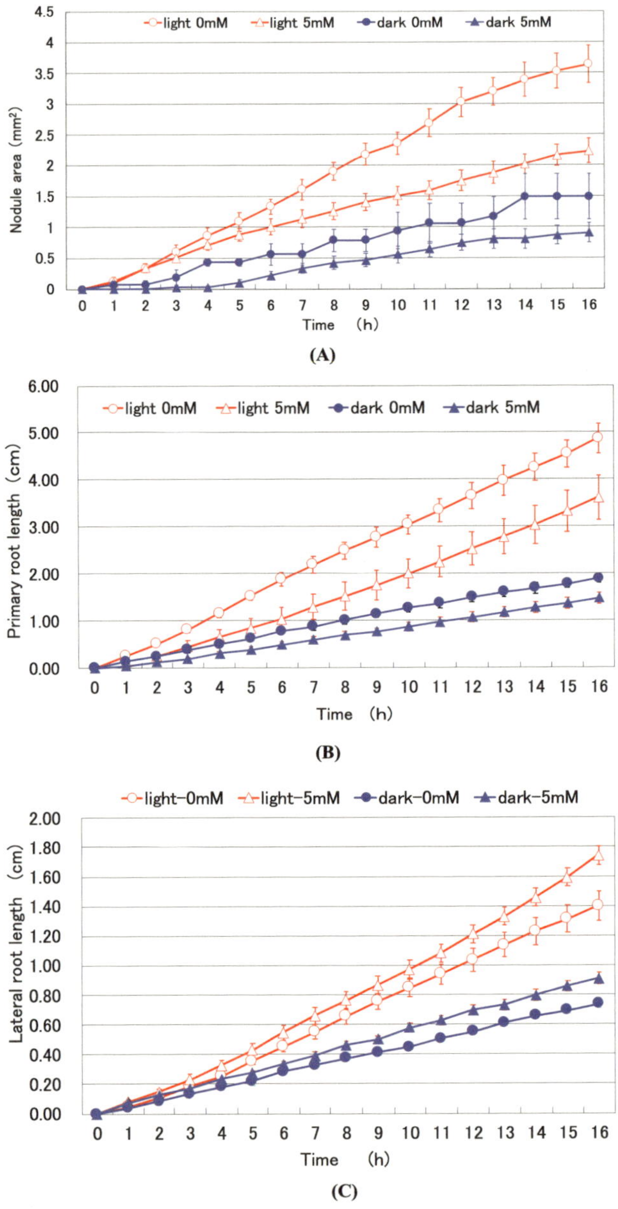

Figure 4.
(A) Increase in nodule area with 0 mM or 5 mM NO_3^- under light or dark conditions, (B) Increase in primary root length with 0 mM or 5 mM NO_3^- under light or dark conditions, and (C) Increase in lateral root length with 0 mM or 5 mM NO_3^- under light or dark conditions. From Saito et al. [19].

mainly through the decrease in photo-assimilate supply to the nodules because dark conditions repressed the nodule growth, and dark plus nitrate additively repressed the nodule growth.

Similar responses are observed in the increase in primary root growth with 0 mM or 5 mM NO_3^- under light or dark conditions (**Figure 4B**). The primary root growth was the fastest with 0 mM NO_3^- under light conditions, and that with 5 mM NO_3^- repressed

(73% at 16 h), 0 mM NO_3^- under dark conditions (40% at 16 h), and 5 mM NO_3^- under dark conditions (29% at 16 h).

Interestingly, the opposite responses to nitrate were observed for the growth of lateral roots (**Figure 4C**). Under light conditions, the increase in the lateral roots with 5 mM NO_3^- was promoted to 121% of the 0 mM under light conditions. The increase in the lateral roots was repressed in 0 mM NO_3^- under dark conditions (51% at 16 h), but 5 mM NO_3^- still increased the lateral root growth under dark conditions compared with 0 mM NO_3^-.

Initial transport of photo-assimilate labeled with $^{11}CO_2$ exposed to the matured leaf showed that ^{11}C was leached to the roots and nodules within 1 h, and the distribution of photo-assimilate was higher in the root parts in contact with 5 mM NO_3^- solution compared with 0 mM NO_3^- solution [13]. Quantitative analysis of photo-assimilate transport was conducted using $^{14}CO_2$ as a tracer, which was supplied to the whole shoot for 2 h, then the distribution of ^{14}C was investigated among the organs of soybean plants supplied with 0 mM or 5 mM NO_3^-. The percentage distribution of ^{14}C in roots and nodules was 5.2 and 9.1% of total fixed ^{14}C in the plants with 0 mM NO_3^-, while those in the roots and nodules changed to 9.1 and 4.3%, respectively [13]. The increase in ^{14}C was mainly in the lateral roots after supplying 5 mM NO_3^-, while ^{14}C distribution in the primary roots was not changed between 0 mM and 5 mM NO_3^- [13].

The inhibitory effect of nitrate on nodule growth was shown to be reversible, and when 5 mM NO_3^- was changed to 0 mM NO_3^-, the nodule growth recovered in a day, when 5 mM NO_3^- treatment continued for 14 days [12]. This means the physiological function of nodules may be maintained under 5 mM NO_3^-, while the nodule growth was almost completely stopped. **Figure 5** shows the 2D-PAGE of soybean nodule extract cultivated with 0 mM (A), and 5 mM NO_3^- (B) treatment from 10 DAP to 34 DAP. The patterns and intensities of the protein spots were similar between the 0 mM and 5 mM NO_3^- treatments after 24 days of treatment. Therefore, the functions of nodule were maintained under 5 mM NO_3^- conditions, although nodule growth and nitrogen fixation activity were strongly repressed. This might suggest that nodule function is not disintegrated in the nodule in direct contact with NO_3^-, whereas carbohydrate deficiency temporarily retard the

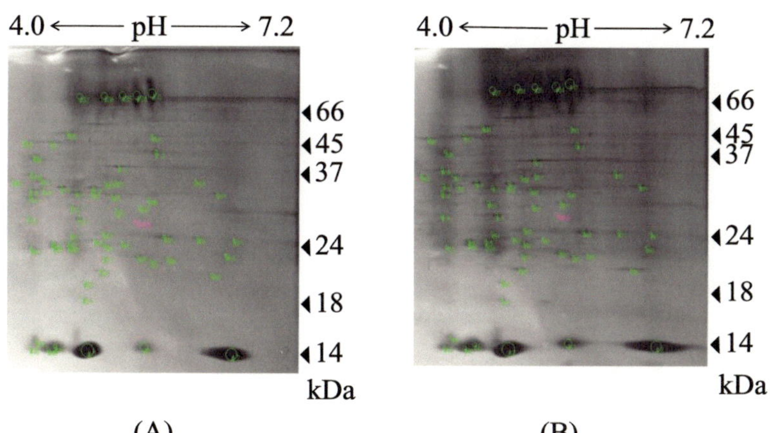

Figure 5.
2D-Page of nodule soluble proteins with 0 mM (A) or 5 mM (B) nitrate. From Saito et al. [19].

nodule growth and nitrogen fixation activity. Therefore, after removal of NO_3^- from the culture solution quickly recovers the nodule growth and nitrogen fixation activity [12, 13].

3. Transcriptome and metabolome analyses of nitrate inhibition

The effect of nitrate supply on the gene expression and metabolite changes in nodules and roots were investigated by transcriptome and metabolome analysis [22]. After supplying 5 mM NO_3^- for 24 h to the nodulated soybean plants at 19 DAP, mRNA was extracted from the nodules and roots, and the cDNAs were hybridized with soybean oligo DNA microarray [22]. The results were compared with control plants supplied with 0 mM NO_3^-. **Figure 6** shows the numbers of probe sets in the roots and nodules, which were up-regulated (4-fold, 2-fold) or down-regulated (4-fold, 2-fold) following 5 mM NO_3^- supply for 24 h compared with the control plants with 0 mM NO_3^-. The number in the 4-fold up-regulated probe sets in roots was 142 and higher than that in the nodules (78). The number of common probe sets 4-fold up-regulated both in roots and nodules was 9. On the other hand, the number of 4-fold down-regulated probe sets in the roots was 92 and lower than that in nodules (116). Similar trends were observed for the 2-fold up-regulation (C), and 2-fold down-regulation (D).

The results showed that the 5 mM NO_3^- supply highly enhanced the gene expression in the roots related to nitrate transporter and metabolisms such as nitrate

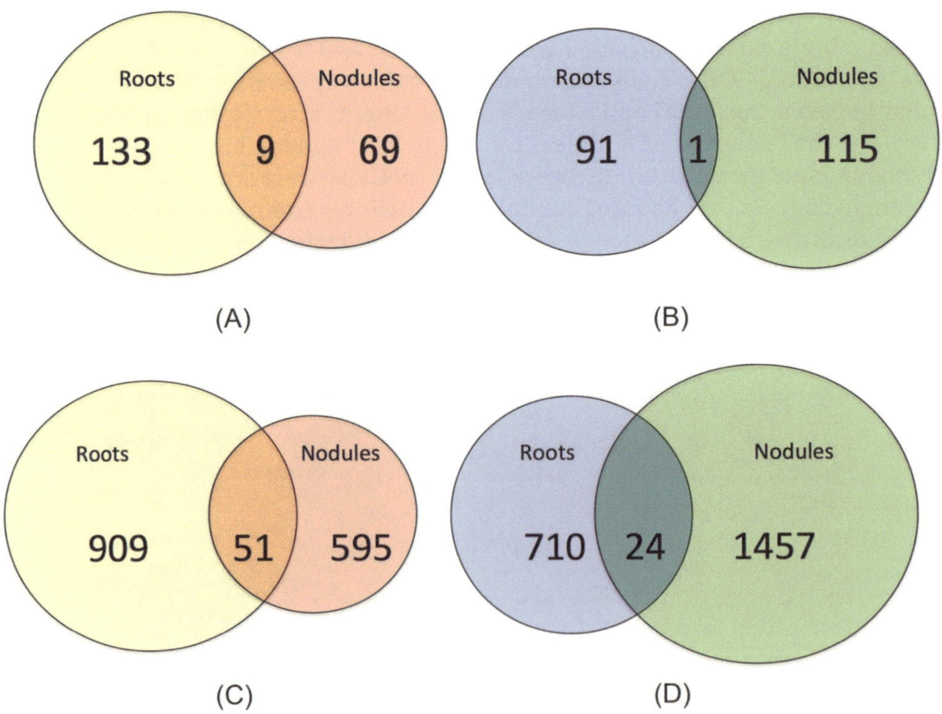

Figure 6.
Numbers of probe sets in the roots and nodules up-regulated or down-regulated by more than 4-fold or 2-fold following the addition of 5 mM nitrate to medium at the level of $p < 0.05$. From Ishikawa et al. [22].

reductase (NR), nitrite reductase (NiR), glutamine synthetase (GS), glutamate synthase (GOGAT), and Asparagine synthetase (AS) (**Figure 7**). The genes for ureide synthesis through purine synthesis and degradation in the roots were slightly increased. The gene expression related to nitrate transport and metabolism in nodules was also slightly promoted after NO_3^- supply, but not so high as in the roots. **Figure 8** shows the ratios of gene expression related to C metabolism, glycolysis, and TCA cycle in roots (left) and nodules (right). The gene expression of the enzymes in glycolysis and TCA cycle in the roots were promoted by NO_3^-, while those in the nodules were mostly depressed.

Metabolome analysis was conducted using the same plants for the transcriptome analysis as above. The ratios of the concentrations of nitrogen compounds, phosphorus compounds, and organic acids concentrations in roots and nodules after 5 mM NO_3^- treatment were compared with 0 mM NO_3^- treatment as shown in **Figure 9**. Most of the nitrogen compounds in the roots were increased by NO_3^- treatment, especially allantoic acid (3.1-fold), glutamine (2.4-fold), and asparagine (2.1-fold). On the other hand, asparagine and N-acetylglucosamine in nodules were increased, but some amino acids were decreased such as alanine. For the phosphorous compounds, 5 mM NO_3^- treatment increased ATP concentration in the roots (3.9-fold) but decreased in nodules (0.7-fold). Similarly, most of the ratios of phosphorous compounds show over 1-fold in the roots, but less than 1-fold in nodules. The same was true in organic acids in roots and nodules.

Both transcriptome and metabolome analysis indicated that NO_3^- treatment caused the promotion of C and N metabolism in the roots, while depressed C metabolism in the nodules. This agrees with the changes in the photo-assimilate supply from nodules to the roots [13].

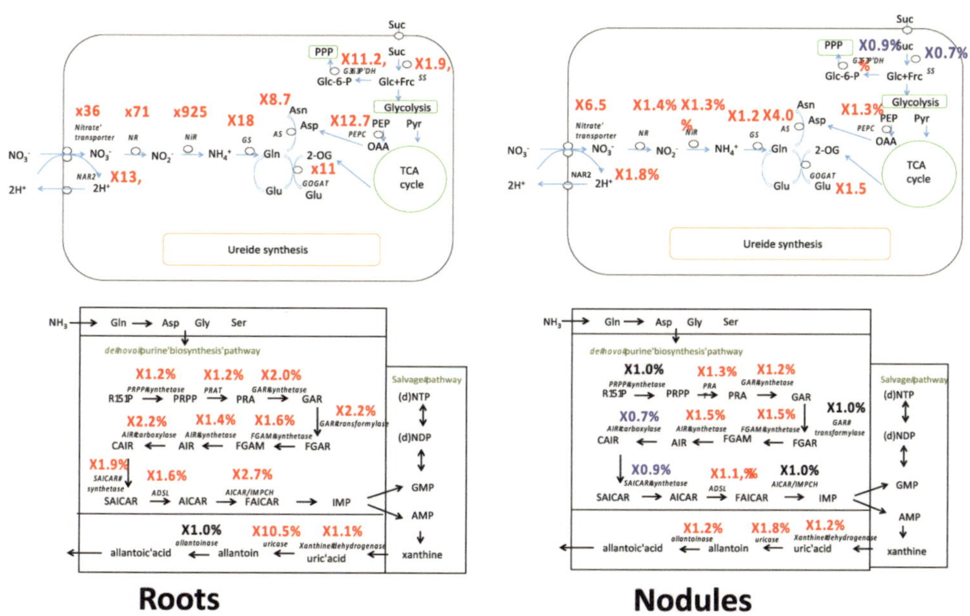

Figure 7.
Comparison of the ratios of gene expression related to N metabolism in roots or root nodules with 5 mM or 0 mM nitrate (+N/−N). From Ishikawa et al. [22].

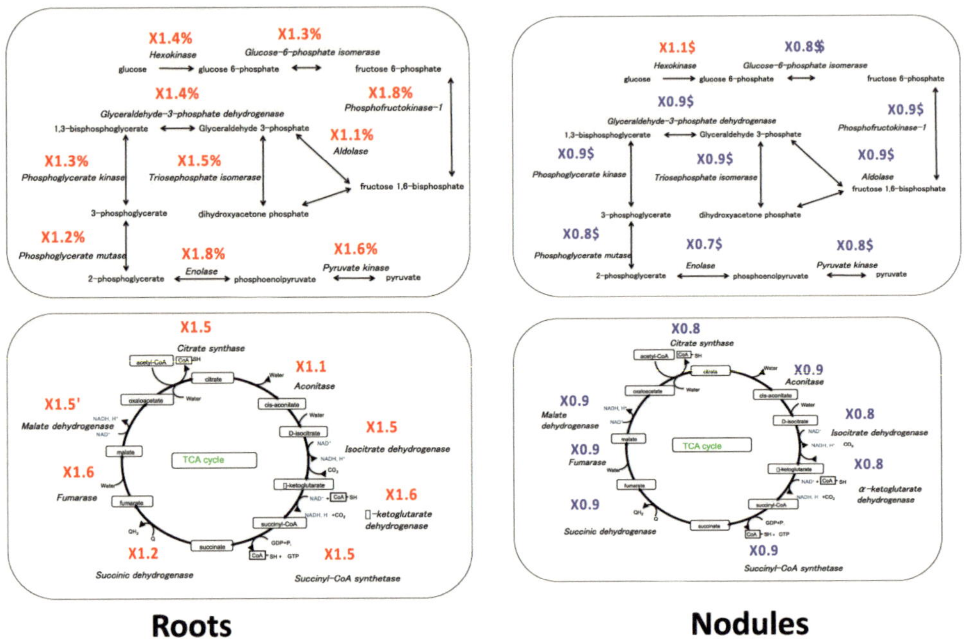

Figure 8.
Ratios of gene expression related to C metabolism (Glycolysis and TCA cycle in roots and nodules with 5 mM or 0 mM nitrate (+N/−N). From Ishikawa et al. [22].

Table 1A Nitrogen compounds

Nitrogen compound	Roots	Nodules
Gln	2.4	1.6
Asn	2.1	3.3
Ala	2.0	0.9
Glu	1.8	1.1
Asp	1.8	1.4
Homoserine	2.8	0.8
2-Aminobutyric acid	1.9	0.8
Allantoin	1.0	1.9
Allantoic acid	3.1	1.4
N-Acetylglucosamine	1.1	3.6

Ratio (+N/−N): <3; 2–3; 1–2; 1<

Table 1B Phosphorous compounds

P- compound	Roots	Nodules
ATP	3.9	0.7
ADP	2.9	0.9
GMP	2.8	N.A.
CDP	2.4	0.8
UDP	2.1	0.9
CTP	2.1	N.A.
AMP	1.6	1.3
UMP	1.6	0.8
CMP	1.5	1.0
Sedoheptulose 7-phosphate	2.5	0.7
Ribulose 5-phosphate	2.2	0.8
NAD^+	2.2	0.9
Ribose 5-phosphate	2.0	0.8
6-Phosphogluconic acid	1.9	0.8
UDP-glucuronic acid	1.4	0.7
Phosphoenolpyruvic acid	1.3	0.9

Ratio (+N/−N): <3; 2–3; 1–2; 1<

Table 1C Organic acid and others

Organic acid	Roots	Nodules
2-Isopropylmalic acid	2.4	1.0
2-Aminoadipic acid	2.1	1.1
N-Acetylglutamic acid	1.9	1.1
Citramalic acid	1.9	0.9
Fumaric acid	1.7	1.0
Succinic acid	1.3	0.9
Malic acid	0.8	0.9

Other	Roots	Nodules
Spermidine	1.8	1.2
O-Acetylserine	1.7	N.A.

Ratio (+N/−N): <3; 2–3; 1–2; 1<

Figure 9.
Ratios of N and P metabolite concentrations in roots and nodules with 5 mM or 0 mM nitrate (+N/−N). From Ishikawa et al. [22].

4. Effects of nitrate, ammonium, urea, and glutamine on nodule growth and nitrogen fixation activity

In addition to the inhibitory effect of nitrate on nodule growth and nitrogenase activity, the repressive effect of ammonium is also reported [14], whereas urea did not reduce nodule dry weight and nitrogen fixation activity in hydroponically grown soybean [17, 18]. The difference in inhibitory effects by the forms of N compounds is not fully understood yet. Yamashita et al. [23] investigated the effect of various forms of nitrogen; nitrate, ammonium, urea, and glutamine on the quick and reversible inhibition of nodule growth and nitrogen fixation activity of soybean plants.

Soybean plants were cultivated in a nitrogen-free nutrient solution in a glass bottle, and nitrate, ammonium, glutamine, or urea (1 mM-N) were supplied from 12 days after planting (DAP) to 17 DAP. The increase in individual nodule growth expressed by nodule volume (mm^3) was shown in **Figure 10**. The inhibitory effects on the nodule growth were severe in nitrate and ammonium treatments, and those by urea and glutamine were smaller than in nitrate and ammonium. On 17 DAP, the increase in nodule dry volume was low in nitrate and ammonium medium in urea and glutamine compared with the control with N-free solution. After N-based solutions were replaced by N-free solutions from 17 DAP to 24 DAP, the nodule growth in all treatments showed the recoveries. These results indicated that the rapid and reversible inhibition is not only by nitrate but also by ammonium, urea, and glutamine. **Figure 11A** shows the dry weight of nodules per plant on 17 DAP. The application of

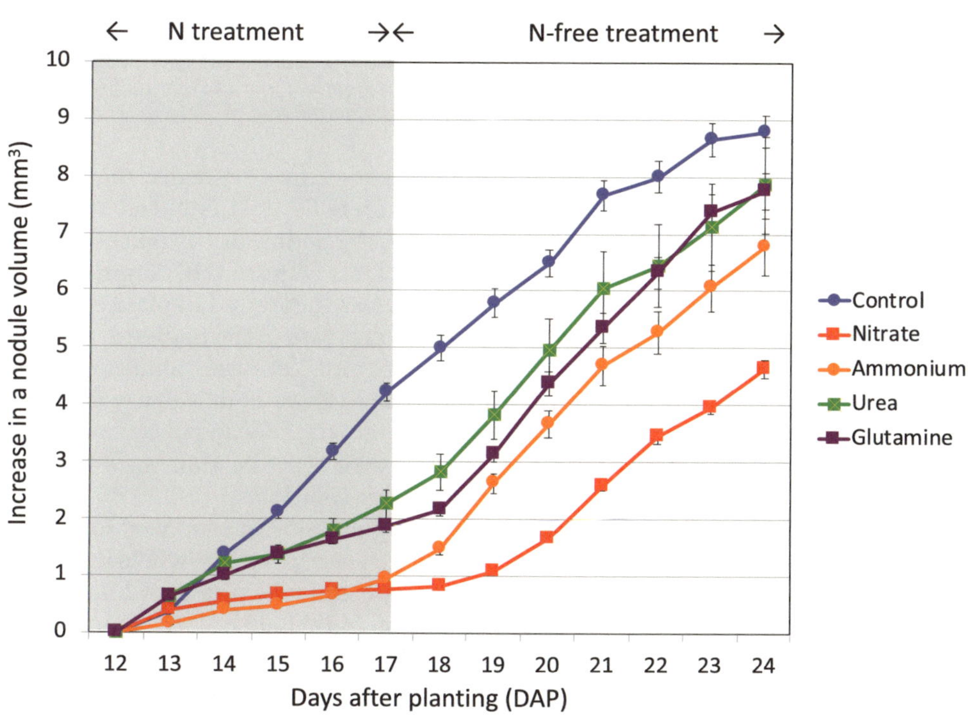

Figure 10.
Changes in nodule volume from 12 to 24 DAP for treatments with control (N-free), nitrate, ammonium, urea, or glutamine from 12 to 17 DAP, thereafter cultivated with a N-free culture solution (Experiment 2). Shaded background indicates N treatment period, and white background indicates cultivation with N-free medium. Average and standard error are shown (n = 5). From Yamashita et al. [23].

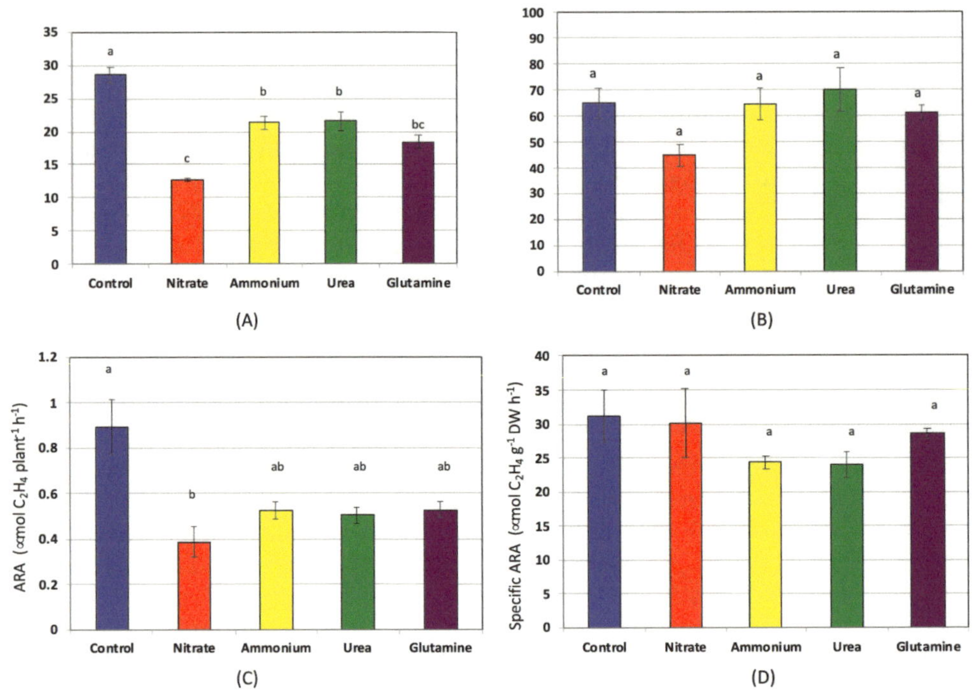

Figure 11.
Comparison of the dry weight of nodules at 17 DAP (A) and at 24 DAP (B), and acetylene reduction activity per plant (C), and specific acetylene reduction activity per nodule g dry weight (D) on 17 DAP of soybean plants supplied with control (N-free), nitrate, ammonium, urea, or glutamine from 12 to 17 DAP, followed by supplying N-free solution from 17 DAP to 24 DAP. Averages and standard errors are shown (n = 4). Different letters above the column indicate significant differences at <0.05 by Tukey's test. From Yamashita et al. [23].

1 mM-N of nitrate, ammonium, urea, and glutamine for 5 days depressed the nodule weight to 45, 75, 76, and 65% of the control nodules on 17 DAP. After N-free solution was supplied for 6 days following 5-day N treatment, the nodule dry weight on 24 DAP increased in all N treatments similar to the control plants (**Figure 11B**). **Figure 11C** shows the ARA per a single plant, and the activities were depressed in nitrate, ammonium, urea, and glutamine treatment like the decrease in the nodule dry weight (**Figure 11A**). The specific ARA per g DW of nodules on 17DAP were almost the same (**Figure 11D**), so the decrease in ARA per plant is due to lower nodule dry weight and not by a decline in the specific ARA. In this experiment, the DW and N concentrations of roots, stems, and leaves increased after 5 days of N application on 24 DAP compared with the control plants supplied with N-free solution.

^{15}N-labeled 1 mM-N nitrate, ammonium, urea, or glutamine was supplied for 3 days from 21 to 24 DAP, in which the solutions were renewed every day. The whole shoot was enclosed in a plastic bag on 23 DAP, and $^{13}CO_2$ was exposed to the plants for 1 h. Then the plants supplied with ^{15}N and ^{13}C were harvested on 24 DAP, 26 h after $^{13}CO_2$ exposure. The labeling of ^{15}N and ^{13}C in each organ was determined by Mass spectrometry. The amount of ^{15}N in nodules (**Figure 12A**) was 0.14 mg from nitrate-^{15}N, 0.26 mg from ammonium-^{15}N, 0.14 mg from urea-^{15}N, and 0.29 mg from glutamine-^{15}N. It is interesting to note that the amount of ^{15}N in nodules is not related to the decrease in nodule DW (**Figure 11A**) and ARA (**Figure 11C**). **Figure 12B** shows the amount of ^{13}C in nodules, and it was 85 mg in control, 27 mg in nitrate, 34 mg in ammonium, 54 mg in urea, and 34 mg in glutamine treatment. The decrease in the amount of ^{13}C in nodules was similar

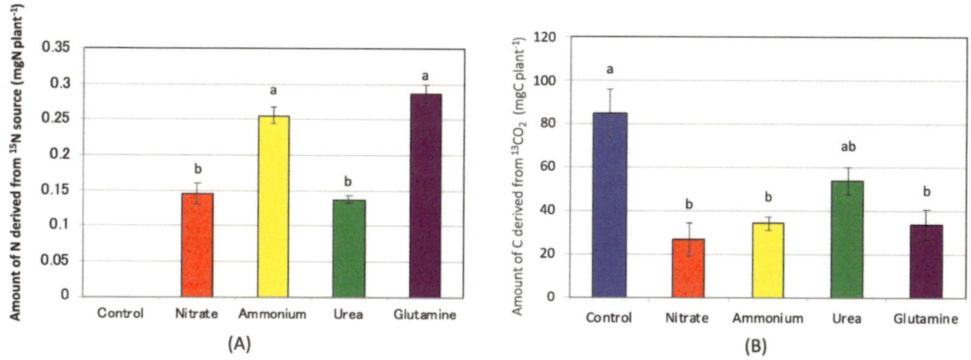

Figure 12.
(A) Amounts of N derived from ^{15}N-labeled source in each tissue of soybean plants on 15 DAP supplied for 3 days from 21 to 24 DAP. (B) amounts of C derived from ^{13}C-labeled CO_2 in each tissue of soybean plants on 24 DAP supplied for 1 hour at 23 DAP. From Yamashita et al. [23].

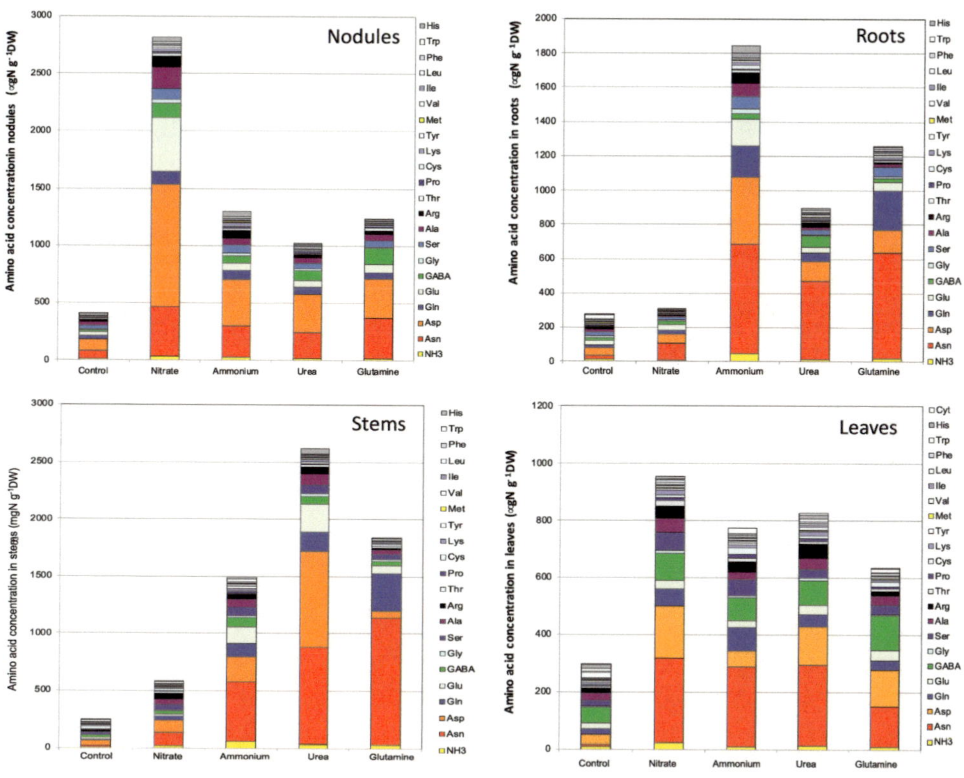

Figure 13.
Free amino acid concentrations in each tissue of soybean plants on 24 DAP supplied with various N compounds from 21 to 24 DAP. From Yamashita et al. [23].

to the nodule DW (**Figure 11A**) and ARA (**Figure 11C**) among treatments. This result supports the hypothesis that the depression of nodule growth and nitrogen fixation activity is related to the decline in photoassimilates partitioning but not N supply.

Changes in the free amino acid concentrations in nodules, roots, stems, and leaves were shown in **Figure 13**. The nitrate treatment significantly increased the concentration of the amino acids, especially Asp, Asn, and Glu in the nodules.

The application of ammonia, urea, or glutamine also increased the concentrations of Asp and Asn in nodules compared with control plants although the increases were not higher than nitrate treatment. On the other hand, the amino acid concentrations in the roots show that ammonium treatment remarkably increased the Asn and Asp concentrations, the urea and glutamine treatments also increased the Asn concentration in the roots, but the increase in Asn in the roots treated with nitrate was relatively low. The increase in the Asn and Asp concentrations was the highest in the stems of plants supplied with urea, followed by glutamine, ammonium, and nitrate treatments. The increase in Asn and Asp was observed from nitrate, ammonium, urea, and glutamine treatments.

Figure 14A shows the effects of nitrogen compounds on total root length on 34 DAP after long-term nitrogen treatment for 2 weeks. The application of nitrate promoted the total root length by over 2-fold compared with the control plants. On the other hand, the application of ammonium inhibited the root length by only a half of the control. The application of urea and glutamine slightly increased the total root length. Similar trends were observed for the root dry weight (**Figure 14B**). **Figure 14C** and **D** show the increase in primary root and lateral roots for the first week of N application from 20 to 27 DAP, and the second week from 27 to 34 DAP. By NO_3^- application, the growth of the lateral roots was promoted, but the primary root was not. On the other hand, the inhibitory effects of NH_4^+ were evident both for the

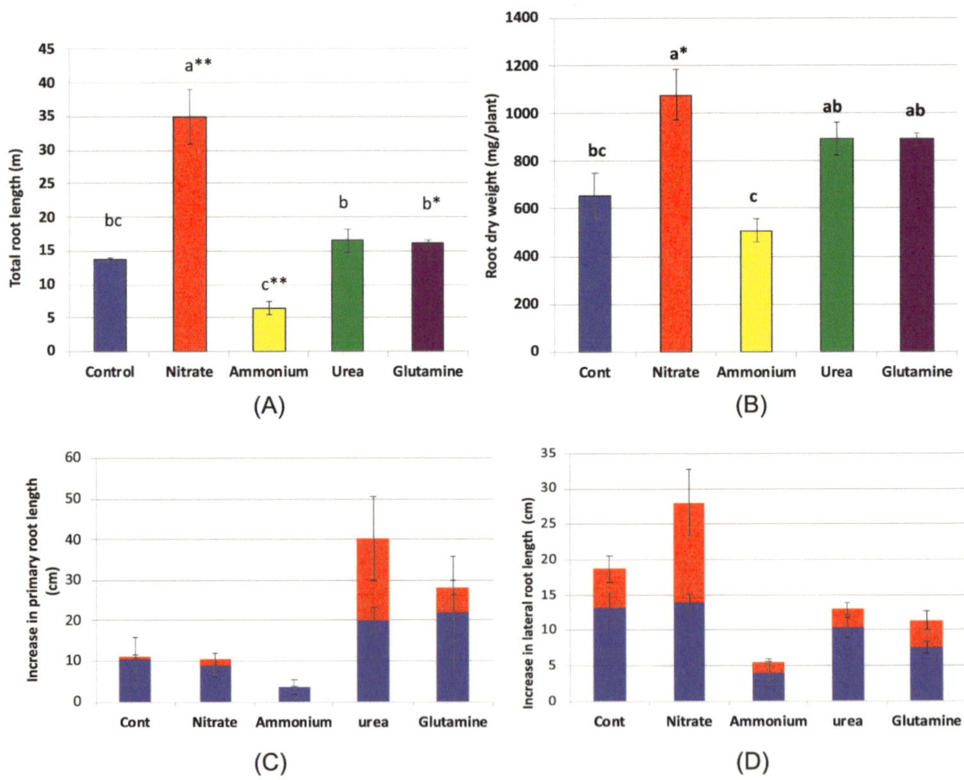

Figure 14.
Total root length (A) and the dry weight of roots (B) on 34 DAP after two weeks of the treatments with various forms of N compounds, and the increase in the primary root length (C), and lateral root length (D) for the first week (blue bar) and the second week (red bar). (A) Total root length, (B) Root dry weight, (C) Increase in primary root length, and (D) Increase in lateral root length. From Yamashita et al. [23].

primary root and the lateral roots. The promotive effects on the length of the primary root and not on the lateral root length were observed by urea and glutamine applications like nitrate.

5. The N composition transported in xylem sap of soybean plants cultivated with nitrate, ammonium, urea

Soybean seeds were inoculated with *Bradyrhizobium diazoefficiens* (strain USDA110), and the seedlings were cultivated in an N-free culture solution until 20 DAP were treated with 5 mM-N of nitrate, ammonium, urea for 3 days until 23 DAP [24]. Control plants were cultivated continuously with N-free solution from 20 to 23 DAP. Another group of soybean plants was not inoculated, and seedlings were cultivated with 5 mM NO_3^- from 5 to 23 DAP. The xylem sap was collected for 1 h from the cut basal part of the stem on 23 DAP. **Figure 15** shows the concentration of N compounds in xylem sap of soybean plants treated with 3 days of nitrate, ammonium, or urea. Non-nodulated plants were cultivated with NO_3^- due to the lack of nodules.

The concentration of ureides (sum of allantoin and allantoic acid) was the highest in control plants depending on nitrogen fixation, followed by ammonium, and urea treatments. The concentration of ureides was very low in nodulated plants treated with nitrate. A small amount of ureides was present in non-nodulated plants. This result is in accordance with the effect of nitrogen compounds on nitrogen fixation activity because most of the ureides in xylem sap originated from fixed nitrogen

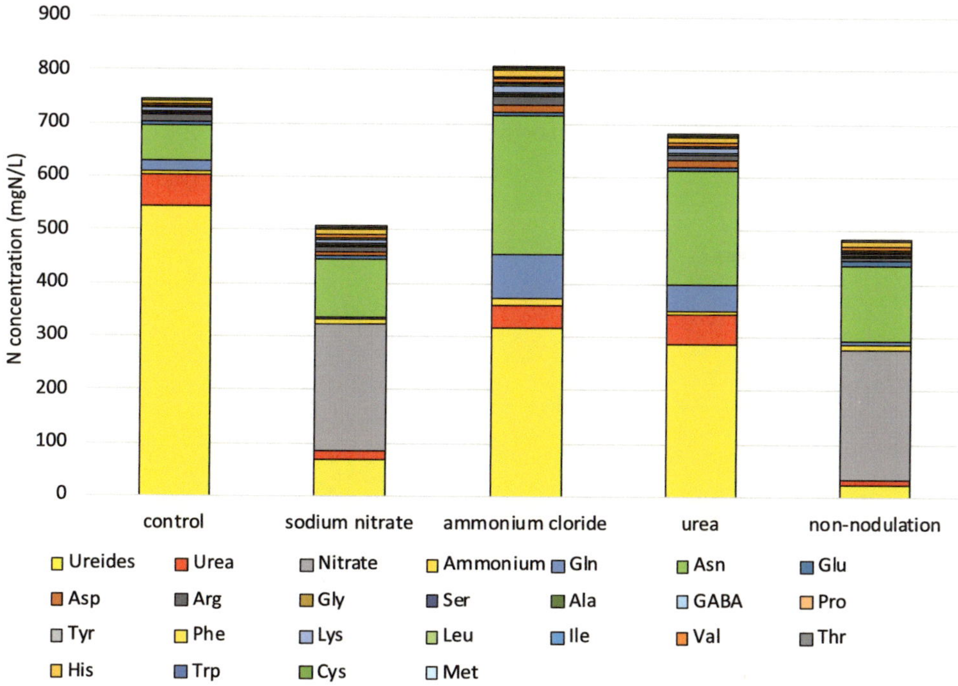

Figure 15.
Nitrogen concentrations of N compounds in xylem sap of nodulated soybean plants treated with control (N-free), nitrate, ammonium, urea, and non-nodulated plants with nitrate. From Ono et al. [24].

in nodules, although a small amount is produced in the roots. Nitrate was detected only in the xylem sap of soybean treated with nitrate, and the concentration of nitrate accounted for about 50% of total N in xylem sap either in nodulated or non-nodulated plants supplied with NO_3^-. The asparagine concentration was the lowest in control plants totally depended on nitrogen fixation, and it was higher in ammonium and urea treatments. The concentration of glutamine was also higher in ammonium and urea treatments, but very low in nitrate treatment. It is interesting that urea is always present in xylem sap and it was higher in control, ammonium, and urea treatments.

The concentrations of ureides and urea or arginine and urea in xylem sap were plotted in **Figure 16**. The concentration of urea was positively correlated with ureides concentration (**Figure 16A**), but the correlation was not observed between urea and arginine in xylem sap (**Figure 16B**) which is the alternative precursor of urea production [25]. Appreciable amounts of urea were present in all the organs for all treatments, and the positive correlations between urea and ureides were observed in the nodules, roots, stems, and leaves. This may indicate that some urea originated from ureides in soybean plants, especially in the roots [24].

The ureides, allantoin, and allantoate, are universal metabolites in all organisms including plants, animals, and microorganisms generated by the degradation of futile purines. Soybean plants transport the fixed nitrogen in the nodules mainly in the form of ureides (ca. 80–90% of total N) supplemented with amides and amino acids [26, 27]. On the other hand, nitrate and asparagine are the principal forms of N transport in the xylem sap of the non-nodulated soybean plants [26]. A small percentage of N about 10% was transported in the form of ureides from the non-nodulated roots, which means some ureides can be synthesized in the roots as well as nodules. The concentrations of ureides, nitrate, and amide N transported through xylem sap could be used to evaluate the percentage dependence of N derived from nitrogen fixation [27].

Figure 17 shows a model of ureide synthesis in the nodules and ureide degradation in soybean. The fixed ammonia from N_2 in the bacteroid, a symbiotic state of

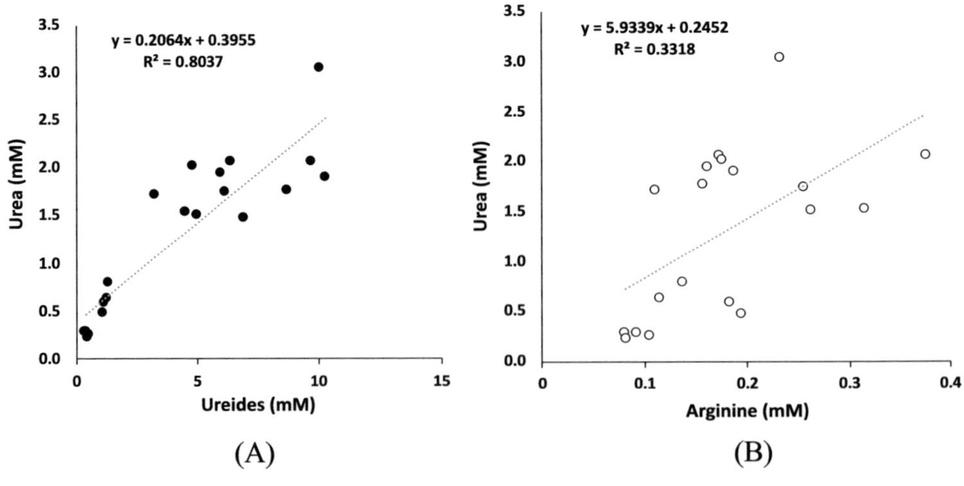

Figure 16.
Correlations between the concentrations of ureides and urea (A), and arginine and urea (B) in the soybean xylem sap. (A) Ureide vs. Urea (B) Arginine vs. Urea. From Ono et al. [24].

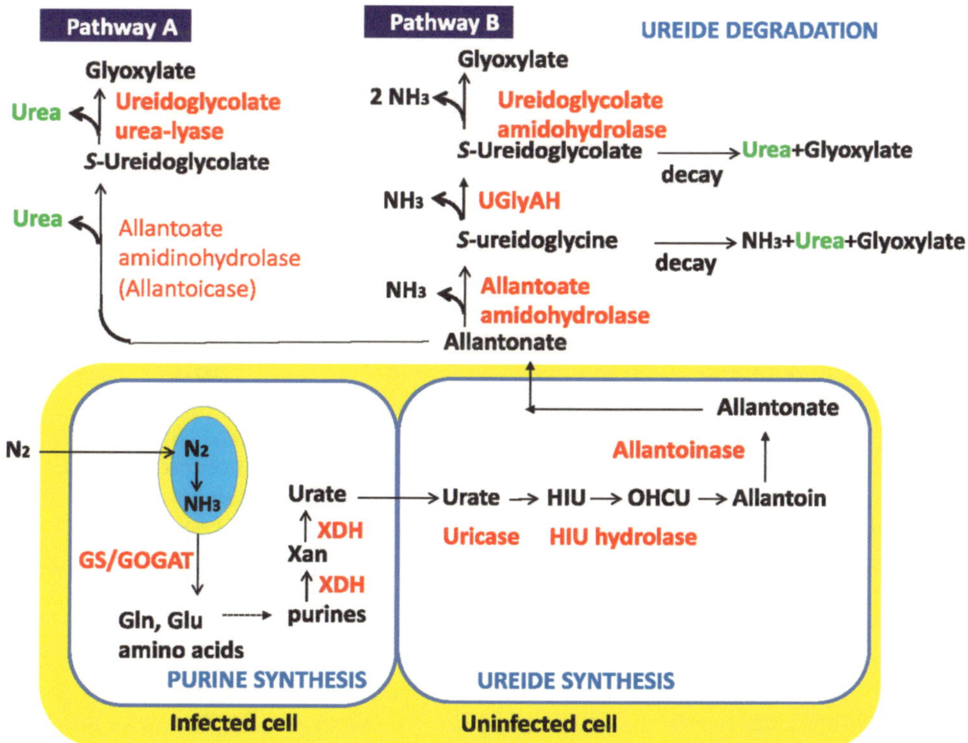

Figure 17.
A model of ureide synthesis in the nodules and ureide degradation in soybean. GS: glutamine synthetase; GOGAT: glutamate synthase; XDH: xanthine dehydrogenase; Xan: xanthine; HIU: hydroxyisourate; OHCU: 2-oxo-4-hydroxy-4-carboxy-5-ureidoimidazoline; UGlyAH: ureidoglycine aminohydrolase. From Ono et al. [24].

rhizobia, is rapidly excreted to the cytosol of the infected cells, then the ammonium is assimilated by glutamine synthetase (GS)/glutamate synthase (GOGAT) pathway to glutamine [28–31]. *De novo* synthesis of purine occurs in the infected cells, and urate is transported to the adjacent uninfected cells and hydrolyzed to allantoin. Some allantoin is further decomposed to allantoate in nodules, then allantoate and allantoin are transported through the xylem. There are two purine degradation pathways in microorganisms, allanotate amidinohydrolase (Pathway A in **Figure 17**) and allantoate amidohydrolase (Pathway B in **Figure 17**). In pathway A, allantoate is hydrolyzed and produces two molecules of urea and one molecule of glyoxylate. On the other hand, in pathway B-left no urea is released but 4 ammonium are sequentially released with one molecule of glyoxylate. Studies of ureide degradation in soybean have been done in the leaves, the sink organ of ureides. Shelp and Ireland [32] reported that allantoate degradation is via pathway A, however, Winkler et al. [33, 34] reported that both the leaf extracts and intact leaves of soybean directly liberate ammonium without releasing urea (Pathway B-left). Werner et al. suggested that allantoate degradation intermediates, ureidoglycine and uredoglycolate are non-enzymatically decayed and release urea and glyoxylate (Pathway B-right). In our results relatively high concentrations of urea were presented in xylem sap and all the parts of soybean, and the concentration was correlated with ureide concentration, suggesting that some part of urea may be derived from ureide degradation, especially in the roots [24].

6. Conclusion

Rapid and reversible repression of nodule growth and nitrogen fixation activity of nodulated soybean were observed when nitrate was supplied in the culture solution. This may be caused by the decrease in the photo-assimilate partitioning to the nodules and not by the transport of N compounds from applied N in the solution. Transcriptome and metabolome analysis supported the above hypothesis. Conversely, the C and N metabolism in the roots was promoted by the application of nitrate. A similar rapid and reversible repression of nodule growth and nitrogen fixation activity was also observed when ammonium, urea, or glutamine was supplied as same as nitrate, however, the inhibitory effect was stronger in nitrate compared with ammonium, urea, or glutamine. Urea was detected in xylem sap and all parts of soybean, and some of the urea originated from ureide degradation.

The plant shoots and roots exchange C and N through the xylem and phloem transport systems [35]. The C and N metabolism is regulated by complex mechanisms to optimize plant organ development and growth. So, understanding the CN metabolism can be related to the agricultural crop production, and maintenance of the agroecosystem.

Acknowledgements

This research was partially supported by the Grants-in-Aid for Scientific Research (No. 18380049, No. 26292036) from Japanese Society for the Promotion of Science.

Conflict of interest

The authors declare no conflict of interest.

Author details

Takuji Ohyama[1]*, Sayuri Tanabata[2], Norikuni Ohtake[3], Takashi Sato[4], Kuni Sueyoshi[3], Yoshihiko Takahashi[3], Shinji Ishikawa[3], Yuki Ono[3], Natsumi Yamashita[3] and Akinori Saito[3]

1 Department of Agricultural Chemistry, Tokyo University of Agriculture, Tokyo, Japan

2 College of Agriculture, Ibaraki University, Mito, Japan

3 Graduate School of Science and Technology, Niigata University, Niigata, Japan

4 Faculty of Bioresource Sciences, Akita Prefectural University, Akita, Japan

*Address all correspondence to: to206474@nodai.ac.jp

References

[1] FAOSTAT. Available from: https://www.fao.org/faostat/en/#data

[2] Ohyama T. The role of legume-rhizobium symbiosis in sustainable agriculture. In: Sulieman S, Tran LP, editors. Legume Nitrogen Fixation in Soils with Low Phosphorus Availability. Berlin: Springer; 2017. pp. 1-20. DOI: 10.1007/978-3-319-55729-8_1

[3] Ohyama T. Nutrition of soybean seeds. In: Ohyama et al., editors. Traditional and Modern Japanese Soy Foods-Manufacturing, Nutrition and Cuisine of a Variety of Soy Foods for Health and Joy of Taste. New York: NOVA Publishers; 2013. pp. 1-10

[4] Ohyama T et al. Soybean seed production and nitrogen nutrition. In: Board JE, editor. A Comprehensive Survey of International Soybean Research-Genetics, Physiology, Agronomy and Nitrogen Relationships. London: InTech; 2013. pp. 115-157

[5] Ohyama T et al. Role of nitrogen on growth and seed yield of soybean and a new fertilization technique to promote nitrogen fixation and seed yield. In: Kasai M, editor. Soybean, The Basis of Yield, Biomass and Productivity. London: InTech; 2017. pp. 153-185

[6] Sato T, Onoma N, Fujikake H, Ohtake N, Sueyoshi K, Ohyama T. Changes in four leghemoglobin components in nodules of hypernodulating soybean (*Glycine max* [L] Merr.) mutant and its parent in the early nodule developmental stage. Plant and Soil. 2001;**237**:129-135

[7] Harper JE. Soil and symbiotic nitrogen requirements for optimum soybean production. Crop Science. 1974;**14**(2):255

[8] Rabie RK, Kumazawa K. Effect of nitrate application and shade treatment on the nitrogen fixation and yield of soybean plant. Soil Science & Plant Nutrition. 1979;**25**(4):467

[9] Fred EB, Graul EJ. The effect of soluble nitrogenous salts on nodule formation. Journal of American Society of Agronomy. 1916;**8**(5):316-328

[10] Streeter J. Inhibition of legume nodule formation and N_2 fixation by nitrate. CRC Critical Reviews in Plant Sciences. 1988;**7**(1):1-23

[11] Ohyama T et al. Effect of nitrate on nodulation and nitrogen fixation of soybean. In: El-Shemy HA, editor. Soybean, Physiology and Biochemistry. London: InTech; 2011. pp. 333-364

[12] Fujikake H, Yashima H, Sato T, Ohtake N, Sueyoshi K, Ohyama T. Rapid and reversible nitrate inhibition of nodule growth and N_2 fixation activity in soybean (*Glycine max* (L.) Merr.). Soil Science & Plant Nutrition. 2002;**48**(2):211-217

[13] Fujikake H, Yamazaki A, Ohtake N, et al. Quick and reversible inhibition of soybean root nodule growth by nitrate involves a decrease in sucrose supply to nodules. Journal of Experimental Botany. 2003;**54**(386):1379-1388

[14] Imsande J. Inhibition of nodule development in soybean by nitrate or reduced nitrogen. Journal of Experimental Botany. 1986;**37**(3):348-355

[15] Tanaka A, Fujita K, Terasawa H. Growth and dinitrogen fixation of soybean root system affected by partial exposure to nitrate. Soil Science & Plant Nutrition. 1985;**31**(4):637-645

[16] Yashima H et al. Long-term effect of nitrate application from lower part of roots on nodulation and N_2 fixation in upper part of roots of soybean (*Glycine max* (L.) Merr.) in two-layered pot experiment. Soil Science & Plant Nutrition. 2005;**51**(7):981-990

[17] Vigue JT, Harper JE, Hageman RH, Peters DB. Nodulation of soybeans grown hydroponically on urea. Crop Science. 1977;**17**(1):169-172

[18] Ohyama T, Nicholas JC, Harper JE. Assimilation of $^{15}N_2$ and $^{15}NO_3^-$ by partially nitrate tolerant nodulation mutants of soybean. Journal of Experimental Botany. 1993;**44**(269):1739-1747

[19] Saito A et al. Effect of nitrate on nodule and root growth of soybean (Glycine max (L.) Merr.). International Journal of Molecular Sciences. 2014:4464-4480. DOI: 10.3390/ijms15034464

[20] Available from: https://www.google.com/search?client=firefox-b-d&q=NODAME

[21] Tanabata S, Tanabata T, Saito A, et al. Computational image analysis method for measuring size of nodule growth in soybean. Japanese Society of Soil Science and Plant Nutrition. 2014;**85**(1):43-47. (in Japanese)

[22] Ishikawa S, Ono Y, Ohtake N, Sueyoshi K, Tanabata S, Ohyama T. Transcriptome and metabolome analyses reveal that nitrate strongly promotes nitrogen and carbon metabolism in soybean roots, but tends to repress it in nodules. Plants. 2018;**7**(32). DOI: 10.3390/plants7020032

[23] Yamashita N, Tanabata S, Ohtake N, Sueyoshi K, Sato T, Highchi K, et al. Effects of different chemical forms of nitrogen on the quick and reversible inhibition of soybean nodule growth and nitrogen fixation activity. Frontiers in Plant Science. 2019;**10**(131). DOI: 10.3389/fpls.2019.00131

[24] Ono Y, Fukasawa M, Sueyoshi K, Ohtake N, Sato T, Tanabata S, et al. Application of nitrate, ammonium, or urea changes the concentrations of ureides, urea, amino acids and other metabolites in Xylem Sap and in the organs of soybean plants (*Glycine max* (L.) Merr.). International Journal of Molecular Sciences. 2021;**22**(4593):4573. DOI: 10.3390/ijms22094573

[25] Blume C, Ost J, Mühlenbruch M, Peterhänsel C, Laxa M. Low CO_2 indices urea cycle intermediate accumulation in Arabidopsis thaliana. PLoS One. 2019;**14**(1):e0210342

[26] Takahashi Y, Chinushi T, Nagumo Y, Nakano T, Ohyama T. Relative concentration of ureides-N in root-bleeding sap of nodulated and non-nodulated soybean isoline. Japanese Society of Soil Science and Plant Nutrition. 1991;**62**(4):431-433

[27] Takahashi Y, Chinushi T, Ohyama T. Quantitative estimation of N_2 fixation and N absorption rate in field grown soybean plants by relative ureide method. Bullut in Faculty of Agriculture Niigata University. 1993;**45**:91-105

[28] Ohyama T, Kumazawa K. Incorporation of ^{15}N into various nitrogenous compounds in intact soybean nodules after exposure to $^{15}N_2$ gas. Soil Science & Plant Nutrition. 1978;**24**(4):525-533

[29] Ohyama T, Kumazawa K. Assimilation and transport of nitrogenenous compounds originated from $^{15}N_2$ fixation and $^{15}NO_3$ absorption.

Soil Science & Plant Nutrition. 1979;**25**(1):9-19

[30] Ohyama T, Kumazawa K. Nitrogen assimilation in soybean nodules. I. The role of GS/GOGAT system in the assimilation of ammonia produced by N_2 fixation. Soil Science & Plant Nutrition. 1980;**26**(1):109-115

[31] Ohyama T, Kumazawa K. Nitrogen assimilation in soybean nodules. II. $^{15}N_2$ assimilation in bacteroid and cytosol fractions of soybean nodules. Soil Science & Plant Nutrition. 1980;**26**(2):205-213

[32] Shelp BJ, Ireland RJ. Ureide metabolism in leaves of nitrogen-fixing soybean plants. Plant Physiology. 1985;**77**(3):779-783. DOI: 10.1104/pp.77.3.779

[33] Winkler RG, Polacco JC, Blevins DG, Randall DD. Enzymatic degradation of allantoate in developing soybeans. Plant Physiology. 1985;**79**(3):787-793. DOI: 10.1104/pp.79.3.787

[34] Winkler RG, Blevins DG, Polacco JC, Randall DD. Ureide catabolism of soybeans. II. Pathway of catabolism in intact leaf tissue. Plant Physiology. 1987;**83**(3):585-591. DOI: 10.1104/pp.83.3.585

[35] Marouane B, Mitsui T, Sueyoshi K, Ohyama T. Recent advances in carbon and nitrogen metabolismin C3 plants. International Journal of Molecular Sciences. 2021;**22**(1):318. DOI: 10.3390/ijms22010318

Chapter 3

Physiological and Biochemical Basis of Stress Tolerance in Soybean

Md. Mannan, Ismot Rima and Abdul Karim

Abstract

Soybean is considered as a species sensitive to several abiotic stresses, such as drought, salinity, and waterlogging, when compared with other legumes, and these abiotic stresses have a negative effect on soybean plants' growth and crop productivity. Clearing the conception on the physiological and biochemical responses to drought is essential for an overall understanding of the mechanism of plant resistance to water-restricted conditions and for developing drought resistance screening techniques that can be used for plant breeding. Plants can adapt in response to water scarcity situations by altering cell metabolism and activating various defense mechanisms. Higher salt tolerance in resistant soybean genotypes was associated with better water relation, salt dilution by juiciness, and better osmotic adaptation with an accumulation of more amino acids, sugars, and proline. In addition, less damaging chlorophylls, higher photosynthetic efficiency and cell membrane stability, and higher calcium content contributed to the higher salt tolerance of soybean genotypes. Plants adapted to flooded conditions have mechanisms to cope with this stress. Aerenchyma formation increased availability of soluble carbohydrates, greater activity of glycolytic pathways and fermenting enzymes, and involvement of antioxidant defense mechanisms to cope with post-hypoxic/post-anoxic oxidative stress. Ethylene, a gaseous plant hormone, plays an important role in altering a plant's response to oxygen deficiency.

Keywords: physiology, biochemical, mechanism, drought, salinity, waterlogged, tolerance, soybean

1. Introduction

Soybean (*Glycine max* L.) belongs to the family Leguminosae, was domesticated in China, which has the first records of soybean [1]. It is known as "golden bean" and "miracle crop" of the 20th century for its versatile and fascinating crop characteristics. It is an important grain legume because of its high protein (40–42%), oil content (18–22%) as well as Fe, Cu, Mn, Ca, Mg, Zn, Co, P, K, Vitamin B_1 and B_2 [2]. Besides oil, soybean is used as soy milk, soy flour, etc., and as ingredients of animal and poultry feed. In addition to being a source of macronutrients and minerals, soybeans contain secondary metabolites, such as isoflavones [3], saponins, phytic acid, oligosaccharides, goitrogens [4], and phytoestrogens [5]. Soybean is also viewed as an attractive crop for the production of biodiesel [6]. It also has the ability to fix atmospheric nitrogen (17–127 kg N ha^{-1} $year^{-1}$) [7] and therefore requires minimal input of nitrogen fertilizer that often accounts for the single largest energy input in agriculture.

Soybean is inherently more stress tolerant [8] than other legume crops, but it still suffers considerable damage due to different abiotic stress. Drought effects negatively on soybean growth, physiology, and yield and yield reduction was observed 40% or even more due to drought. Drought-tolerant soybeans maintain higher proline and other osmoticums as well as higher chlorophyll content and water status in their leaves. Osmotic adjustment in plants subjected to salt stress can occur with the accumulation of high concentrations of either inorganic ions or low-molecular-weight organic solutes. Compatible solute accumulation in the cytoplasm is considered a mechanism to impart salt tolerance [9, 10]. The osmolytes generally found in higher plants are of low-molecular-weight sugars, organic acids, amino acids, proteins, and quaternary ammonium compounds. According to Cram [11], among the various organic osmotica, sugars contribute up to 50% of the total osmotic potential in glycophytes subjected to exposure to salt stress. Higher content of soluble proteins was observed in salt-tolerant than in salt-sensitive cultivars of barley, sunflower [12], and rice [13, 14]. It has also been reported that amino acids (alanine, arginine, glycine, serine, leucine, valine, etc.) and amides (glutamine, asparagine, etc.) accumulate in plants exposed to salt stress [15]. Total free amino acids in the leaves have been reported to be higher in salt-tolerant than in salt-sensitive lines of sunflower [12], safflower [16], *Eruca sativa* [17], and *Lens culinaris* [18]. Salt tolerance of plants related to ion transport through cation channel and governed by membrane protein plays a significant role in the redistribution of Na^+ and K^+ ions. Proline is widely distributed in higher plants, accumulates more than other amino acids [19], and regulates the accumulation of available N. Proline accumulation usually occurs in the cytosol and contributes significantly to cytoplasmic osmoregulation [20]. It is osmotically very active, contributes to membrane stability, and reduces the effect of NaCl on cell membrane destruction [21].

Waterlogging occurs whenever the soil is so wet that there is insufficient oxygen in the pore space for plant roots to be able to adequately respire. Lack of oxygen in the rhizosphere of plants causes their root tissues to rot. This usually happens from the root tips, making the roots look like they have been trimmed. As a result, plant growth and development are blocked. If the anaerobic condition persists, the plant will eventually die. Floods and inundation are abiotic and hierarchical stresses that, together with water scarcity, salinity, and extreme temperatures, are among the major determinants of the worldwide distribution of plant species. During waterlogging or submersion, plants are exposed to a reduced oxygen supply due to the low rate of oxygen diffusion in water and its limited solubility [22]. Turbid floodwaters can become anaerobic, especially overnight. Growth is greatly inhibited in the deficiency (hypoxia) or complete absence (anoxia) of oxygen. The mechanisms for different stress tolerance are complex and depend upon anatomical, biochemical, and physiological changes occurring in the whole plant rather than in a single cell. This chapter mostly discusses physiological and biochemical parameters that are related to stress tolerance in soybean.

1.1 Physiological and biochemical basis of drought tolerance in soybean

Water deficit or drought is one of the major abiotic stresses that negatively affect crop production worldwide. Intensification of the global water cycle [23] will be extreme events of drought and humidity increasing its frequency of occurrence in different areas of the globe, including the tropical and subtropical areas [24]. Soybean is a legume of great economic importance, but its production is highly dependent on optimal rainfall or abundant irrigation. In addition, during dry periods, additional irrigation may be required for drought-sensitive soybean varieties. Effects of water stress on soybeans, including osmotic regulation, reduce leaf surface area, plant height, decrease branching, reduce chlorophylls, low stomatal conductance and transpiration; fresh and dry matter reduction and finally yield loss have been well documented. Water stress, as a key abiotic limiting factor for soybean production, can cause soybean yield reduction up to 40% or even more [25]. The mechanisms of water stress tolerance, especially at low water stress levels, involve processes at the cellular level, most importantly osmotic regulation and protection of the membrane system. Osmotic regulation is a decrease in osmotic potential due to the active accumulation of organic and inorganic solutes in the cell. High concentrations of inorganic ions become detrimental to cell metabolism and must be sequestered in the vacuole. To maintain osmotic balance, specific types of organic molecules (such as soluble sugars, betaines, and proline) accumulate in the cytoplasm. Those compounds protect plants against stresses by cellular adjustment through the protection of membranes integrity and enzymes stability [26] are termed as compatible solutes, because they can be accumulated in high concentrations without impairing normal physiological function. Water-deficit stress adversely affects many physiological processes related to water use efficiency in soybean, thus leading to a decrease in plant productivity [27]. Relative water content (RWC) is used extensively to determine the water status of plants related to their fully turgid condition. According to Beltrano et al. [28] plants that are able to maintain high levels of RWC under water-deficit conditions are less affected by stress and are able to maintain normal growth and yield. Leaf water potential is considered to be a reliable parameter for quantifying plant water stress response. The effects of water stress on photosynthetic rates of soybean leaves are readily detectable at leaf water potentials about −1.0 to −1.2 MPa [29]. Siddique et al. [30] reported that changes in plant water potential might be attributed to a change in osmotic pressure—the osmotic component of water potential. Water stress significantly reduced the leaf water potential of soybean plants and the potentials fell from −0.88 MPa in unstressed leaves to −1.18 MPa in drought-stressed leaves [31]. Such observation was also reported by Ohashi et al. [32] in soybean. Leaf water potential in all the genotypes was higher under control conditions than that in stress conditions. Raper et al. [29] also reported that the effect of drought stress on photosynthetic rates of soybean declined rapidly with further reductions in leaf water potential to about −1.8 MPa, and then continue to decline gradually with decreasing water potential.

Drought-tolerant soybean cultivars have been investigated for revealing the mechanisms of tolerance and survival. Drought-tolerant soybeans try to adapt in water-deficit conditions through an increase in total sugar, proline, betains, sugar alcohols, and organic acids in their cell. The proline accumulation is a metabolic response characteristic of plants under abiotic stresses, it being showed the increase in the drought-tolerant soybean genotypes [33] because the free proline work as an osmotic adjustor that reduces the negative effects provoked in the plants under adverse conditions [34], besides promoting higher resistance in cells under these circumstances [35]. The proline is synthesized from glutamate and ornitine, in which the production of this organic solute, under conditions of the water shortage, occurs at the major part from glutamate [36]. The relative melondialdehyde (MDA) content was significantly higher (111%) in water stress conditions than control in drought-tolerant soybean variety

Bina soybean1 [33], the lower relative values of MDA in Bina soybean1 indicate that at the cellular level this genotype is better equipped with efficient free-radical quenching system that offers protection against oxidative stress. The soybean variety Bina1 showed relatively higher tolerance to water stress in terms of yield compared with other genotypes. Higher water content, leaf proline and sugar accumulation, and lower MDA accumulation contributed to the higher drought tolerance of Bina soybean1 compared with other genotypes [33]. Rima et al. [37] found that higher water-deficit stress tolerance in soybean genotype G00081 was associated with higher water content in leaf, higher accumulation of proline, and less reduction of leaf chlorophyll.

1.2 Physiological and biochemical basis of salinity tolerance in soybean

Soil salt is one of the major problems of crop production in the arid and semi-arid regions of the world. Salt affects plant growth and development by causing a lack of water, reduced uptake and accumulation of essential nutrients, and increased accumulation of toxic ions, such as Na^+ and Cl^- in plant cells. All of these factors cause changes in various physiological and biochemical processes, such as photosynthesis, protein synthesis, and nucleic acid metabolism [38, 39]. Even in well-hydrated soil, salt causes water scarcity by reducing the osmotic potential of dissolved soil material, making it more difficult for roots to extract water from the surrounding medium [10]. Excess sodium inhibits the growth of many salt-sensitive plants, which includes most of the crop plants. The osmotic adjustment is considered as one of the important mechanisms of water-deficit tolerance of plants [40], which promotes the protection of the plant cell structures including membranes and chloroplasts [41].

Ashraf and Harris [42] reported a considerable variation in the accumulation of soluble sugars in response to salt stress between tolerant and susceptible plants of both inter-specific and/or intra-specific genotypes. Regulation of ion transport is one of the important factors responsible for the salt tolerance of plants. Membrane proteins play a significant role in the selective distribution of ions within the plant or cell. According to DuPont [43], membrane proteins are involved in cation selectivity and redistribution of Na^+ and K^+. It is well established that Na^+ moves passively through a general cation channel from the saline growth medium into the cytoplasm of plant cells [44, 45], and the active transport of Na^+ is also occurred through Na^+/H^+ antiports in plant cells [46]. Salt tolerance in the plant is generally associated with low uptake and accumulation of Na^+, which is mediated through the control of influx and/or by active efflux from the cytoplasm to the vacuoles and also back to the growth medium [44].

Some basic structural components of the membrane are affected by salinity. Vascularization of the plasma membrane is reported to be associated with salt tolerance in halophytes [47]. There can also be a significant increase in the endoplasmic reticulum. The increase in vesicles and endoplasmic reticulum may be a mechanism of compartmentalizing or exporting Na^+ ions [48]. Under saline conditions, plasma membrane leakage increase in glycophytes, and there is a linear relationship between external salinity and membrane leakage [49].

The mechanisms for salt tolerance are complex and depend upon anatomical, biochemical, and physiological changes occurring in the whole plant rather than in a single cell. Mannan et al. [50] opined that the relatively high salt tolerance of AGS 313 was associated with the limited accumulation of sodium and high accumulation of different mineral ions in different plant parts, as well as the maintenance of better water relations under salinity than in the case of susceptible variety Shohag.

Such variation in the response of both genotypes to salt-induced water deficit was attributed to the genetic ability of the resistant trait to undergo certain modifications in their metabolic pathway, thus declining their osmotic and water potentials with a concomitant preliminary decrease in their RWC. White and Izquierdo [51] reported that under severe stress conditions plant cells accumulate metabolites and make the osmotic potential of the cell more negative to maintain turgor pressure. The osmotic potential may be regulated through shifts in concentration of some osmoprotectants, such as proline and sugar. This mechanism is considered to be an important adaptation of plants to stress conditions. Relative protein content was higher in tolerant genotype AGS 313 than that of the susceptible genotype Shohag [52]. The decreased soluble protein due to salinity as found in their study is in agreement with the results of Subbarao et al. [53]. The enhanced protease activity with increasing concentration of NaCl supports the results of Sheoran et al. [54]. Ashraf and Tufail [12] observed a higher content of soluble protein in salt-tolerant cultivars than in salt-sensitive ones of barley and sunflower. Mansour [15] advocated that under salt stress conditions proteins might play a role in osmotic adjustment and thus it is accumulated in response to heat, cold, drought, waterlogging, and high and low mineral nutrient stress. Proteins that are accumulated in plants grown under saline conditions might provide a storage form of nitrogen that would be neutralized when stress is over [55]. In the salinity stress period, proline accumulation was twice in tolerant genotype AGS 313 than that of susceptible one Shohag [52]. Accumulation of proline in response to salinity was also observed by Weimberg et al. and Reddy and Vora [56, 57]. Enhanced proline accumulation with increased salinity levels was also observed by Khawale et al. [58] in different grape cultivars. Accumulation of proline content in plants grown in saline conditions might be due to hydrolysis of storage proteins needed for osmoregulation [59] and determining resistant capacity.

1.3 Physiological and biochemical basis of waterlogging tolerance in soybean

Waterlogging is defined as ponding of water over an area of crop land [60]. Waterlogging occurs whenever the soil is so wet that there is insufficient oxygen in the pore space for plant roots to be able to adequately respire. A lack of oxygen in the root area of a plant causes the root tissue to decompose. This usually comes from the tip of the root, so it looks like the root has been cut. The result of this is to stop the growth and development of the plant. In most cases, submersion does not last long enough for the plant to die. After a period of submersion, the plant begins to respire again. As long as the soil is moist, old roots close to the surface allow the plant to survive. However, additional root cuttings due to submersion and/or dry conditions can weaken plants to the point where they can become very unproductive and eventually die. Waterlogging is a widespread phenomenon drastically reducing the growth and production of soybean in many regions of the world [61], mostly due to the occurrence of flat topography [62], high water tables, and poor drainage of clay-like soils [63]. Effect of waterlogging in soybean plants may include leaf yellowing, reduced root growth, reduced nodulation, stunted growth, defoliation, reduced yields, and plant death [64]. Waterlogging treatment caused a reduction in plant growth in terms of leaf area and growth rate in all the genotypes, and the level of reduction was more pronounced in sensitive genotypes. Solaiman et al. [65–67] stated that waterlogging induced several physiological disturbances, including a reduction in growth, dry matter, photosynthesis, and pod formation that resulted in a low yield of soybean similar to that in other beans. Submersion causes energy starvation in plants as a result of root respiration difficulties due to O_2 deficiency [68]. Flooding and ultimately anaerobic metabolic energy limitation, accumulation of toxic products (e.g., lactic acid), and carbon loss (due to ethanol loss from the roots) can result in severe stunting and death in most crops [69]. Soybeans accumulate alanine [70], an amino acid produced by the enzyme alanine aminotransferase (AlaAT) under hypoxic conditions. Alanine synthesis plays an important role in the regulation of glycolysis, preventing excessive accumulation of pyruvic acid while maintaining intracellular carbon and nitrogen resources [71]. The production of lactic acid and ethanol, the accumulation of alanine has no harmful side effects on the cells.

2. Conclusions

Water-deficit stress exerted inhibitory consequences on plant morphology, physiological and biochemical parameters that sooner or later decreased the yield of soybean. Drought additionally decreased water content and chlorophyll. On the alternative hand, water-deficit stress improved proline and malondialdehyde content material in soybean leaves. Drought tolerance of soybean became located related to better water content, higher proline and much less malondialdehyde content material, and much less degradation of chlorophyll in the leaf. Salt tolerance in soybean became related to higher water relations, higher osmotic adjustment maintained with the aid of using collecting extra amino acid, sugar, and proline, much less chlorophylls degradation and better photosynthetic efficiency. Waterlogging tolerant soybean plant life capin a position to build up the better quantity of soluble sugars, boom fermentation enzymes and antioxidant protection mechanism under oxygen deficiency.

Conflict of interest

There is no conflict of interest.

Author details

Md. Mannan*, Ismot Rima and Abdul Karim
Department of Agronomy, Bangabandhu Sheikh Mujibur Rahman Agricultural University, Gazipur, Bangladesh

*Address all correspondence to: mannanagr@bsmrau.edu.bd

References

[1] Singh G. The Soybean Botany, Production, and Uses. CPI Anthony Rowe; 2010. pp. 2-13

[2] Messina MJ. Soyfoods: Their role in disease prevention and treatment. In: Soybeans. Boston, MA: Springer; 1997. pp. 442-477

[3] Sakai T, Kogiso M. Soy isoflavones and immunity. The Journal of Medical Investigation. 2008;**55**(3-4):167-173

[4] Liener IE. Implications of antinutritional components in soybean foods. Critical Reviews in Food Science & Nutrition. 1994;**34**(1):31-67

[5] Ososki AL, Kennelly EJ. Phytoestrogens: A review of the present state of research. Phytotherapy Research: An International Journal Devoted to Pharmacological and Toxicological Evaluation of Natural Product Derivatives. 2003;**17**(8):845-869

[6] Pimentel D, Patzek T. Ethanol production using corn, switchgrass and wood; biodiesel production using soybean. In: Biofuels, Solar and Wind as Renewable Energy Systems. Dordrecht: Springer; 2008. pp. 373-394

[7] Burris RH, Roberts GP. Biological nitrogen fixation. Annual Review of Nutrition. 1993;**13**(1):317-335

[8] Singh BB, Hartmann P, Fatokun C, Tamo M, Tarawali S, Ortiz R. Recent progress in cowpea improvement. Chronica Horticulturae. 2003;**43**(2): 8-12

[9] Hare PD, Cress WA, Van Staden J. Dissecting the roles of osmolyte accumulation during stress. Plant, Cell & Environment. 1998;**21**(6):535-553

[10] Jaleel CA, Gopi R, Manivannan P, Panneerselvam R. Antioxidative potentials as a protective mechanism in Catharanthusroseus (L.) G. Don. Plants under salinity stress. Turkish Journal of Botany. 2007;**31**(3):245-251

[11] Cram WJ. Negative feedback regulation of transport in cells. The maintenance of turgor, volume and nutrient supply. In: Transport in Plants II. Berlin, Heidelberg: Springer; 1976. pp. 284-316

[12] Ashraf M, Tufail M. Variation in salinity tolerance in sunflower (*Helianthus annuus* L.). Journal of Agronomy and Crop Science. 1995;**174**(5):351-362

[13] Lutts S, Kinet JM, Bouharmont J. Effects of salt stress on growth, mineral nutrition and proline accumulation in relation to osmotic adjustment in rice (*Oryza sativa* L.) cultivars differing in salinity resistance. Plant Growth Regulation. 1996;**19**(3):207-218

[14] Pareek A, Singla SL, Grover A. Salt responsive proteins/genes in crop plants. In: Jaiwal PK, Singh RP, Gulati A, editors. Strategies for Improving Salt Tolerance in Higher Plants. New Delhi: Oxford and IBH Publication Co.; 1997. pp. 365-391

[15] Mansour MMF. Nitrogen containing compounds and adaptation of plants to salinity stress. Biologia Plantarum. 2000;**43**(4):491-500

[16] Ashraf M, Fatima H. Responses of salt-tolerant and salt-sensitive lines of safflower [*Carthamustinctorius* L.] to salt stress. Acta Physiologiae Plantarum. 1995;**17**(1):61-70

[17] Ashraf MY, Wu L. Breeding for salinity tolerance in plants.

Critical Reviews in Plant Sciences. 1994;**13**(1):17-42

[18] Hurkman WJ, Tao HP, Tanaka CK. Germin-like polypeptides increase in barley roots during salt stress. Plant Physiology. 1991;**97**(1):366-374

[19] Abrahám E, Rigó G, Székely G, Nagy R, Koncz C, Szabados L. Light-dependent induction of proline biosynthesis by abscisic acid and salt stress is inhibited by brassinosteroid in Arabidopsis. Plant Molecular Biology. 2003;**51**(3):363-372

[20] Ketchum RE, Warren RS, Klima LJ, Lopez-Gutiérrez F, Nabors MW. The mechanism and regulation of proline accumulation in suspension cell cultures of the halophytic grass Distichlisspicata L. Journal of Plant Physiology. 1991;**137**(3):368-374

[21] Mansour MMF. Protection of plasma membrane of onion epidermal cells by glycinebetaine and proline against NaCl stress. Plant Physiology and Biochemistry. 1998;**36**(10):767-772

[22] Armstrong AC. The effect of drainage treatments on cerealyields: Results from experiments on clay lands. Journal of Agricultural Science. 1978;**91**:229-235

[23] Huntington TG. Evidence for intensification of the global water cycle: Review and synthesis. Journal of Hydrology. 2006;**319**(1-4):83-95

[24] Lintner BR, Biasutti M, Diffenbaugh NS, Lee JE, Niznik MJ, Findell KL. Amplification of wet and dry month occurrence over tropical land regions in response to global warming. Journal of Geophysical Research: Atmospheres. 2012;**117**:D11106. DOI: 10.1029/2012JD017499

[25] Pathan MS, Lee JD, Shannon JG, Nguyen HT. Recent advances in breeding for drought and salt stress tolerance in soybean. Advances in Molecular Breeding toward Drought and Salt Tolerant Crops. 2007;**17**(10):739-773

[26] Farooq M, Wahid A, Kobayashi N, Fujita D, Basra SMA. Plant drought stress: Effects, mechanisms and management. In: Lichtfouse E, Navarrete M, Debaeke P, Véronique S, Alberola C, editors. Sustainable Agriculture. Dordrecht: Springer; 2009. DOI: 10.1007/978-90-481-2666-8_12

[27] Hamayun M, Khan SA, Shinwari ZK, Khan AL, Ahmad N, Lee IJ. Effect of polyethylene glycol induced drought stress on physio-hormonal attributes of soybean. Pakistan Journal of Botany. 2010;**42**(2):977-986

[28] Beltrano J, Ronco MG, Arango MC. Soil drying and rewatering applied at three grain developmental stages affect differentially growth and grain protein deposition in wheat (*Triticumaestivum*L.). Brazilian Journal of Plant Physiology. 2006;**18**:341-350

[29] Raper CD, Kramer PJ, Wilcox JR. Soybeans: Improvement, production and uses. Stress Physiology. 1987;**16**:589-641

[30] Siddique MRB, Hamid AIMS, Islam MS. Drought stress effects on water relations of wheat. Botanical Bulletin of Academia Sinica. 2000;**41**(1):35-39

[31] Makbul S, Guler NS, Durmus N, Guven S. Changes in anatomical and physiological parameters of soybean under drought stress. Turkish Journal of Botany. 2011;**35**(4):369-377

[32] Ohashi Y, Saneoka H, Fujita K. Effect of water stress on growth, photosynthesis, and photoassimilate translocation in soybean and tropical

pasture legume siratro. Soil Science and Plant Nutrition. 2000;**46**(2):417-425

[33] Sarkar KK, Mannan MA, Haque MM, Ahmed JU. Physiological basis of water stress tolerance in soybean. Bangladesh Agronomy Journal. 2015;**18**(2):71-78

[34] Kishor K, Polavarapu B. Role of proline in cell wall synthesis and plant development and its implications in plant ontogeny. Frontiers in Plant Science. 2015;**6**:544. DOI: 10.3389/fpls.2015.00544

[35] Zhu JK. Salt and drought stress signal transduction in plants. Annual Review of Plant Biology. 2002;**53**(1):247-273

[36] Delauney AJ, Verma DPS. Proline biosynthesis and osmoregulation in plants. The Plant Journal. 1993;*4*(2):215-223

[37] Rima IA, Mannan MA, Mamun MAA, Kamal ZU. (2019). morphophysiological traits of soybean as affected by drought. Bangladesh Agronomy Journal. 2019;**22**(2):41-54

[38] Zhang HX, Blumwald E. Transgenic salt-tolerant tomato plants accumulate salt in foliage but not in fruit. Nature Biotechnology. 2001;**19**(8):765-768

[39] Munns R. Genes and salt tolerance: Bringing them together. New Phytologist. 2005;**167**(3):645-663

[40] Costa RCL. Nitrogen assimilation and osmotic adjustment in nodulated plants of stringed beans *Vigna unguiculata (L.)* Walp. under water stress [doctoral dissertation, Ph. D. Thesis]. Brasil: Universidade Federal do Ceará; 1999

[41] Martinez JP, Lutts S, Schanck A, Bajji M, Kinet JM. Is osmotic adjustment required for water stress resistance in the Mediterranean shrub Atriplexhalimus L? Journal of Plant Physiology. 2004;**161**(9):1041-1051

[42] Ashraf MPJC, Harris PJC. Potential biochemical indicators of salinity tolerance in plants. Plant Science. 2004;**166**(1):3-16

[43] DuPont FM. Salt-induced changes in ion transport: Regulation of primary pumps and secondary transporters. In: Transport and Receptor Proteins of Plant Membranes. Boston, MA: Springer; 1992. pp. 91-100

[44] Jacoby B. Mechanisms involved in salt tolerance of plants. Handbook of Plant and Crop Stress. 1999;**2**:97-123

[45] Mansour MMF, Salama KHA, Al-Mutawa MM. Transport proteins and salt tolerance in plants. Plant Science. 2003;**164**(6):891-900

[46] Shi H, Lee BH, Wu SJ, Zhu JK. Overexpression of a plasma membrane Na+/H+ antiporter gene improves salt tolerance in *Arabidopsis thaliana*. Nature Biotechnology. 2003;**21**(1):81-85

[47] Flowers TJ, Troke PF, Yeo AR. The mechanism of salt tolerance in halophytes. Annual Review of Plant Physiology. 1977;**28**(1):89-121

[48] Hagemeyer J. Salt. In: Prasad MNV, editor. Plant Ecophysiology. New York, Toronto Singapore: John Wiley & Sons. Inc; 1997. pp. 174-205

[49] Leopold AC. Evidence for toxicity effects of salt on membranes. In: Salinity Tolerance in Plants: Strategies for Crop Improvement. 1984. pp. 67-76

[50] Mannan MA, Karim MA, Khaliq QA, Haque MM, Mian MAK, Ahmed JU. Effect of salinity on photosynthesis, cell membrane stability and water

retention capacity of two soybean genotypes differing in salinity tolerance. Bangladesh Agronomy Journal. 2010;**13**(1&2):18-24

[51] White JW, Izquierdo J. Physiology of yield potential and stress tolerance. In: van Schoonhoven A, Voysest O, editors. Common Beans: Research for Crop Improvement. Wallingford, U. K.: C. A. B. Int. and Cali, Colombia: CIAT; 1991. pp. 287-382

[52] Mannan MA, Karim MA, Khaliq QA, Ahmed JU. Salinity effect on dry matter accumulation in different vegetative parts of soybean. Journal of the Bangladesh Society for Agricultural Science and Technology. 2009;**6** (1 & 2):35-38

[53] Subbarao GV, Johansen C, Jana MK, Rao JK. Physiological basis of differences in salinity tolerance of pigeon pea and its related wild species. Journal of Plant Physiology. 1990;**137**(1):64-71

[54] Sheoran IS, Garg OP. Effect of salinity on the activities of RNase, DNase and protease during germination and early seedling growth of mung bean. Physiologia Plantarum. 1978;**44**(3):171-174

[55] Singh NK, Bracker CA, Hasegawa PM, Handa AK, Buckel S, Hermodson MA, et al. Characterization of osmotin: A thaumatin-like protein associated with osmotic adaptation in plant cells. Plant Physiology. 1987;**85**(2):529-536

[56] Weimberg R, Lerner HR, Poljakoff Mayber A. A relationship between potassium and proline accumulation in salt stressed *Sorghum bicolor*. Physiologia Plantarum. 1982;**55**(1):5-10

[57] Reddy MP, Vora AB. Effect of chloride and sulphate of sodium and potassium salinity on germination and free proline content of bajra. INSA Proceedings Part B. 1983;**49**:702-705

[58] Khawale RN, Singh SK, Patel VB, Singh SP. Changes due to in vitro sodium chloride induced salinity in grape (*Vitis vinifera* L.). Indian Journal of Plant Physiology. 2003;**8**(1):378

[59] Trinchant JC, Yang YS, Rigaud J. Proline accumulation inside symbiosomes of faba bean nodules under salt stress. Physiologia Plantarum. 1998;**104**(1):38-49

[60] Scott HD, Batchelor JT. Dry weight and leaf area production rates of irrigated determinate soybeans 1. Agronomy Journal. 1979;**71**(5):776-782

[61] Van Nguyen L, Takahashi R, Githiri SM, Rodriguez TO, Tsutsumi N, Kajihara S, et al. Mapping quantitative trait loci for rootdevelopment under hypoxia conditions in soybean (*G. max* L. Merr.). Theoretical and Applied Genetics. 2017;**130**:743-755

[62] Collaku A, Harrison SA. Losses in wheat due to waterlogging. Crop Science. 2002;**42**(2):444-450

[63] Jitsuyama Y. Hypoxia-responsive root hydraulic conductivity influences soybean cultivar-specific waterlogging tolerance. American Journal of Plant Sciences. 2017;**8**:770-790

[64] Linkemer G, Board JE, Musgrave ME. Waterlogging effects on growth and yield components in late-planted soybean. Crop Science. 1998;**38**(6):1576-1584

[65] Solaiman Z, Colmer TD, Loss SP, Thomson BD, Siddique KHM. Growth responses of cool-season grain legumes to transient waterlogging. Australian Journal of Agricultural Research. 2007;**58**(5):406-412

[66] Pociecha E, Kościelniak J, Filek W. Effects of root flooding and stage of development on the growth and photosynthesis of field bean (*Vicia faba L. minor*). Acta Physiologiae Plantarum. 2008;**30**(4):529-535

[67] Celik G, Turhan E. Genotypic variation in growth and physiological responses of common bean (*Phaseolus vulgaris* L.) seedlings to flooding. African Journal of Biotechnology. 2011;**10**(38):7372-7380

[68] van Dongeng JT, Licausi F. Oxygen sensing and signaling. Annual Review of Plant Biology. 2015;**66**:345-367

[69] Tamang BG, Magliozzi JO, Maroof MAS, Fukao T. Physiological and transcriptomic characterization of submergence and reoxygenation responses in soybean seedlings. Plant, Cell and Environment. 2014;**37**:2350-2365

[70] Borella J, Oliveira HC, de Oliveira DSC, Braga EJB, de Oliveira ACB, Sodek L, et al. Hypoxia-driven changes in glycolytic and tricarboxylic acid cycle metabolites of two nodulated soybean genotypes. Environmental and Experimental Botany. 2017;**133**:118-127

[71] Rocha M, Licausi F, Araujo WL, Nunes-Nesi A, Sodek L, Fernie AR, et al. Glycolysis and the tricarboxylic acid cycle are linked by alanine aminotransferase during hypoxia induced by waterlogging of *Lotus japonicus*. Plant Physiology. 2010;**152**:1501-1513

Section 3

Cultivation and Productivity of Soybean

Chapter 4

Nitrogen Budget in a Paddy-Upland Rotation Field with Soybean Cultivation

Fumiaki Takakai, Takemi Kikuchi, Tomomi Sato, Masato Takeda, Saki Kanamaru, Yasuhiro Aono, Shinpei Nakagawa, Kentaro Yasuda, Takashi Sato and Yoshihiro Kaneta

Abstract

To reduce the over-production of rice, the paddy-upland rotation system, which alternates every few years between paddy rice cultivation and upland crop cultivation in drained (converted) paddy fields, is now commonly practiced in Japan. Recently, depletion of available soil nitrogen (N) and a subsequent decline in soybean yield in converted upland fields with repeated rotation have been reported in northern Japan. To evaluate the N budget in the paddy-upland rotation field with soybean and rice, a 6-year lysimeter experiment was conducted. In the rotation system, a considerable loss of N occurred in both the upland soybean and paddy rice cultivation periods (−11.9 and − 2.3 g N m^{-2} y^{-1}, respectively). To mitigate the N loss in the rotation system, N supply from organic matter application is required. The effects of applying different types of organic matter (leguminous green manure, hairy vetch, and livestock manure compost) on the N budget in soybean cultivated fields were investigated. Compared to the N loss in the control plot without organic matter application, the N loss was mitigated in the hairy vetch plot, and N accumulation occurred in the livestock manure compost plot (−13.7, −3.5, and +11.8 g N m^{-2} y^{-1}, respectively).

Keywords: flooded paddy rice, hairy vetch, livestock manure compost, nitrogen budget, organic matter application, paddy-upland rotation, upland soybean

1. Introduction

In Japan, rice production has been restricted for more than 40 years due to declining rice consumption. The area planted to paddy rice during summer, which was more than 3 million ha in the 1960s, has continued to decline since 1970, reaching 1.58 million ha in 2019 (**Figure 1**) [1, 2]. As a countermeasure, crop rotation in shifting the cultivation of paddy fields to crops other than staple food rice (crop rotation) has been implemented in earnest since 1970. As of 2019, 18% of the total paddy area was planted with crops other than paddy rice in the summer (**Figure 1**).

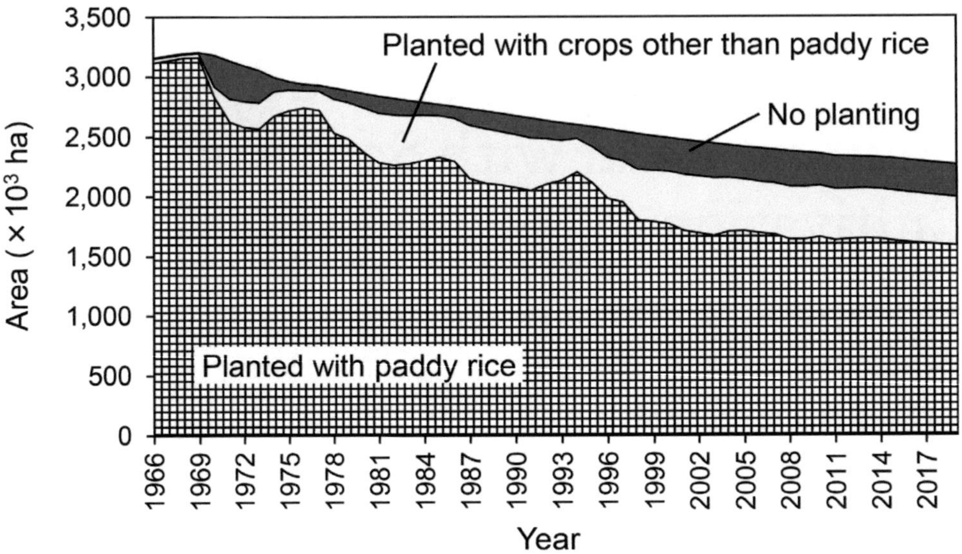

Figure 1.
Trends in crop cultivation in paddy fields during the summer season in Japan. Source: [2].

One of the systems of crop rotation is "paddy-upland rotation," in which paddy fields are planted with rice and upland crops alternately for one to several years. Although there are no statistics on how much of the total area is under the paddy-upland rotation, the rotations have become a major cultivation system for crop rotation. In Japan, a three-crop in two-year rotation system: paddy rice in summer, followed by wheat or barley from autumn to next early summer, and then soybean cultivation in summer has been conducted. On the other hand, in northern Japan, where it is relatively cold, a rotation system with annual cropping of paddy rice and upland crops such as soybean has also been conducted.

As mentioned above, soybean is an important rotational crop cultivated in paddy fields in Japan. Of the total area under soybean cultivation in Japan, about 80% is planted in paddy fields, and 90% in the Tohoku region of northern Japan, a major paddy field area (**Table 1**) [2]. The area of soybean cultivated in paddy fields is 115,900 ha, or 29% of the total area of cultivation with crops other than paddy rice shown in **Figure 1** (403,000 ha).

While the average yield of the world's major soybean-producing countries is approaching 3 Mg ha^{-1}, the yield in Japan has remained low at 1.55–1.65 Mg ha^{-1} [3, 4]. Shimada [4] pointed out that there are many factors contributing to the low soybean yield in Japan, but one of the main factors is the inhibition of N_2 fixation due to wet damage and drought stress in paddy-upland rotation fields.

Soybean assimilates N from atmospheric N_2 by symbiotic N_2 fixation in root nodules [3]. The contribution of atmospheric N_2 to the N accumulation of soybean is highly variable and depends largely on the surrounding environment such as oxygen and moisture. Yoneyama et al. [5] reported that the average percentage of soybean N accumulation derived from N_2 fixation in Japan was 50%. Ohyama et al. [3] reported that the percentages of soybean N accumulation derived from N_2 fixation in rotated paddy fields in Niigata, Japan ranged from 59 to 75%, whereas soybean plants require a large amount of N compared to other crops because of the large protein

Region	Soybean cultivation area (×10^3 ha)			Percentage of paddy field (%)
	Paddy	Upland	Total	
Hokkaido	18.4	20.7	39.1	47
Tohoku	32.7	2.4	35.1	93
Hokuriku	11.7	0.7	12.4	94
Kanto-Tozan	7.7	2.2	9.9	78
Tokai	11.4	0.5	11.9	96
Kinki	9.3	0.2	9.4	98
Chugoku	4.0	0.4	4.3	92
Shikoku	0.5	0.0	0.5	94
Kyushu-Okinawa	20.3	0.7	21.0	97
Total	115.9	27.6	143.5	81

Table 1.
Soybean cultivation area by agricultural region in Japan (2019). Source [2].

accumulation in their seeds (about 35–40%). In order to meet this high N requirement, N derived from N_2 fixation in root nodules alone is not sufficient; soybean should also absorb significant amounts of N from the soil. Then, most of the accumulated N in soybean could be removed from the field as harvested grain. Therefore, there is a possibility that N output from the soybean cultivated field exceeds the N input to the field, and thus the N loss could occur. Therefore, the N budget of a converted paddy field with soybean cultivation could be negative, indicating N loss from the field.

Recently, depletion of available soil N followed by a decline in soybean yield in a repeated paddy-upland rotation field has been reported in northern Japan [6]. Nishida et al. [7, 8] reported a decrease in available soil N with an increase in upland frequency (i.e., the number of years in soybean cultivation per total cultivation years) in fields with paddy-upland rotation in Akita, Tohoku region, northern Japan (**Figure 2**). This indicates that soybean cultivation reduces the soil N fertility of paddy-upland rotation fields. They also reported that when the upland frequency exceeded 60%, the amount of available soil N was less than the minimum value of suitable concentrations of available soil N in the paddy field (80 mg kg^{-1}) [9]. And that the concentration of available soil N could be increased by repeated cattle manure compost application at the rate of 2–3 kg m^{-2}. A similar trend of declining available soil N has been reported in various parts of Japan in recent years [10–13].

Takahashi et al. [14] reported that soil N fertility of a converted paddy field could be a major controlling factor for soybean yield when moisture injury is not severe. In maintaining soil N fertility in paddy-upland rotation fields with soybean cultivation, it is necessary to manage N during soybean cultivation based on the N budget. In general, the N budget in a rice paddy field is considered to be neutral [15] or slightly positive (accumulation) [16], suggesting that the N loss from the rotated paddy field occurs mainly during soybean cultivation. However, the N budget in rice paddy fields could vary widely depending on field management practices [17]. Therefore, in the paddy-upland rotation fields, the N budget including that during paddy rice cultivation should be evaluated and measures to improve the budget should be considered.

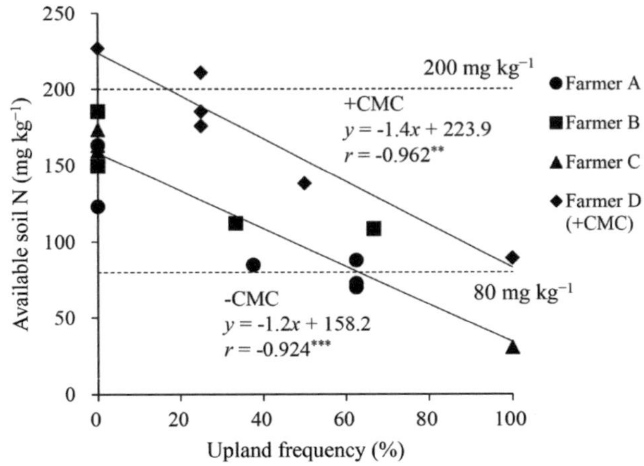

Figure 2.
*Relationship between upland frequency and available soil nitrogen (N). ***P < 0.001, **P < 0.01. CMC, cattle manure compost. Broken lines indicate the minimum (80 mg kg^{-1}) and maximum (200 mg kg^{-1}) values of the suitable range of available soil N in paddy fields [9]. Modified from [8].*

2. Detailed nitrogen budget measurement in a paddy-upland rotation field with soybean cultivation

Detailed N budget in a paddy-upland rotation field was evaluated for 6 years (3 years for upland soybean, then 3 years for flooded paddy rice) in a lysimeter plot at the Akita Prefecture Agricultural Experiment Station, located in the Tohoku region, northern Japan [18, 19]. The lysimeter was filled with soil collected from a rice paddy field on gray lowland soil, which is a major paddy soil in this region (soil texture: clay loam). No organic matter other than soybean and paddy rice residues was applied, and the crop was grown according to the guidelines for soybean and paddy rice cultivation in Akita Prefecture [20, 21]. Soybean (cv. Ryuho) was cultivated from early June to early October. Paddy rice (cv. Yumeobako or Akitakomachi) was cultivated from late May to mid-September. The major N flows of inputs (fertilizer, bulk N deposition, irrigation, and symbiotic N_2 fixation in soybean) and outputs (harvested grain, leaching, surface drainage, and N_2O emission) were measured (**Figure 3**). Symbiotic N_2 fixation in soybean was measured using the relative ureide method [22, 23]. Other N flows were estimated from literature values. The N budget was calculated by the difference between the total input N flow and the total output N flow. Positive and negative values indicate net N accumulation and loss in the field, respectively. Nitrogen budgets were measured for 3 years in soybean and rice cultivation, respectively, and averaged to give annual values.

The average yields of soybean and rice were 341 and 519 g m^{-2}, respectively. The yield of soybean was much higher than the average of Akita Prefecture (about 140 g m^{-2}) [20], whereas the yield of paddy rice was lower than the target value of Akita Prefecture (570 g m^{-2}) [21]. It could be due to severe damage by insects in the third year (422 g m^{-2}) [19].

Among the N inputs during soybean cultivation, symbiotic N_2 fixation by nodule accounted for the majority of the inputs, about over 80% (**Figure 4**). The percentages of N accumulation derived from N_2 fixation for 3 years ranged from 60 to 69% [19]. On the other hand, the N input from fertilizer was about 6% of the total.

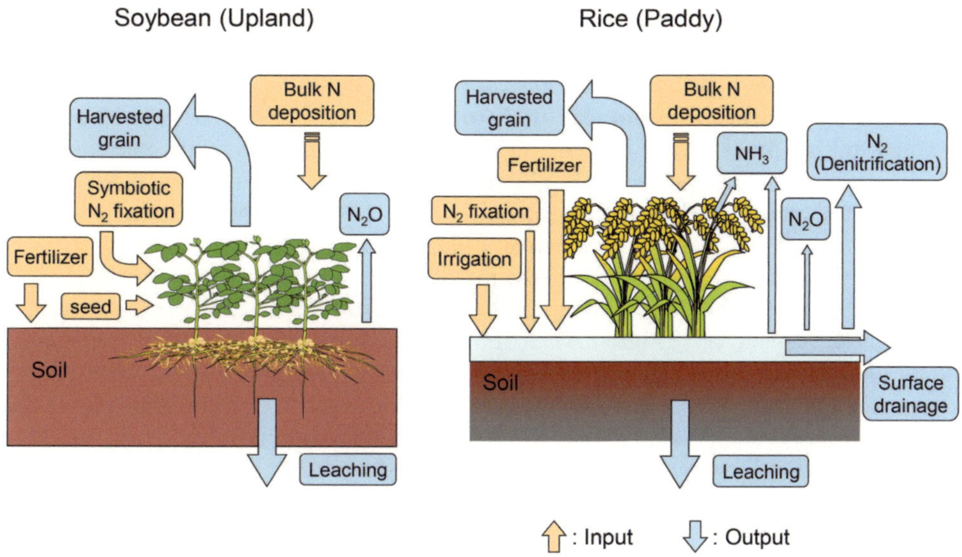

Figure 3.
Outline of major nitrogen (N) flows in soybean upland field and rice paddy field. NH_3, ammonia volatilization; N_2, dinitrogen; N_2O, nitrous oxide. N_2 emission via denitrification in upland was not considered in this study. Modified from [19].

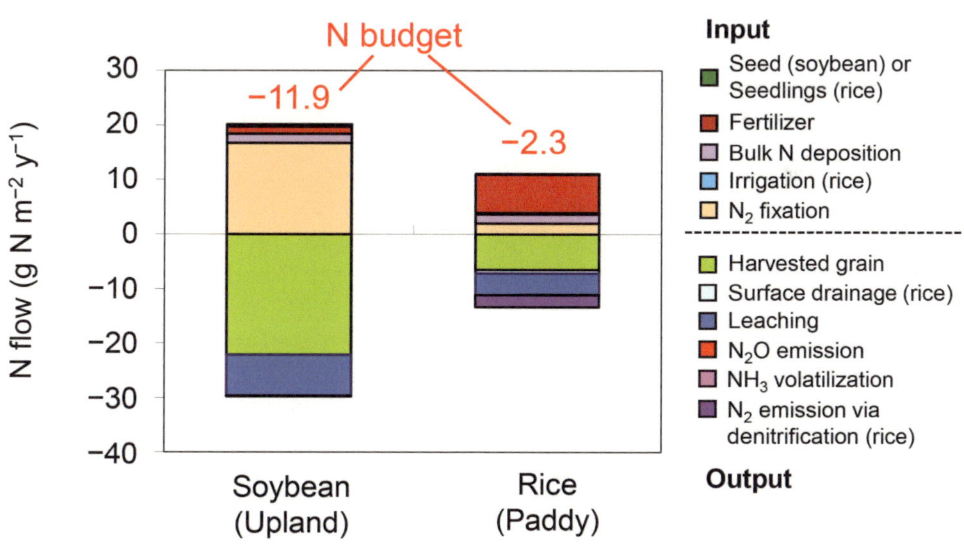

Figure 4.
Comparison of the nitrogen (N) flows and budgets in soybean and rice cultivated fields. Positive and negative values indicated N input and output, respectively. The N budget was calculated by subtracting N output from input. NH_3, ammonia; N_2, dinitrogen; N_2O, nitrous oxide. Modified from [19].

This is due to the fact that in soybean cultivation in converted paddy fields, the amount of N fertilizer applied is set as low as 0–2 g N m^{-2} to prevent over-luxuriant growth [20]. The major components of N output were harvested grain and leaching (74 and 25%, respectively). During the soybean cultivation under upland conditions, the annual N budget was negative (−11.9 g N m^{-2} y^{-1}), indicating net N loss from the field.

A review of N balance in soybean cultivation reported that the mean value of partial N balance, which is calculated from the difference between the input due to N_2 fixation and the output due to harvest, is close to neutral (−0.4 g N m^{-2} growing $season^{-1}$) [24]. However, the value is likely to be negative if a detailed N balance is obtained, taking into account N losses such as leaching, as in this study. Similar to the present study, the N budget including N flow due to water movement such as leaching in a converted paddy field with soybean cultivation in Shiga, central Japan, was negative (−5.4 to −4.0 g N m^{-2} growing $season^{-1}$) [25].

During the paddy rice cultivation, the major input N flow was fertilizer application (63%), whereas the major output N flows were harvested grain and leaching (49 and 29%, respectively; **Figure 4**). Although less than soybean cultivation, the N budget during paddy rice cultivation was also negative (−2.3 g N m^{-2} y^{-1}), indicating N loss from the field. The N loss during the paddy rice cultivation could be due to the limitation of N fertilization to paddy fields converted from upland fields. In Akita Prefecture, to avoid over-luxuriant growth and lodging due to the increased N uptake, it is recommended that basal N fertilization decreases by 100% and by 50–70% in the first and second years after conversion, respectively [21].

In paddy-upland rotation fields including soybean cultivation, the field N budget in both upland soybean and paddy rice is negative, and soil N fertility is likely to decrease due to repeated rotation [6]. It will be essential to take measures to improve the N budget in paddy-upland rotation fields to maintain soil N fertility and crop productivity.

3. Cultivation managements for the mitigation of N loss from paddy-upland rotation fields with soybean cultivation

As mentioned above, soybean productivity in Japan is low and needs to be improved in the future. As soybean yields increase, a corresponding increase in N loss from the field is expected [24]. Hence, measures to improve the N budget to maintain the soil N fertility become more important. To improve the N budget in the field, N inputs need to be increased.

To increase N_2 fixation, which is a major N input in soybean cultivation fields, control of groundwater level in converted paddy fields has been reported to be effective [26, 27]. Deep placement application of slow-release fertilizers [3, 28–35] has been proposed as a fertilization method that does not inhibit N_2 fixation in soybean. Nitrogen supply by application of organic matter is also effective in improving the N budget. Organic matter not only supplies nutrients but also affects the physical, chemical, and biological properties of the soil to promote soybean growth. The application of livestock manure compost (LMC) has been reported to increase soybean production by improving N availability and N_2 fixation in soybean nodules [11, 36–39]. Green manure cultivation of N-fixing legumes is also effective in improving the N budget. In a converted paddy field, cultivation of the leguminous green manure hairy vetch (HV) before soybean cultivation has been shown to promote soybean growth by increasing N supply to the crop and improving soil physical properties [11, 40–42]. A standard application rate of 2 kg m^{-2} of LMC [20] is expected to supply about 20 g N m^{-2} of N to the field. Nitrogen supply by HV cultivated before soybean cultivation ranges from 10 to 20 g N m^{-2} [43, 44]. These N supplies are sufficient to compensate for the N losses during soybean cultivation described above (−11.9 g N m^{-2} y^{-1}, [19]). However, increased N supply to the field may alter other N flows, such as inhibition of N_2 fixation in nodules and increased leaching and N_2O emission. Therefore, there is a need to quantitatively evaluate the effect of organic matter application on the N budget in soybean cultivation in converted paddy fields.

The preliminary results of an experiment for different types of organic matter (HV and LMC; **Figure 5**) application conducted in lysimeter plots at the Center of Field Education and Research, Faculty of Bioresource Sciences, Akita Prefectural University are reported. The results of this study are for a single year, the first year of conversion from paddy. Three lysimeter plots were filled with soil collected from a rice paddy field on gley lowland soil (Fluvic Gleysols in WRB), one of the major paddy soils in this region (soil texture: heavy clay). The three plots were designated as a control plot with no organic matter application except for crop residue, HV, and LMC plots. In the HV plot, HV was sown and cultivated after paddy rice cultivation in the autumn of the previous year, cultivated until before soybean sowing in early June, and then plowed into the soil. In the LMC plot, 2 kg m^{-2} of cow dung-based LMC was applied and incorporated into the soil before soybean sowing. Soybean (cv. Ryuho) was cultivated from early June to early October. No chemical N fertilizer was applied to all plots. Soybean cultivation and N budget measurements were conducted basically as in Section 2 [18, 19].

The effect of organic matter application on the N flows and budgets in a converted soybean field is shown in **Figure 6**. The soybean yields in the HV and LMC plots (358 and 343 g m^{-2}, respectively) were higher than that in the control plot (314 g m^{-2}) (data not shown). Although soybean in the HV plots grew more vigorously than in the other two plots, damage to harvested grain by insects was significant, and the difference in grain yield excluding damaged grains among the plots was small. The total grain yield including damaged grains, which was used in the calculation of N budget, in the control, HV, and LMC plots was 414, 550, and 443 g m^{-2}, respectively.

The N inputs from HV and LMC application were 18.0 and 25.1 g N m^{-2}, which were higher than the respective N input from soybean N_2 fixation. Symbiotic N_2 fixation by soybean nodules was lower in the HV plot (7.9 g N m^{-2}) and higher in the LMC plot (12.1 g N m^{-2}) compared to the control plot (10.2 g N m^{-2}). The percentages of N accumulation derived from N_2 fixation for the control, HV, and LMC plots were 30, 19, and 34%, respectively. Hairy vetch has a low C/N ratio of around 10 [43, 44], indicating rapid mineralization of its N after incorporating into the soil. Therefore, inorganic N could inhibit soybean nodule growth and N_2 fixation activity [45]. On the other hand, LMC may not inhibit N_2 fixation because of its slow decomposition

Figure 5.
Hairy vetch (left) and livestock manure compost (right).

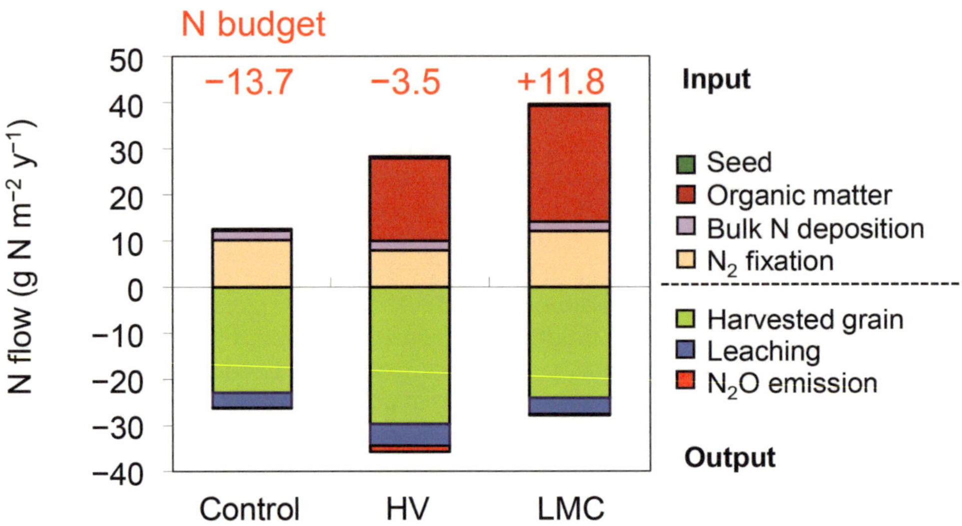

Figure 6.
Effect of organic matter application on the nitrogen (N) flows and budgets in a converted soybean field. Positive and negative values indicated N input and output, respectively. The N budget was calculated by subtracting N output from input. HV, hairy vetch; LMC, livestock manure compost; N_2, dinitrogen; N_2O, nitrous oxide.

(mineralization). The lower amount and percentages of N accumulation derived from N_2 fixation than the values shown in Section 2 (**Figure 4**, [19]) may be due to the fact that the study site was located in a reclaimed land and soil N fertility was high [46]. Therefore, soil N uptake via soybean roots could be high and dependence on symbiotic N_2 fixation could be low.

The major components of N output were harvested grain and leaching (83–87% and 12–13%, respectively). The N output by harvested grain in the HV plot (29.7 g N m^{-2}) was higher than that of the control and LMC plots (23.0 and 24.1 g N m^{-2}, respectively) because of the higher total grain yield. The increase in leached N by organic matter application (1.5 and 0.2 g N m^{-2} y^{-1} for the HV and LMC plots, respectively) was much smaller than that of the amount of applied N. The leaching may have been low because the texture of studied soil is heavy clay and its drainage is poor.

The N budget in the control plot without organic matter application was negative (−13.7 g N m^{-2} y^{-1}; **Figure 6**), and was similar to the value reported previously (−11.9 g N m^{-2} y^{-1}; **Figure 4**) [19]. The N loss during soybean cultivation was mitigated by HV application, and N accumulation occurred by LMC application (−3.5 and +11.8 g N m^{-2} y^{-1}, respectively). The application of HV increased N inputs, but only mitigated the N loss because it suppressed symbiotic N_2 fixation and increased N output by harvested grain and leaching. The application of LMC did not suppress symbiotic N_2 fixation, but rather promoted it. It was shown that different types of applied organic matter had different effects on the N budgets in converted paddy fields.

Application of organic matter during soybean cultivation in paddy-upland rotation fields can improve the N budgets. For effective use of hairy vetch, it may be necessary to consider management practice to avoid inhibition of N_2 fixation and to reduce the environmental load such as leaching and N_2O emission. Unlike chemical fertilizers, the N supplied to plants by organic matter such as LMC continues for several years [47]. Nitrogen derived from organic matter applied during upland crop (soybean) cultivation is expected to affect the N budget during subsequent paddy rice cultivation. In the future, it will be necessary to evaluate the effect of organic matter application on the N budget in the entire paddy-upland rotation system, including paddy rice cultivation. Furthermore, accumulation of soil N due to continuous application of organic matter [6, 48–50] is considered to affect the field N budget. Therefore, it is necessary to evaluate the effects of organic matter application on the N dynamics and budget in the field and on crops on a long-term basis.

4. Conclusions

In paddy-upland rotation fields including soybean cultivation, the field N budget in both upland soybean and paddy rice is negative, and soil N fertility is likely to decrease due to repeated rotation. Application of organic matter is an effective measure to improve the N budget in paddy-upland rotation fields to maintain soil N fertility and crop productivity. The effect of organic matter application on the N budget depends on the type of organic matter. Further evaluation of organic matter management practices and their impact on the N budget is needed.

Acknowledgements

We are deeply grateful to the staff members of the Akita Prefectural Agricultural Experiment Station (especially Mr. Kazuki Sekiguchi and Mr. Keiji Sasaki) and the Center of Field Education and the Research, Faculty of Bioresource Sciences (present: Agri-Innovation Education and Research Center), Akita Prefectural University for their support in the management of the experimental fields. We also thank Ms. Emiko Sato, Ms. Keiko Hatakeyama, Ms. Tomoko Suzuki, and the students of the Laboratory of Soil Science, Faculty of Bioresource Sciences, Akita Prefectural University for their great help with the field survey and laboratory analyses. We also thank Ms. Hiroko Sato (Akita Prefectural Livestock Experiment Station) for providing the livestock manure compost.

Conflict of interest

The authors declare no conflict of interest.

Author details

Fumiaki Takakai[1*], Takemi Kikuchi[1], Tomomi Sato[1], Masato Takeda[1], Saki Kanamaru[1], Yasuhiro Aono[1], Shinpei Nakagawa[2], Kentaro Yasuda[3], Takashi Sato[1] and Yoshihiro Kaneta[1]

1 Faculty of Bioresource Sciences, Akita Prefectural University, Akita, Japan

2 Akita Prefectural Agricultural Experiment Station, Akita, Japan

3 Agri-Innovation Education and Research Center, Akita Prefectural University, Ogata, Japan

*Address all correspondence to: takakai@akita-pu.ac.jp

References

[1] Takahashi T, Sumida H, Nira R. A new framework for study of irrigated paddy rice and upland crops rotation farming and its relation to soil and plant nutrition science. 1. Advances and perspectives in irrigated paddy rice and upland crops rotation farming. Japanese Journal of Soil Science and Plant Nutrition. 2013;**84**:202-207 (In Japanese)

[2] Ministry of Agriculture, Forestry and Fisheries. Statistics of Agriculture, Forestry and Fisheries [Internet]. 2019. Available from: https://www.maff.go.jp/j/tokei/index.html (In Japanese) [Accessed: January 01, 2022]

[3] Ohyama T, Tewari K, Ishikawa S, Tanaka K, Kamiyama S, Ono Y, et al. Role of nitrogen on growth and seed yield of soybean and a new fertilization technique to promote nitrogen fixation and seed yield. In: Kasai M, editor. Soybean - the Basis of Yield, Biomass and Productivity. London: IntechOpen; 2017. pp. 153-185. DOI: 10.5772/66743

[4] Shimada H. A new framework for study of irrigated paddy rice and upland crops rotation farming and its relation to soil and plant nutrition science. 2.Expectations of integrated manuring technology for productive soybean cultivation on drained paddy fields. Japanese Journal of Soil Science and Plant Nutrition. 2013;**84**:208-214 (In Japanese)

[5] Yoneyama T, Nakano H, Kuwahra M, Takahashi T, Kambayashi I, Ishizuka J. Natural ^{15}N abundance of field grown soybean grains harvested in various locations in Japan and estimate of the fractional contribution of nitrogen fixation. Soil Science & Plant Nutrition. 1986;**32**:443-449

[6] Sumida H, Kato N, Nishida M. Depletion of soil fertility and crop productivity in succession of paddy rice-soybean rotation. Bulletin of the National Agricultural Research Center for the Tohoku Region. 2005;**103**:39-52 (In Japanese with English summary)

[7] Nishida M, Sekiya H, Yoshida K. Status of paddy soils as affected by paddy rice and upland soybean rotation in Northeast Japan, with special reference to nitrogen fertility. Soil Science & Plant Nutrition. 2013;**59**:208-217. DOI: 10.1080/00380768.2012.762588

[8] Nishida M. Decline in fertility of paddy soils induced by paddy rice and upland soybean rotation, and measures against the decline. JARQ. 2016;**50**:87-94

[9] Ministry of Agriculture, Forestry and Fisheries. Basic direction for improvement of soil fertility [Internet]. 2009. Available from: http://www.maff.go.jp/j/seisan/kankyo/hozen_type/h_dozyo/pdf/chi4.pdf (In Japanese) [Accessed: January 01, 2022]

[10] Hattori M, Nagumo Y, Sato T, Fujita Y, Higuchi Y, Ohyama T, et al. Effect of continuous cropping and longterm paddy-upland rotation on yield reduction of soybean in Niigata prefecture. Japanese Journal of Crop Science. 2013;**82**:11-17 (In Japanese with English summary)

[11] Hirokawa T, Inahara M, Koike J. Declines in soil nitrogen fertility after rotation of medium and coarse-textured gray lowland soil and restoration methods with green manure and cattle manure compost. Bulletin of the Agricultural Research Institute, Toyama Prefectural Agricultural, Forestry & Fisheries Research Center. 2011;**2**:11-26 (In Japanese with English summary)

[12] Matsumoto S, Yoshikawa M. Influence of continuous cropping on

yield of black soybean and chemical properties of soils in the field converted from paddy. Japanese Journal of Crop Science. 2010;**79**:268-274 (In Japanese with English summary)

[13] Odahara K, Fukushima Y, Araki M, Kaneko A, Aramaki K. The soil fertility status and soybean productivity in paddy-upland rotation fields in Japan's Chikugo River basin. Japanese Journal of Soil Science and Plant Nutrition. 2012;**83**:405-411 (In Japanese with English summary). DOI: 10.20710/dojo.83.4_405

[14] Takahashi T, Matsuzaki N, Shioya Y, Hosokawa H. Influence of soil properties on the yield of soybean in upland fields converted from rice paddies—A case study in Joetsu region, Niigata. Bulletin of the National Agricultural Research Center. 2005;**6**:51-58 (In Japanese with English summary)

[15] Nishida M. Changes in nitrogen fertility of soils in paddy-upland rotation fields and strategies for its fertility management. In: In Japanese Society of Soil Science and Plant Nutrition, editor. Tokyo: Hakuyu-sha: Fertility and Management of Soil in Paddy-Upland Rotation Fields in Japan-Factors of Fertility Change and Approaches for Its Control; 2010. pp. 27-52 (In Japanese)

[16] Katayanagi N, Ono K, Fumoto T, Mano M, Miyata A, Hayashi K. Validation of the DNDC-Rice model to discover problems in evaluating the nitrogen balance at a paddy-field scale for single-cropping of rice. Nutrient Cycling in Agroecosystems. 2013;**95**:255-268. DOI: 10.1007/s10705-013-9561-1

[17] Koyama T, App A. Nitrogen balance in flooded rice soils. In: Rockwood WG, editor. Nitrogen and Rice. Los Baños: International Rice Research Institute; 1979. pp. 95-104

[18] Takakai F, Takeda M, Kon K, Inoue K, Nakagawa S, Sasaki K, et al. Effects of preceding compost application on the nitrogen budget in an upland soybean field converted from a rice paddy field on gray lowland soil in Akita. Soil Science & Plant Nutrition. 2010;**56**:760-772. DOI: 10.1111/j.1747-0765.2010.00503.x

[19] Takakai F, Kikuchi T, Sato T, Takeda M, Sato K, Nakagawa S, et al. Changes in the nitrogen budget and soil nitrogen in a field with paddy–upland rotation with different histories of manure application. Agriculture. 2017;7:39. DOI: 10.3390/agriculture7050039

[20] Department of Agriculture, Forestry and Fisheries, Akita Prefecture. Guidelines for Soybean Cultivation. Akita, Japan: Department of Agriculture, Forestry and Fisheries, Akita Prefecture; 2015. pp. 26-29 (In Japanese)

[21] Department of Agriculture, Forestry and Fisheries, Akita Prefecture. Guidelines for Rice Cultivation. Akita, Japan: Department of Agriculture, Forestry and Fisheries, Akita Prefecture; 2014. pp. 43-51 (In Japanese)

[22] Takahashi Y, Chinushi T, Nakano T, Ohyama T. Evaluation of N_2 fixation activity and N absorption activity by relative ureide method in field-grown soybean plants with deep placement of coated urea. Soil Science & Plant Nutrition. 1992;**38**:699-708

[23] Takahashi Y, Chinushi T, Ohyama T. Quantitative estimation of N_2 fixation activity and N absorption rate in field grown soybean plants by relative ureide method. Bulletin of the Faculty of Agriculture, Niigata University. 1993;**45**:91-105

[24] Salvagiotti F, Cassman KG, Sprecht JE, Walters DT, Weiss A, Dobermann A.

Nitrogen uptake, fixation and response to fertilizer N in soybeans: A review. Field Crops Research. 2008;**108**:1-13

[25] Hasukawa H, Takahashi Y, Toritsuka S, Kawamura H, Yamada Y. Field–scale environmental impact assessment and investigation of nitrogen balance for immediate soybean cultivation after construction of new soil water control system. Japanese Journal of Soil Science and Plant Nutrition. 2014;**85**:509-514 (In Japanese)

[26] Shimada S, Kokubun M, Matsui S. Effects of water table on physiological traits and yield of soybean (*Glycine max*). I. Effects of water table and rainfall on leaf chlorophyll content, root growth and yield. Japanese Journal of Crop Science. 1995;**64**:294-303

[27] Shimada S, Hamaguchi H, Kim Y, Matsuura K, Kato M, Kokuryu T, et al. Effects of water table control by farm-oriented enhancing aquatic system on photosynthesis, nodule nitrogen fixation, and yield of soybeans. Plant Production Science. 2012;**15**:132-143. DOI: 10.1626/pps.15.132

[28] Takahashi Y, Chinushi T, Nagumo Y, Nakano T, Ohyama T. Effect of deep placement of controlled release nitrogen fertilizer (coated urea) on growth, yield and nitrogen fixation of soybean plants. Soil Science & Plant Nutrition. 1991;**37**:223-231

[29] Takahashi Y, Chinushi T, Nakano T, Hagino K, Ohyama T. Effect of placement of coated urea on root growth and rubidium uptake activity in soybean plants. Soil Science & Plant Nutrition. 1991;**37**:735-739

[30] Takahashi Y, Ohyama T. Technique for deep placement of coated urea fertilizer in soybean cultivation. JARQ. 1999;**33**:235-242

[31] Tewari K, Suganuma T, Fujikake H, Ohtake N, Sueyoshi K, Takahashi Y, et al. Effect of deep placement of N fertilizers and different inoculation methods of Bradyrhizobia on growth, N_2 fixation activity and N absorption rate of field-grown soybean plants. Journal of Agronomy and Crop Science. 2004;**190**:46-58

[32] Tewari K, Onda M, Ito S, Yamazaki A, Fujikake H, Ohtake N, et al. Comparison of the depth of placement of lime nitrogen on growth, N_2 fixation activity, seed yield and quality of soybean (*Glycine max* [L.] Merr.) plants. Soil Science & Plant Nutrition. 2006;**52**:453-463

[33] Tewari K, Onda M, Ito S, Yamazaki A, Fujikake H, Ohtake N, et al. Effect of deep placement of slow-release fertilizer (lime nitrogen) applied at different rates on growth, N2 fixation and yield of soya bean (*Glycine max* [L.] Merr.). Journal of Agronomy and Crop Science. 2006;**192**:417-426

[34] Tewari K, Sato T, Abiko M, Ohtake N, Sueyoshi K, Takahashi Y, et al. Analysis of the nitrogen nutrition of soybean plants with deep placement of coated urea and lime nitrogen. Soil Science & Plant Nutrition. 2007;**53**:772-781

[35] Salvagiotti F, Specht JE, Cassman KG, Walters DT, Weiss A, Dobermann A. Growth and nitrogen fixation in high–yielding soybean: Impact of nitrogen fertilization. Agronomy Journal. 2009;**101**:958-970

[36] Nira R, Hamaguchi H. Nitrogen accumulation in soybean [*Glycine max* (L.) Merr.] is increased by manure compost application in drained paddy fields as a result of increased soil nitrogen mineralization. Soil Science & Plant Nutrition. 2012;**58**:764-771. DOI: 10.1080/00380768.2012.749524

[37] Nira R, Hamaguchi H, Ishitsuka A, Sekiguchi T. Soil nitrogen fertility and soybean growth responses to no-tillage and manure compost application on paddy-upland rotation fields. Communications in Soil Science and Plant Analysis. 2020;**51**:2569-2580. DOI: 10.1080/00103624.2020.1845349

[38] Miura K, Togami K, Yoshizumi K, Kudo K, Aoki K. Causes of a decrease in soybean yields by continuous cropping and effects of successive application of cattle manure compost on soybean yields and soil physicochemical properties in terms of soil types on the Pacific seaboard of the Tohoku region. Bulletin of the National Agricultural Research Center. 2017;**119**:59-78

[39] Yoshida S. Effect of farmyard manure on the nitrogen nutrition of soybean. Japanese Journal of Crop Science. 1979;**48**:17-24 (in Japanese with English summary). DOI: 10.1626/jcs.48.17

[40] Sato T, Yoshimoto Y, Watanabe S, Kaneta Y, Sato A. Effect of hairy vetch planting on changes in soil physical properties and soybean early growth in a heavy clayey soil field. Japanese Journal of Soil Science and Plant Nutrition. 2007;**78**:53-60 (In Japanese with English summary)

[41] Sato T, Yoshimoto S, Nakamura Y, Sato E, Takakai F, Shibuya T, et al. Effect of planting with leguminous green manure plant hairy vetch on growth and yield of succeeding crop soybean in a heavy clay soil field converted from paddy field. Japanese Journal of Soil Science and Plant Nutrition. 2011;**82**:123-130 (In Japanese with English summary)

[42] Sato T, Sato E, Takakai F, Yokoyama T, Kaneta Y. Effects of hairy vetch foliage application on nodulation and nitrogen fixation in soybean cultivated in three soil types. Soil Science & Plant Nutrition. 2011;**57**:313-319. DOI: 10.1080/00380768.2011.569375

[43] Sato T. Nitrogen fixation and assimilation in green manure crops. In: Ohyama T, Sueyoshi K, editors. Nitrogen Assimilation in Plants. Kerala: Research Signpost; 2010. pp. 233-241

[44] Sato T. Effects of *rhizobium* inoculation on nitrogen fixation and growth of leguminous green manure crop hairy vetch (*Vicia villosa* Roth). In: Ohyama T, editor. Advances in Biology and Ecology of Nitrogen Fixation. London: IntechOpen; 2014. pp. 225-236. DOI: 10.5772/56992

[45] Streeter JG. Synthesis and accumulation of nitrate in soybean nodules supplied with nitrate. Plant Physiology. 1982;**69**:1429-1434

[46] Takakai F, Goto M, Watanabe H, Hatakeyama K, Yasuda K, Sato T, et al. Effects of the autumn incorporation of rice straw and application of lime nitrogen on methane and nitrous oxide emissions and rice growth of a high-yielding paddy field in a cool-temperate region in Japan. Agriculture. 2021;**11**:1298. DOI: 10.3390/agriculture11121298

[47] Nishida M, Sumida H, Kato N. Fate of nitrogen derived from 15N-labeled cattle manure compost applied to a paddy field in the cool climate region of Japan. Soil Science & Plant Nutrition. 2008;**54**:459-466. DOI: 10.1111/j.1747-0765.2008.00255.x

[48] Cheng W, Padre AT, Sato C, Shiono H, Hattori S, Kajihara A, et al. Changes in the soil C and N contents, C decomposition and N mineralization potentials in a rice paddy after long-term application of inorganic fertilizers and organic matter. Soil Science & Plant Nutrition. 2016;**62**:212-219. DOI: 10.1080/00380768.2016.1155169

[49] Takakai F, Kominami Y, Ohno S, Nagata O. Effect of the long-term application of organic matter on soil carbon accumulation and GHG emissions from a rice paddy field in a cool-temperate region, Japan -I. comparison of rice straw and rice straw compost. Soil Science & Plant Nutrition. 2020;**66**:84-95. DOI: 10.1080/00380768.2019.1609335

[50] Takakai F, Hatakeyama K, Nishida M, Nagata O, Sato T, Kaneta Y. Effect of the long-term application of organic matter on soil carbon accumulation and GHG emissions from a rice paddy field in a cool-temperate region, Japan -II. Effect of different compost applications. Soil Science & Plant Nutrition. 2020;**66**:96-105. DOI: 10.1080/00380768.2019.1681881

Chapter 5

Inocculant Formulation and Application Determine Nitrogen Availability and Water Use Efficiency in Soybean Production

Canon E.N. Savala, David Chikoye and Stephen Kyei-Boahen

Abstract

Inoculation of suitable rhizobia enhances biological nitrogen fixation in soybean production and are economically viable for use among smallholder farmers due to its low price over inorganic commercial fertilizer blends. In Mozambique, inoculants are available in liquid or solid form (powder/peat or granular). Field studies were conducted in 2017 and 2018 seasons in three agroecologies (Angonia, Nampula and Ruace) in Mozambique to evaluate the performance of inoculants when applied directly to soil and on seed before planting. Data on nodulation, plant growth, nitrogen fixed, ^{13}C isotope discrimination related water use efficiency, yield and yield components were analyzed in Statistical Analysis System® 9.4. Nodulation, yield, and yield components were significant for the different application methods, and solid form tended to be better than liquid form. The nitrogen derived from atmosphere (%Ndfa) were 45.3%, 44.2% and 43.6% with a yield of 2672, 1752 and 2246 kg ha^{-1} for Angonia, Nampula and Ruace, respectively. Overall, inoculants applied on soil or seed increase the amount of biologically fixed nitrogen and has the potential of improving soybean productivity in Mozambique.

Keywords: carbon isotope, nodulation, promiscuous, soybean, rhizobia, water use efficiency, yield

1. Introduction

1.1 Inoculation history in Africa (Mozambique)

Soybean production in Mozambique is gaining pace through land area expansion at the expense of other crops mainly driven by lucrative prices and the unsatisfied market demand particularly the poultry industry [1]. However, climate change effects, low soil fertility and poor crop management keep yield below the world average. Some farmers are seeking solutions to these challenges by adopting region adapted improved varieties, use of soil amendments such as organic manures and inoculant application to improve nitrogen availability. Nitrogen is the most limiting nutrient in soybean production due to its high uptake by plants, vulnerability to leaching, denitrification and removal through crop harvest [2]. Inoculation of rhizobia enhances biological nitrogen fixation (BNF) in soybean production and is economically viable for use among smallholder farmers due to its low price over inorganic commercial fertilizer blends [3, 4]. Likewise, soybean producers have the quest to improve yield which necessitates inoculation with effective rhizobial strains [5–7]. Inoculation improves soybean yield and increases crop resilience to climatic changes effects across Africa such as drought incidences experienced in Mozambique through better water use efficiency (WUE) [8]. Although many African countries currently produce inoculant that is effective for both promiscuous and non-promiscuous soybean varieties and other legumes like beans, cowpea, and groundnuts [9], Mozambique as a country lacks the capacity and facilities for local production. However, production volumes of these inoculants seldom satisfy in-country or regional demand warranting importation of supplementary stocks from as far as south America [10]. Unfortunately, produced inoculants fail to reach smallholder farmers in Africa on time due to logistic constrains linking production and distribution. Development of promiscuous soybean varieties, capable of fixing nitrogen with indigenous rhizobia [11] offer a promising solution to improving BNF. In addition, advancement in research has led to isolation of promising indigenous rhizobia that establish symbiotic association with soybean [12, 13]. The research was based on the notion that African soils have indigenous rhizobia strains capable of colonizing soybean root. Unfortunately, isolated indigenous rhizobia strains are yet to be commercialized despite performing better than or like the well-known USDA 110 strain. Commercial production in solid or liquid form of identified indigenous rhizobia strains is necessary to improve their efficiency since naturally they occur in low populations in the soil coupled with low efficacy as effective nitrogen fixers.

Inoculants can be packaged in liquid, peat, or granular forms. Only the liquid or peat/powder forms of inoculants are found in Mozambique with the latter being more abundant and easier to handle among producers. Both forms of inoculants can be applied on seed or directly on soil before planting. Although both forms of inoculants improve yield, variations in the amount tend to occur due to other factors such as viability, storage and environment especially soil moisture in a specific site [14]. In many cases, seed yield inoculated with liquid formula seldom gives better than the peat inoculants. Liquid inoculants offer limited protection to the rhizobia hence survivability can be a challenge in sub-optimal conditions [15–17] while peat carriers provide more protection to the live cells to a limited extend as it is still important to plant the seed or cover the soil soon after application. Bacterial cells survival on the seed or soil in Mozambique could mainly be affected by desiccation and high temperatures [18]. The most common inoculant application method in Mozambique is on seed although there exists a potential for soil application especially among the large-scale commercial soybean producers who have the capacity to mechanize farm operations.

1.2 Plant nitrogen uptake

Soybeans acquire N from either BNF or soil and sometimes inorganic N fertilizer if applied. Maximum N demand in soybean occurs between the R3 and R5 stages of development [19]. Proportions of N absorbed from these sources differ with the cropping system and management. Since BNF is an energy consuming process, soybean will not invest in it where either the soil or fertilizer N is adequate. On the other hand, unavailability of N from any of the sources during plant growth will result in N translocation from other parts of the plant such as leaves to the grain, which diminishes the photosynthesis thus reducing yield potential [20]. Soybean plant N derived from BNF leads to improved productivity. Nitrogen availability in soybean production can be enhanced through inoculation. Inoculating soybean with liquid or peat based effective rhizobia strains promotes nodulation and plant growth that contribute to increased yield. Through BNF, soybean can satisfy between 50% and 60% of its nitrogen requirement [21]. Farmers in Mozambique rarely apply external inorganic fertilizer on soybean. Therefore, the N sources of soybean production is either soil or BNF where inoculants are applied, or effective indigenous rhizobia strains exist in the soil. More so, where inoculants are applied, there exists no means to quantify the amount of N fixed in the fields other than the yield obtained. Benefits of BNF are higher when phosphorus fertilizer is applied in addition to rhizobia inoculation on soybean [5] or cowpea [3] in Mozambique.

1.3 Carbon isotope discrimination, water use efficiency and yield

Carbon is released from the plant through the leaves as CO_2 during transpiration. Likewise, water is lost from the plant by the same process through the stomata. Transpiration is important in plants as it facilitates mass-flow movement of nutrients from the roots to the above ground parts. This process is inversely correlated to availability of soil moisture content hence affecting plant WUE [22]. WUE is the ratio of plant dry matter production against the water used over a period. It can also be defined at a point in time as the ratio between the rate of carbon fixation and the rate of transpiration. ^{13}C isotope discrimination is used to determine a fraction of carbon isotope during CO_2 uptake and fixation and related to WUE that is an important physiological character as an indicator of plant adaptability to drought conditions through the functioning of the stomata [23]. It is strongly linked to the ratio of the intercellular and atmospheric concentration of CO_2 (C_i/C_a) associated with stomatal conductance and chloroplast affinity for CO_2 [24]. Therefore, the intercellular and atmospheric CO_2 ratio theoretically links WUE to ^{13}C isotope discrimination. These relationship is useful in breeding for selection of high transpiration efficiency, and increased and grain yield in soybean as demonstrated with wheat [25]. Kumar et al. [26] demonstrated a positive relationship between grain yield and ^{13}C isotope discrimination and a negative one to transpiration efficiency. Since transpiration is inverse to WUE the increase in ^{13}C isotope discrimination and WUE lead to increase in grain yield. In essence, ^{13}C isotope discrimination offer a promise to selection of criterion for high yielding drought adapted varieties. Therefore, in our study, we sort to understand how liquid or solid inoculant affect soybean WUE and yield. Earlier studies have reported that inoculation improves yield as it leads to more available N from the BNF process. However, the yield increase varies with soybean varieties and type of inoculant especially nitrogen availability even if similar strains are used [8]. The objective of this study was to evaluate soybean WUE and yield response to liquid or solid inoculants applied to soil and on seed before planting.

2. Materials and methods

2.1 Site selection and description

Field studies using soybean variety '*Safari*' (SeedCo. material) were conducted in 2017 and 2018 growing seasons at three locations, Nampula 15.2741° S, 39.3150° E,

365 m above sea level (m a.s.l.), Angonia 14.5473° S, 34.1873° E, 1224 m a.s.l. and Ruace 15.2345° S, 36.6887° E, 772 m a.s.l. in Mozambique. New fields previously under maize for two growing periods were used for each season. According to the Soils Atlas of Africa, the predominant soil type at the sites in Nampula is Haplic Lixisols while in Angonia and Ruace are Chromic Luvisols [27]. Ten soil samples were taken from 0 to 30 cm soil layer using a soil auger in a W pattern across the field for the trial before plowing or harrowing. Soils from each site were combined into a composite sample and four subsamples drawn for chemical and particle-size analysis (**Table 1**). The pH was determined using a high impedance voltmeter on 1:2 soil–water suspension. Total N was determined using The Kjeldahl method, P by Olsen's method, and K plus other bases by ICP-OES after extraction with Mehlich 3.

2.2 Inoculant sources and preparation

Two inoculants were sourced from Novozymes BioAg (Cell-Tech® liquid and Cell-Tech® peat) in Saskatoon, SK Canada and Soygro (Soyflo-liquid and Soycap-powder) in Potchefstroom South Africa. According to the manufacturers' specifications, the inoculants contained 2×10^9 cells/ml or cells/g of *Bradyrhizobium diazoefficiens* formerly known as *Bradyrhizobium japonicum* [28] USDA110 strain for Cell-Tech® and USDA122 strain for Soygro. The Cell-Tech® liquid inoculant was applied at the rate of 1900 ml/ha (3.8 ml/20 m^2 plot) while the Cell-Tech® peat was 2.32 kg/681 kg seed (170.5 g/50 kg seed/ha). On the other hand, the Soyflo-liquid and Soycap-powder were applied at 2000 ml/ha (4.0 ml/20 m^2 plot) and (250 g/50 kg seed/ha) respectively.

2.2.1 Seed application

Liquid inoculants required for 2 kg soybean seed were weighed and diluted with 100 ml of distilled water before applying on seed in a plastic bag. The seeds were then mixed well for the surfaces to be fully coated with the inoculant. For the solid-based

Location	Angonia	Gurue	Nampula
pH	6.4	6.2	6.6
P (ppm)	25.0	44.8	0.3
K (ppm)	122.8	421.0	90.4
Ca (ppm)	772.8	1755.0	800.5
Mg (ppm)	165.5	301.8	113.0
Na (ppm)	29.4	17.9	29.3
EC (dS/cm)	0.059	0.057	0.050
CEC ($cmol_c$/kg)	6.6	15.0	6.0
N (%)	0.09	0.15	0.13
Sand (%)	64.0	56.8	63.2
Silt (%)	6.6	12.1	2.0
Clay (%)	29.4	31.1	34.8

Table 1.
Soil characteristics at the experimental sites' soils.

inoculants, the seeds were weighed into a plastic bag then moist with water for Cell-Tech® peat or Mollyflo for the Soycap-powder. Seeds were then mixed well in the plastic bag until all the surfaces were coated with a film of water or Mollyflo. Then respective quantities of solid-based inoculants added and mixed well to cover the surfaces of all the seeds. All the preparations were done under shade and the seeds planted within 2 h of mixing with the inoculant.

2.2.2 Soil application

Volumes of inoculants to be applied on soil per plot were measured using a syringe into 2 l hand sprayers before adding 1 l of distilled water. The mixture was then agitated gently to equally distribute the inoculant cells in the water. Later the mixture was sprayed into open seed furrows followed immediately with seed placement and covering with soil. To apply the solid-based inoculants onto soil, quantities of respective plot inoculants were weight and mixed with 100 g moist fine sieved (1 mm sieve) soil in a wide mouth plastic container with a lid. Then soil and inoculant were mixed thoroughly by shaking. The lid was then perforated using a hot nail to open many holes like a saltshaker. This mixture was then applied in open furrow followed by immediate planting of seeds and covering with soil. To avoid scorching of the rhizobia strains to death in the sun, immediately planting the seeds and covering with soil is recommended.

2.3 Experimental layout

A disc plow was used for land preparation followed by two passes of harrowing. Both seasons' experiments were planted between 16 and 24 December depending on the onset of rains in each site. Experimental treatments were formulated by combining the two inoculants, their formula (liquid or solid) and place of application (seed or soil) plus a control (no amendment). These resulted in nine treatments that were layered out in a Randomized Complete Block Design (RCBD). A non-promiscuous soybean variety Safari was planted in plots of 20 m^2 in four replications. Plots consisted of seven rows of 8 m in length, 0.50 m row-spacing and 0.1 m between plants within rows. During establishment of the trials, similar treatments were planted by one person for all the four replicates to avoid contamination. Planting and weeding (twice) were done by hand at site-specific scheduling. The experiment was conducted under rainfed conditions for both seasons with no external water supply through irrigation. Pests were controlled once at beginning of flowering using 100 ml of Cypermethrin (200 g active ingredient/l) and 50 ml of Lambda Cyhalothrin (50 g active ingredient/l) applied using 15 l knapsack sprayer.

2.4 Data collection

Data on nodulation, plant growth, nitrogen fixed, ^{13}C related WUE, yield and yield components were collected. At R3 (flowering to podding) growth stage when pods had reached 10–12 mm long at one of the four uppermost nodes on main stem, five randomly selected soybean plants were excavated using a hoe and spade from each plot ensuring that all the roots were recovered. All the soil was washed out of the roots and all nodules plucked out carefully by hand. The nodules were counted and later placed in envelopes before drying in an oven at 60°C for 48 h to determine nodule dry weight. Plant biomass were also dried in an oven at 60°C until constant dry weight was achieved. Later the biomass was ground to pass through a 2-mm mesh sieve for plant tissue N analysis stable light isotope ratio mass spectrometer. At maturity, 10 plants were randomly selected and harvested for determination of pod density and seed weight. Pods from each plot were threshed manually and grain yield was determined. The moisture content of grain samples from each plot was measured using Farmex MT-16 grain moisture Tester (AgraTronix LLC, Streetsboro, Ohio, USA) and grain yield in kg ha^{-1} was adjusted to 13% moisture content. Above-ground plant biomass from whole plots were sun-dried to 10% moisture content for 10 days to determined harvest biomass weight.

2.4.1 Measurement of shoot N and C isotopes

The isotopic analyses of ^{15}N and ^{13}C were performed at the Mammal Research Institute, University of Pretoria, Pretoria, South Africa using a Stable Light Isotope Laboratory on a Flash EA 1112 Series coupled to a Delta V Plus stable light isotope ratio mass spectrometer via a ConFlo IV system (Thermo Fischer, Bremen, Germany). Aliquots of 1.2 mg were weighed into toluene pre-cleaned tin capsules. During the analysis, a standard (Merck Gel: $\delta^{13}C = -20.57‰$, $\delta^{15}N = 6.8‰$, C% = 43.83, N% = 14.64) and a blank sample were run after every 12 samples. The air nitrogen was used as the reference isotope values for nitrogen. The ^{15}N natural abundance expressed as the δ (delta) notation is the ‰ deviation of the ^{15}N natural abundance of the sample from atmospheric N_2 (0.36637 atom % ^{15}N) was calculated [29] with the analytical precision values used being <0.2‰ for $\delta^{13}C$ and < 0.2‰ for $\delta^{15}N$.

$$\delta^{15}N = \left[\left(\left({}^{15}N/{}^{14}N\right)_{plant} - \left({}^{15}N/{}^{14}N\right)_{atm}\right)/\left({}^{15}N/{}^{14}N\right)_{atm}\right] \times 1000 \qquad (1)$$

The percentage *N* derived from the legume (%Ndfa) was determined using [30]:

$$\%Ndfa = \left(\left(\delta^{15}N_{ref} - \delta^{15}N_{plant}\right)/\left(\delta^{15}N_{ref} - B_{value}\right)\right) \times 100 \qquad (2)$$

Where, $\delta^{15}N_{ref}$ is the mean ^{15}N natural abundance of the collected reference plants (maize), $^{15}N_{leg}$ is the ^{15}N natural abundance of soybean, and the *B* value is the ^{15}N natural abundance of the test legume wholly dependent on N_2 fixation for its N nutrition. The B_{value} replaces atmospheric N_2 as it incorporates the isotopic fractionation associated with N_2 fixation. The *B* value used for estimating %Ndfa in this study was −0.72‰ [29, 31, 32]. The amount of N-fixed was calculated based on the method established by [33].

$$N - \text{fixed} = (\%Ndfa/100) \times \text{legume biomass N} \qquad (3)$$

Where legume biomass N refers to the N content of plants shoots.

2.4.2 Carbon assimilation and water use efficiency

To perform the $^{13}C/^{12}C$ isotopic analysis, the plants shoots were weighed (subsampled) into tin capsules and analyzed on a mass spectrometer as described for the $^{15}N/^{14}N$ isotopic analysis. Shoot C content was calculated by relating plant %*C* to the biomass of the plant.

$$\text{Shoot C content} = \%C \times \text{shoot biomass per plant} \qquad (4)$$

Reference carbon isotope values were the Vienna Pee-Dee Belemnite (PDB). Change in $_{13}C$ (Δ13C) was calculated as follows

$$\Delta^{13}C = \left(\delta^{13}C_{atm} - \delta^{13}C_{plant}\right) / \left(1 + \delta^{13}C_{plant}\right). \tag{5}$$

Where $\delta^{13}C_{atm}$ is ^{13}C change in atmospheric CO_2 (−8) and $\delta^{13}C_{plant}$ in plant material.

The relationship between carbon fixation and stomatal conductance in soybean at R3 stage was determined based on the model linking the isotope discrimination ($\Delta^{13}C$) to plant and atmospheric ^{13}C [34]. A linear relationship was used to relate the isotope discrimination to plant physiological properties.

$$\Delta^{13}C = a + (b\text{–}a)/(C_i/C_a). \tag{6}$$

Where a is the discrimination against $^{13}CO_2$ during CO_2 diffusion through the stomata (a = 4·4‰), b is the discrimination associated with carboxylation (b = 27‰), and C_i and C_a are the intercellular and atmospheric ambient CO_2 concentrations respectively. According to Fick's law (1855) that states 'the rate of diffusion of a substance across unit area (such as a surface or membrane) is proportional to the concentration gradient'. Then Movement of CO_2 can be expressed as;

$$A = g_{CO2}(C_i/C_a) \tag{7}$$

Since the ratio of leaf conductance to water vapor is 1.6 g CO_2, and therefor change in ^{13}C can be related to the A/gH_2O ratio as follows:

$$\Delta^{13}C = a + (b\text{–}a)\left(1 - 1.6\left(A/C_a \text{ g } gH_2O\right)\right. \tag{8}$$

WUE defined as the ratio of the fluxes of net photosynthesis and conductance for water vapor (A/E) which indicates carbon assimilated per unit of water umol mol^{-1}) [35]. Therefore, water-use efficiency at growth level (WUE_g) – biomass accumulated over water transpired (g C kgH_2O^{-1}) was calculated as:

$$WUE_g = C_a\left[\left(b - \Delta^{13}C\right)/1.6(b\text{–}a)\right] \tag{9}$$

2.5 Data analysis

Analyses of variance (ANOVAs) were performed using PROC GLM in Statistical Analysis System (SAS)® 9.4 [36]. First a combined analysis across locations and cropping seasons was performed. Since location and season effects were dominant, the two variables were combined to form environment. Secondly, a factorial ANOVA was performed, to evaluate the effects of environment, treatment, and their interactions. Environments effects were considered random and were significant for all the variables [37] while the treatments factors were fixed effects for each environment. Means were determined for treatments, and comparisons done using Tukey adjustment at $p \leq 0.05$ significance level based on the standard error of means (SEM) [36].

3. Results

3.1 Nodulation

Formation of nodules is an indicator of BNF through the symbiotic relationship of soybean plant and the inoculant strains. Data on nodule count and dry weight per plant were collected for both crown and lateral nodules. There were no significant differences ($p \leq 0.05$) in the nodule count and dry weight between treatments, sites and their interactions for both crown and lateral nodules. It was however evident that crown nodules of inoculated soybean averaging at 20.4 nodules plant^{-1} were more than lateral nodules at 18.6 nodules plant^{-1} against the check of 3.4 nodules plant^{-1} and 3.2 nodules plant^{-1} respectively. Total nodule counts, and weight combined both crown and lateral nodules were significant between treatments at Angonia in 2017, Ruace in 2017 and Ruace in 2018 (**Tables 2** and **3**). Angonia and Ruace sites are in well suited high potential soybean production agroecologies while Nampula site is in a low to marginal production region.

In Angonia and Ruace in 2017, nodule counts were lowest for the uninoculated soybean and the nodule count per plant was observed to be the highest from seed inoculated soybean with Soycap-powder (**Table 2**). Comparable nodules were formed for inoculated soybean at Ruace in 2018 except for Soycap-powder soil application. A common trend was observed between manufacturers/source liquid and solid inoculants regardless of the application on soil or seed. The liquid inoculants had numerically lower nodules formed than the solid (peat or powder) based. Generally, liquid based inoculants averaged at 36.5, 37.5 and 41.2 versus 56.3, 56.2 and 45.8 nodules plant^{-1} for Angonia 2017, Ruace 2017 and Ruace 2018 respectively. Except for Ruace 2018 with 50.2 and 36.8 nodules plant^{-1} for seed and soil inoculant application, mean number of nodules formed between the two inoculation methods were not different for the other environments. The total number of nodules formed per plant were significantly higher ($p \leq 0.05$) for the inoculated soybean in all the sites at 46.4, 46.9 and 43.5 than the uninoculated plants at 9.0, 8.5 and 11.1 nodules plant^{-1} (**Table 2**).

Similar trends of nodules plant^{-1} were also observed for the nodule dry weight (mg plant^{-1}). Inoculated soybean had heavier nodules than the uninoculated ones averaging at 206.9, 218.8 and 249.7 mg plant^{-1} versus 33.5, 36.6 and 69.9 mg plant^{-1} for Angonia 2017, Ruace 2017 and Ruace 2018 respectively (**Table 3**). It was also noted that the dry weight per nodule at Ruace in 2018 was higher than at Angonia and

Treatment (inoculant application)	Angonia 2017	Ruace 2017	Ruace 2018
Control	9.0^{d}	8.5^{b}	11.1^{c}
Seed Cell-Tech liquid	36.6^{bc}	42.6^{a}	48.1^{ab}
Seed Cell-Tech peat	52.6^{abc}	57.5^{a}	60.5^{a}
Seed Soyflo-liquid	38.8^{bc}	38.3^{a}	36.0^{ab}
Seed Soycap-powder	63.9^{a}	58.6^{a}	56.1^{ab}
Soil Cell-Tech liquid	37.9^{bc}	34.9^{a}	42.9^{ab}
Soil Cell-Tech peat	53.4^{abc}	54.4^{a}	37.8^{ab}
Soil Soyflo-liquid	32.8^{c}	34.2^{a}	37.6^{ab}
Soil Soycap-powder	55.4^{ab}	54.4^{a}	28.7^{bc}
%CV	10.3	13.2	14.7
p-Value	<0.0001	<0.0001	0.0001

The subscripts signify statistical differences at p < 0.05. Same letters indicate no differences while different letters show significance in the treatments within the season.

Table 2.
Nodule count per plant of inoculated soybean.

Treatment (inoculant application)	Angonia 2017	Ruace 2017	Ruace 2018
Control	33.5^{d}	36.6^{b}	69.0^{b}
Seed Cell-Tech liquid	134.3^{cd}	174.1^{a}	247.0^{ab}
Seed Cell-Tech peat	259.4^{ab}	247.3_{a}	275.1^{ab}
Seed Soyflo-liquid	155.4^{bc}	176.6^{a}	228.0^{ab}
Seed Soycap-powder	294.3^{a}	295.7^{a}	310.7^{a}
Soil Cell-Tech liquid	147.0^{bc}	165.1^{a}	249.5^{a}
Soil Cell-Tech peat	238.9^{abc}	255.3^{a}	228.3^{ab}
Soil Soyflo-liquid	169.6^{bc}	180.1^{a}	239.0^{a}
Soil Soycap-powder	256.7^{ab}	256.7^{a}	220.5^{ab}
%CV	28.8	26.9	35.6
p-Value	<0.0001	<0.0001	0.0127

The subscripts signify statistical differences at $p < 0.05$. Same letters indicate no differences while different letters show significance in the treatments within the season.

Table 3.
Nodule weight (mg) per plant of inoculated soybean.

Ruace 2017 for all the treatments. The average weight per nodule was Angonia 2017 (4.3 mg nodule^{-1}), Ruace 2017 (4.6 mg nodule^{-1}) and Ruace 2018 (6.0 mg nodule^{-1}). The heaviest weight per nodule was from soybean that were inoculated with Soycap powder applied on the soil at 7.7 mg nodule^{-1} in Ruace 2018. As observed for the nodule counts, significantly heavier nodules ($p \leq 0.05$) were obtained when Soycap-powder inoculant was applied on seed which gave 294.3, 295.7 and 310.7 mg plant^{-1} of dry nodule weight at Angonia 2017, Ruace 2017 and Ruace 2018 respectively (**Table 3**). Application of the inoculants in liquid form had lighter nodules for all the sites at 151.5, 174.0 and 240.9 mg plant^{-1} against using inoculants in solid form with 262.3, 263.7, and 258.6 correspondingly. From the contrast analysis, nodule dry weight had a likelihood of increasing over the uninoculated by 173.4 mg plant^{-1} in Angonia 2017, 181.9 mg plant^{-1} in Ruace 2017 and 180.4 mg plant^{-1} in Ruace 2018 when using inoculant either as liquid or in solid form. There was a strong correlation between number of nodules and dry weight in all the environments with the coefficients ranging between 0.92 and 0.96 (**Table 4**). This suggests that variation in the nodule dry weight attributed to nodule count was between 85.1% at Angonia 2018 to 91.6% at Ruace 2017.

Environment	Correlation coefficient	Significance level
Angonia 2017	0.926	<0.0001
Nampula 2017	0.935	<0.0001
Ruace 2017	0.957	<0.0001
Angonia 2018	0.922	<0.0001
Ruace 2018	0.938	<0.0001

Table 4.
The correlation between nodule count and nodule dry weight of soybean.

3.2 Nitrogen uptake in non-promiscuous soybean safari

Nitrogen is important in soybean production. Soybean has the ability of obtaining nitrogen from the atmosphere through BNF. The proportion of nitrogen derived from the atmosphere denoted as %Ndfa by soybean used as an indicator of nitrogen fixed through BNF. The %Ndfa was as low as 3.8% for control treatment in Angonia 2017 to as high as 69.8% for soybean that were inoculated with Cell-Tech liquid inoculant at Ruace 2018 (**Figure 1**). Our study showed that inoculating soybean seed with Soycap-powder could derive as high as 50.8% of the nitrogen from the atmosphere across the environments compared to 14.1% for the uninoculated soybean. The proportion of N derived from the atmosphere significantly varied with treatment for each environment. Therefore, the highest %Ndfa was 44.0% for soil Cell-Tech peat in Nampula 2017, 46.9% for seed Soycap-powder in Angonia 2017, 66.4% for seed Soyflo-liquid and 69.8% for soil Cell-Tech liquid inoculant at Ruace 2018. In each environment, % Ndfa due to inoculation was significant ($p \leq 0.05$) between the treatments resulting in average %Ndfa of 27.5% for Nampula 2017, 36.4% for Angonia 2017, 59.1% for Angonia 2018 and 47.1% for Ruace 2018 (**Figure 1**). Consequently, the proportion of N derived from the atmosphere was higher at Angonia in 2018.

Nitrogen uptake associated to BNF by the Safari variety per hectare was also calculated across the seasons for each site. Inoculating soybean increased the amount of plant N uptake at all the three sites. Plant N uptake was highest at Angonia with 235 kg N ha^{-1}, followed by Ruace with 150 kg N ha^{-1} and at Nampula with 137 kg N ha^{-1} for the inoculated soybean against the uninoculated counterparts at 113 kg N ha^{-1}, 46 kg N ha^{-1} and 98 kg N ha^{-1} correspondingly for all the sites (**Table 5**). Different treatments had significantly high amount of plant N uptake at

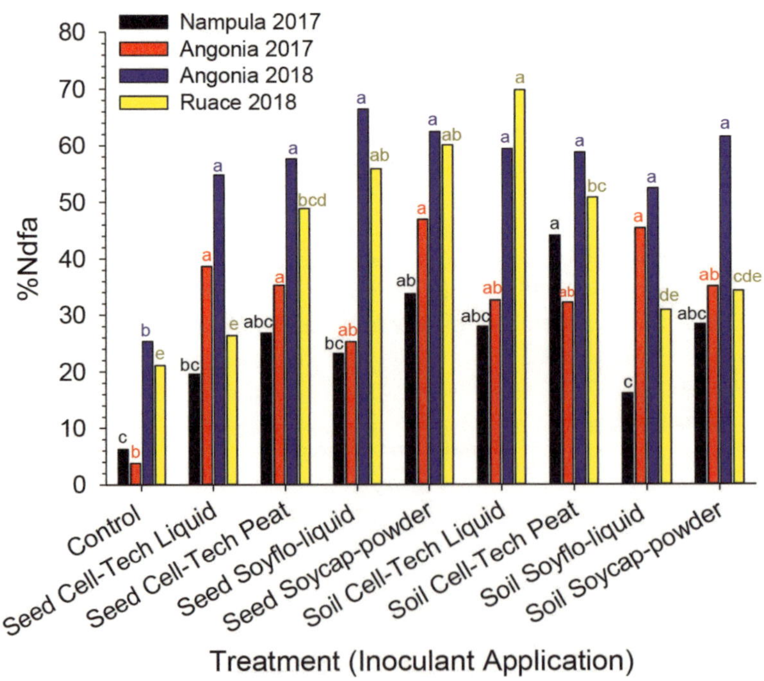

Figure 1.
Proportion of nitrogen derived from the atmosphere (%Ndfa).

Treatment (inoculant application)	Nampula	Angonia	Ruace
Control	98^{b}	113^{c}	46^{d}
Seed Cell-Tech liquid	110^{ab}	181^{abc}	112^{c}
Seed Cell-Tech peat	137^{ab}	313^{a}	129^{bc}
Seed Soyflo-liquid	110^{ab}	261^{ab}	144^{ab}
Seed Soycap-powder	136^{ab}	213^{abc}	144^{ab}
Soil Cell-Tech liquid	149^{ab}	221^{abc}	194^{a}
Soil Cell-Tech peat	154^{ab}	178^{bc}	171^{ab}
Soil Soyflo-liquid	134^{ab}	202^{abc}	120^{bc}
Soil Soycap-powder	158^{a}	307^{ab}	188^{ab}
%CV	24.1	32.1	18.8
p-Value	0.0257	0.0519	<0.0001

The subscripts signify statistical differences at $p < 0.05$. Same letters indicate no differences while different letters show significance in the treatments within the season.

Table 5.
Amount of plant nitrogen derived from BNF (kg ha^{-1}) by soybean in 2018 growing season following inoculant application.

each site. The highest plant N uptake was 158 kg N ha^{-1} at Nampula, 307 kg N ha^{-1} at Angonia for soil Soycap-powder and 194 kg N ha^{-1} for soil Cell-Tech liquid at Ruace when averaged across the seasons. Like the nodulation data, the amount of plant N uptake per ha for liquid based inoculant was numerically lower than the solid form at every application method (seed or soil) at Nampula. Since the form of inoculant also affected the amount of plant N uptake per ha at each site, solid-based inoculants resulted in more N absorbed by the plant than liquid-based at 146 vs. 126, 253 vs. 216 and 158 vs. 143 kg N ha^{-1} for Nampula, Angonia and Ruace respectively (**Table 5**).

3.3 ^{13}C isotope discrimination and water use efficiency

Water-use efficiency at growth level (WUE_g), an indicator of biomass accumulation over water transpired was calculated based on the assimilation of carbon at R3 growth stage. Before the calculations, the C:N ratio of plant biomass was also determined. Our data indicate that no significant differences existed for the C:N ratio values across the treatments with an average of 13.6 (data not presented). Similarly, ^{13}C isotope discrimination (a fraction of carbon isotope of soybean leaves during CO_2 uptake and fixation) was not significant with an average of 20.1‰ across treatments within environments except for Ruace 2018 where seed Cell-Tech peat inoculant had the lowest significant ($p \leq 0.05$) value of 19.7‰ than the other treatment (**Figure 2**). The highest numerical ^{13}C isotope discrimination at Ruace 2018 was soil Soyflo-liquid application with 20.93‰. Like the ^{13}C isotope discrimination, WUE_g was not significant among the treatments within each environment averaging at 11.8 g C kgH_2O^{-1} except at Ruace in 2018 where seed Cell-Tech peat inoculant had the highest significant ($p \leq 0.05$) value of 12.0 g C kgH_2O^{-1} (**Figure 2**). The WUE_g average ranged from 11.6 g C kgH_2O^{-1} at Nampula 2017 to 13.3 g C kgH_2O^{-1} at Angonia 2018. There were also no significant differences in applying either of the inoculants on seed or soil with a mean of 11.0 g C kgH_2O^{-1}. However, application of the inoculant in a solid form

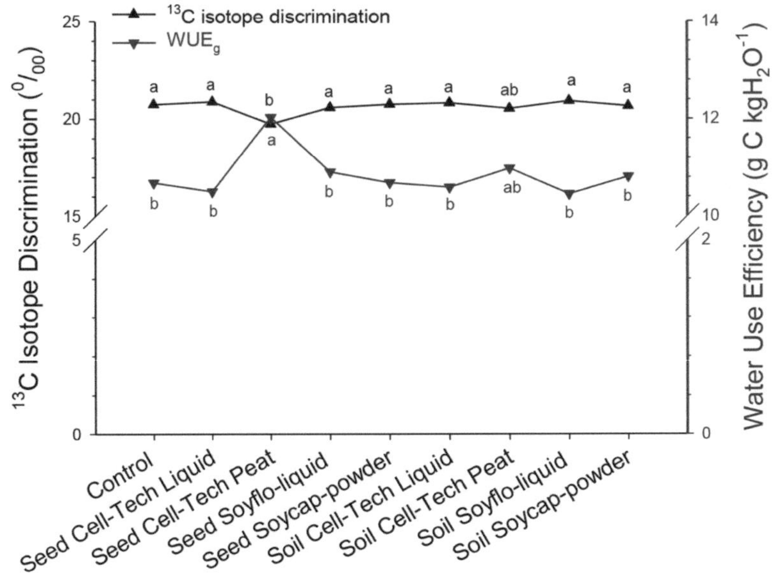

Figure 2.
Relationship between ^{13}C isotope discrimination and WUE in Ruace 2018 growing season.

resulted in a numerically higher WUE_g of 11.1 g C kgH_2O^{-1} against the liquid counterpart with 10.6 g C kgH_2O^{-1}. There is an inverse relationship between the ^{13}C isotope discrimination and WUE_g (**Figure 2**). A treatment with higher isotope discrimination had a corresponding lower WUE_g value. For instance, soybean inoculated with seed Cell-Tech peat inoculant has an isotope discrimination value of 20.89‰ which corresponded to WUE_g of 12.0 g C kgH_2O^{-1}.

3.4 Soybean yield

Inoculation treatment yield was determined within each environment. Significant differences ($p \leq 0.05$) were observed between treatments in all environments except for Ruace 2017 (p-value = 0.9851) with a mean yield of 2186 kg ha^{-1} and Angonia 2018 (p-value = 0.0883) averaging at 2572 kg ha^{-1} (**Figure 3**). However, in these two environments, mean yield of the inoculated soybean 2248 kg ha^{-1} at Ruace 2018 and 2413 kg ha^{-1} at Angonia 2018 were significantly higher than the uninoculated with 1685 and 1756 kg ha^{-1} respectively. In Nampula 2017, seed Soycap powder gave the highest significant yield of 2194 kg ha^{-1} over the uninoculated production of 978 kg ha^{-1} representing over 2.3-fold increase in yield due to inoculation (**Figure 3**). Soybean production increased in the second season at the Nampula site. In Nampula 2018, the highest yield at 2059 kg ha^{-1} was 81% more than the uninoculated treatment with 1140 kg ha^{-1}. Soybean yielded better in Angonia and Ruace sites that are in high soybean production potential agroecologies. For instance, in Angonia 2017 environment, the highest statistical yield was from soil Soycap powder at 3439 kg ha^{-1} against a check of 1646 kg ha^{-1} while in Ruace 2018 was from Cell-tech liquid inoculant applied on the seed before planting with 2684 kg ha^{-1} compared to the uninoculated fields with 1439 kg ha^{-1} (**Figure 3**). From the data, we deduce that applying inoculant in solid form either on seed or soil was better than using the liquid formulation. Mean yield of solid over the liquid inoculants in the different environments were Nampula

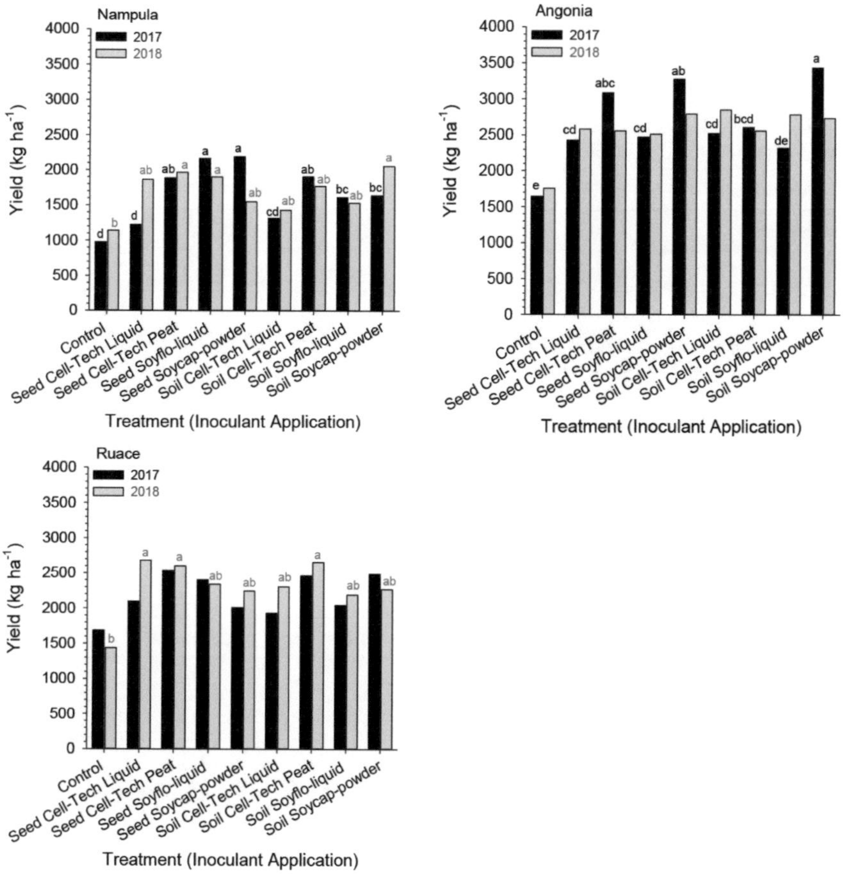

Figure 3.
Yield of inoculated soybean at three experimental sites of Nampula, Angonia and Ruace in 2017 and 2018 growing seasons.

2017 (1907 > 1580 kg ha^{-1}), Angonia 2017 (3103 > 2437 kg ha^{-1}), Nampula 2018 (1838 > 1683 kg ha^{-1}) and Ruace 2018 (2445 > 2380 kg ha^{-1}).

Contrast analysis of yield on whether to apply inoculant or not and using which placement (seed or soil) were conducted at $p \leq 0.05$ (**Table 6**). Inoculation increased yield in all the environments except Angonia 2018. Yield increase in 2017 due to inoculation was 82% in Nampula, 68% in Angonia and 35% in Ruace (**Table 6**). During the second season of 2018 inoculation increased yield from 1140 to 1760 kg ha^{-1} in Nampula and 1439 to 2413 kg ha^{-1} in Ruace. Generally, across all the environments, it was advantageous to apply inoculant on seed than the soil (**Table 6**). The differences in yield due to the inoculant form (liquid or solid), source (Cell-Tech or Soygro) and placement were also determined through contrast analysis (**Table 7**). Soygro inoculants performed statistically better than Cell-Tech counterparts in Nampula and Angonia 2017. In the same locations, solid-based inoculants enhanced yield more than the liquid-based application. Results also show that it is more beneficial to apply inoculants on the seed that the soil directly. For instance, 1868 and 2817 kg ha^{-1} yield obtained from applying inoculant on seed was more than soil placements with 1620 and 2725 kg ha^{-1} in Nampula 2017 and Angonia 2017 respectively (**Table 7**).

Contrasts	Nampula	*p*-Value	Angonia	*p*-Value	Ruace	*p*-Value
			2017			
Control	978		1646		1685	
Control vs. inoculant	1779	<0.0001	2770	<0.0001	2270	0.0251
Control vs. seed	1903	<0.0001	2817	<0.0001	2285	0.0268
Control vs. soil	1655	0.0004	2724	<0.0001	2254	0.0350
Contrasts	**Nampula**	***p*-Value**	**Ruace**	***p*-Value**		
			2018			
Control	1139		1439			
Control vs. inoculant	1753	0.0047	2413	0.0005		
Control vs. seed	1821	0.0029	2469	0.0005		
Control vs. soil	1684	0.0137	2357	0.0014		

Table 6.
Yield gains of inoculation and inoculant application place (seed or soil).

Contrasts	Nampula	*p*-Value	Angonia	*p*-Value
Cell-Tech	1584		2662	
Cell-Tech vs. Soygro	1904	<0.0001	2878	0.0501
Liquid	1580		2437	
Liquid vs. peat	1907	<0.0001	3103	<0.0001
Seed	1868		2817	
Seed vs. soil	1620	0.0002	2725	0.3894

Table 7.
Yield of soybean due to source, grade, and placement of inoculant in 2017 season.

4. Discussions

4.1 Nodulation and plant nitrogen uptake

Inoculation increased the number of nodules and dry weight. Inoculants have been shown to increase the number of nodules per plant in soybean production regardless of the source and stage of plant growth at application ranging from planting time to V6 [38]. Use of the inoculants with compatible rhizobia strain for non-promiscuous varieties [39, 40] and availability of right strain resident rhizobia for promiscuous genotypes [41] leads to formation of more nodules in soybean. In our study, on average, the number of nodules increased by 5.1 times in Angonia 2017, 5.5 times in Ruace 2017 and 3.9 times in Ruace 2018 due to inoculation with liquid and solid inoculants either in seed or direct soil application. Solid based inoculants had high number of nodules and dry weight than the liquid inoculants. Our results corroborate with the findings from a study conducted in the Eastern Region of the south of Vietnam where nodulation of the liquid inoculants was less than the peat-based inoculants for similar rhizobia strains [15, 42]. Solid based inoculants better protect the rhizobia strains from harsh environmental conditions hence leading to increased viability than the liquid inoculants. In addition, solid carrier inoculants attach better onto the seed during inoculation. Also, our data indicated that although crown nodules were fewer in number than the lateral nodules, individual nodules of the former were heavier than the later. It has been reported that crown nodules can account for up to 82% and above of the total nodule count or dry weight in soybean [43]. Crown nodules from our study accounted for 41.7–64.0% of the total nodule dry weight. More crown nodules are formed early in the season following inoculation than the lateral nodules that are formed later after development of lateral roots.

Sources of nitrogen for soybean in our study were either BNF or absorption from soil. The BNF process was enhanced by introduction of compatible rhizobia strain through inoculation. More nitrogen was fixed from the atmosphere for inoculated soybean in Angonia and Ruace relative to Nampula. Nampula lies in a semi-arid region of Mozambique with frequent incidences of drought leading to low soil moisture. High temperatures, drought and low soil moisture has been shown to reduce the effectiveness of rhizobia in BNF process leading to low nodulation hence reduced %Ndfa [44]. In Angonia and Ruace the large share of plant N was from the atmosphere representing as high as 69.8%. Other studies have reported high percentages of plant N in soybean to be associated with atmospheric nitrogen though BNF [45, 46]. As earlier indicated, plant N uptake associated with BNF varies with the biotic factors such as soybean and rhizobia characteristics as well as abiotic factors largely controlled by the environment and management. Due to the differences in the interaction levels of these factors, variations were observed in the amount of N uptake by soybean [47]. For instance, soybean in Angonia a more humid environment, absorbed more N from the atmosphere than Nampula site that is in a drier ecology. A similar trend of N fixed in wet versus drier environment was reported on farmer's fields in humid Dowa (88.9 kg N ha^{-1}) and drier Salima location (47.1 kg N ha^{-1}) in Malawi [48]. Soil moisture that depends on the rainfall amount has been reported to greatly affect amount of N fixed. The amount of N uptake was determined at R3 growth stage in soybean. This growth stage falls within the peak N demand period of flowering to podding in soybean production. Like the amount of N derived from the atmosphere, plant tissue N was enhanced by inoculation [14]. Soybean had accumulated as high as 307 kg N ha^{-1} in Angonia. These findings are like those reported for inoculated TGx 1660-19F soybean with 306 kg N ha^{-1} at Mokwa in the southern Guinea savanna of Nigeria [49]. Although we did not monitor plant N over the growing season, the amount of N in plant tissue varies with the growth stage due to the translocations that occur between plant parts.

4.2 ^{13}C isotope discrimination and water use efficiency

Both ^{13}C isotope discrimination and WUE_g were not significant among the treatments within each environment except at Ruace in 2018. This suggests that these two parameters measured at the R3 stage were not dependent on the application of the inoculant. Like our findings at R3 growth stage in soybean, Zhao et al. [50] also reported that no significant difference existed in C isotope discrimination and corresponding WUE_g at wheat harvest time. Also, Yang et al. [51] reported no clear significance differences in the amount of carbon isotope composition among C3 plants in the Yellow River region in China. For the case of Ruace in 2018 a negative relationship was observed between ^{13}C isotope discrimination and WUE_g. Values of ^{13}C isotope discrimination generally decrease with reductions in water availability. Reduced water availability leads to a decline in transpiration rate hence increased water-use efficiency [22, 35, 52]. Also earlier reported was a negative relationship between ^{13}C isotope discrimination in wheat at tillering stage and WUE of above ground biomass measured over the seedling to tillering period [50]. The change in ^{13}C isotope discrimination in relation to the environment may differ with plant growth stages due to variation in physiological processes within the plant that define its functionality requirements [53]. Since we measured ^{13}C isotope discrimination and WUE_g at one stage for all the treatments, the likelihood of soybean functionality being comparable was high and more dependent on the environment. Therefore, ^{13}C isotope discrimination can be used to determine differences in WUE_g of different soybean growth stages rather than a variation associated to inoculation at a single stage [54].

4.3 Soybean yield

Inoculation increased yield in all the three sites between an average of 602–1124 kg ha^{-1}. Our results agree with a study conducted in 2013 and 2014 in the same locations using storm a non-promiscuous variety that recorded an increase of 523–989 kg ha^{-1} [6]. These results of yield increase due to inoculation also confirms previous report [5, 8] where inoculation alone led to higher soybean yield that uninoculated. Although numerical average increase in yield due to inoculation was higher in Angonia and Ruace than Nampula across the seasons, percent rise in production was higher at Nampula 620–766 kg ha^{-1} (65%) than Angonia 918–1124 (60%) and Ruace 602–974 (52%). Chibeba et al. [6] reported and increased of 47% in yield of inoculated over the uninoculated soybean variety storm. Association between the introduced rhizobia strain and soybean was enhanced in the humid environments of Angonia and Ruace than the drier Nampula. Adequate moisture is required to take full advantage of the BNF process in inoculated soybean. The numerical rise in yield is also a pointer to the earlier reported enhanced nodulation in the inoculated soybean regardless of the placement on seed or soil. Across the sites, average soybean yield of 1440 kg ha^{-1} for the uninoculated fields is above the Mozambique national average of 1216 kg ha^{-1} [55, 56]. Therefore, use of inoculation in this study indicated that soybean yield can be increased by 1052 kg ha^{-1} over the national average figure. Our study observed that inoculant application on seed (2308 kg ha^{-1}) gave higher yield than soil application (2228 kg ha^{-1}) agrees with the report by [57] where seed inoculation 2842 kg ha^{-1} was greater than 2678 kg ha^{-1} for soil inoculation on planting line. Seed inoculation plus good adhesive agent and proper mixing of the seeds in the bag enables better distribution of the rhizobia cells per seed-grain. As a result, the rhizobia cells remain close to the seed and can attach to the root as soon as it germinates leading to better nodulation and BNF process that promote increased yield production. Seed inoculation led to a difference in yield was also noted between the liquid and solid inoculants. Solid inoculants (peat or powder) gave higher yield of 2389 kg ha^{-1} than the liquid inoculant 2147 kg ha^{-1} across the environments. Similar results where solid inoculants gave higher yields than liquid inoculants were reported from a study comparing the two forms of inoculants in Vietnam on promiscuous soybean varieties where identical rhizobia strains of in peat inoculant outyielded the liquid counterparts between 40 and 60 kg ha^{-1} [42]. These results demonstrates that farmers in Mozambique have a basket of inoculation options to choose from in enhancing soybean yield on their fields. However, selection of suitable inoculant should be made with consideration of environmental site conditions especially soil moisture availability over the growing season and the easiness of application.

5. Conclusions

Inoculation improved soybean nodulation by increasing the number of nodule count and its dry weight. Increase in nodulation could be associated to improved soybean productivity through high plant N uptake and yield. Nitrogen uptake and yield increased with application of inoculants. Farmers in Mozambique are likely to produce more soybean through using of solid cased inoculants applied on the seed than the liquid inoculants plus soil application. Although WUE_g related to ^{13}C isotope discrimination at R3 stage did not vary with inoculation, it is recommended that further study be conducted to determine cumulative WUE of the whole plant for the complete growing season while segregating for different growth stages. This could offer information on how to time soybean planting to take advantage of shifting growing seasons characteristics due to climate change. As such, soybean varieties could be selected for adaptability and resilience in specific agroecologies based on carbon assimilation, WUE and plant N uptake that affect yield. Data on inoculation and ^{13}C isotope discrimination could be utilized by breeders in selection of high yielding soybean varieties adapted to drought conditions like those found in Mozambique. The varieties developed would have high transpiration efficiency and WUE.

Acknowledgements

The authors greatly acknowledge financial support from the Consortium of International Agricultural Research Centers (CGIAR) through the Research Program on Grain Legumes and Dryland Cereals (CRP-GLDC) and United States Agency for International Development (USAID) through Feed the Future Mozambique Improved Seeds for Better Agriculture (SEMEAR) project in Mozambique. Thanks to the IITA technical staff at Angonia, Nampula and Ruace stations in Mozambique for managing the trials and collecting of field-related data.

Conflict of interest

The authors declare that the research was conducted in the absence of any commercial or financial benefits that could be construed as a potential conflict of interest.

Author details

Canon E.N. Savala[1*], David Chikoye[2] and Stephen Kyei-Boahen[1]

1 International Institute of Tropical Agriculture (IITA), Nampula, Mozambique

2 IITA-Zambia, Lusaka Province, Zambia

*Address all correspondence to: c.engoke@cgiar.org

References

[1] Santos MS, Nogueira MA, Hungria M. MINI-REVIEW microbial inoculants: Reviewing the past, discussing the present and previewing an outstanding future for the use of beneficial bacteria in agriculture. AMD Express. 2019;**9**(1): 205. DOI: 10.1186/s13568-019-0932-0

[2] Stoorvogel JJ, Smaling EMA, Janssen BH. Calculating soil nutrient balances in Africa at different scales. Supra-national scale. Nutrient Cycling in Agroecosystems. 1993;**35**:227-335. DOI: 10.1007/BF00750641

[3] Kyei-Boahen S, Savala CEN, Chikoye D, Abaidoo R. Growth and yield responses of cowpea to inoculation and phosphorus fertilization in different environments. Frontiers in Plant Science. 2017;**8**:1-13. DOI: 10.3389/fpls.2017.00646

[4] Mutuma SP, Okello JJ, Karanja NK, Woomer PL. Smallholder farmers' use and profitability of legume inoculants in western Kenya. African Crop Science Journal. 2014;**22**(3):205-213. DOI: 10.4314/ACSJ.V22I3

[5] Savala CEN, Kyei-Boahen S. Potential of inoculant and phosphorus application on soybean production in Mozambique. Universal Journal of Agricultural Research. 2020;**8**(2):46-57. DOI: 10.13189/ujar.2020.080204

[6] Chibeba AM, Kyei-Boahen S, de Fátima GM, Nogueira MA, Hungria M. Towards sustainable yield improvement: Field inoculation of soybean with Bradyrhizobium and co-inoculation with Azospirillum in Mozambique. Archives of Microbiology. 2020;**202**(9):2579-2590. DOI: 10.1007/s00203-020-01976-y

[7] Giller KE, Murwira MS, Dhliwayo DKC, Mafongoya PL, Mpepereki S. Soyabeans and sustainable agriculture in southern Africa. International Journal of Agricultural Sustainability. 2011;**9**(1):50-58. DOI: 10.3763/ijas.2010.0548

[8] Savala CEN, Wiredu AN, Okoth JO, Kyei-Boahen S. Inoculant, nitrogen and phosphorus improves photosynthesis and water-use efficiency in soybean production. The Journal of Agricultural Science. 2021;**159**(5–6):349-362. DOI: 10.1017/S0021859621000617

[9] Bala A, Karanja N, Murwira M, Lwimbi L, Abaidoo R, Giller K. Production and use of Rhizobial inoculants in Africa. In: N2Africa: Putting nitrogen fixation to work for smallholder farmers in Africa. 2011. p. 21. Available from: www.N2Africa.org

[10] Woomer PL, Huising J, Giller KE. N2Africa Final Report of the First Phase 2009–2013. 2014. pp. 138. Available from: www.N2Africa.org

[11] Tefera H, Kamara AY, Asafo-Adjei B, Dashiell KE. Breeding progress for grain yield and associated traits in medium and late maturing promiscuous soybeans in Nigeria. Euphytica. 2010; **175**(2):251-260. DOI: 10.1007/s10681-010-0181-4

[12] Nabintu NB, Ndemo OR, Sharwasi NL, Gustave MN, Kazamwali LM, Okoth KS. Demographic factors and perception in rhizobium inoculant adoption among smallholder soybeans (*Glycine max* L . Merryl) farmers of South Kivu Province of Democratic Republic of Congo, African Journal of Agricultural Research. 2020; **16**(11):1562-1572. DOI: 10.5897/AJAR2020.15030

[13] Chibeba AM, Kyei-Boahen S, Guimarães M, Nogueira MA, Hungria M.

Isolation, characterization and selection of indigenous *Bradyrhizobium* strains with outstanding symbiotic performance to increase soybean yields in Mozambique. Agriculture Ecosystems & Environment. 2017;**246**: 291-305. DOI: 10.1016/j.agee.2017. 06.017

[14] Gatabazi A, Vorster BJ, Asanzi Mvondo-She M, Mangwende E, Mangani R, Hassen AI. Nitrogen efficacy of peat and liquid inoculant formulations of *Bradyrhizobium japonicum* strain WB74 on growth, yield and nitrogen concentration of soybean (*Glycine max* L.). Nitrogen. 2021;**2**:332-346. DOI: 10.3390/nitrogen2030023

[15] Albareda M, Rodríguez-Navarro DN, Camacho M, Temprano FJ. Alternatives to peat as a carrier for rhizobia inoculants: Solid and liquid formulations. Soil Biology and Biochemistry. 2008; **40**(11):2771-2779. DOI: 10.1016/j. soilbio.2008.07.021

[16] Tittabutr P, Payakapong W, Teaumroong N, Singleton PW, Boonkerd N. Growth, survival and field performance of bradyrhizobial liquid inoculant formulations with polymeric additives. ScienceAsia. 2007;**33**(1):69-77. DOI: 10.2306/scienceasia1513-1874.2007.33.069

[17] Singleton P, Keyser H, Sande E. Development and evaluation of liquid inoculants. Inoculants and nitrogen fixation of legumes in Vietnam. In: Proceedings of a Workshop; 17–18 February 2001. Vol. 1. Hanoi, Vietnam. Canberra, Australia: Australian Centre for International Agricultural Research (ACAR). 2002. pp. 52-66. ISBN: 18632 03354

[18] Deaker R, Roughley RJ, Kennedy IR. Legume seed inoculation technology - a review. Soil Biology and Biochemistry. 2004;**36**(8):1275-1288. DOI: 10.1016/j. soilbio.2004.04.009

[19] Zapata F, Danso SKA, Hardarson G, Fried M. Time course of nitrogen fixation in field - grown soybean using nitrogen—15 methodology1. Agronomy Journal. 1987;**79**(1):172-176. DOI: 10.2134/AGRONJ1987. 00021962007900010035X

[20] van Kessel C, Hartley C. Agricultural management of grain legumes: Has it led to an increase in nitrogen fixation? Field Crop Research. 2000;**65**(2–3):165-181. DOI: 10.1016/S0378-4290(99)00085-4

[21] Salvagiotti F, Cassman KG, Specht JE, Walters DT, Weiss A, Dobermann A. Nitrogen uptake, fixation and response to fertilizer N in soybeans: A review. Field Crop Research. 2008;**108**(1):1-13. DOI: 10.1016/j.fcr.2008.03.001

[22] Diefendorf AF, Mueller KE, Wing SL, Koch PL, Freeman KH. Global patterns in leaf 13C discrimination and implications for studies of past and future climate. Proceedings of the National Academy of Sciences of the United States of America. 2010;**107**(13): 5738-5743. DOI: 10.1073/ pnas.0910513107

[23] Du B, Zheng J, Ji H, Zhu Y, Yuan J, Wen J, et al. Stable carbon isotope used to estimate water use efficiency can effectively indicate seasonal variation in leaf stoichiometry. Ecological Indicators. 2021;**121**:107250. DOI: 10.1016/j. ecolind.2020.107250

[24] Frank AB, Berdahl JD. Gas exchange and water relations in diploid and tetraploid Russian wildrye. Crop Science. 2001;**41**(1):87-92. DOI: 10.2135/ cropsci2001.41187x

[25] Rebetzke GJ, Richards RA, Condon AG, Farquhar GD. Inheritance

of carbon isotope discrimination in bread wheat (*Triticum aestivum* L.). Euphytica. 2006;**150**(1–2):97-106. DOI: 10.1007/s10681-006-9097-4

[26] Kumar BNA, Azam-Ali SN, Snape JW, Weightman RM, Foulkes MJ. Relationships between carbon isotope discrimination and grain yield in winter wheat under well-watered and drought conditions. Journal of Agricultural Science. 2011;**149**:257-272. DOI: 10.1017/S0021859610000730

[27] Jones A, Breuning-Madsen H, Brossard M, Dampha A, Deckers J, Dewitte O, Gallali T, et al. editors. Soil Atlas of Africa. Luxembourg: Publications Office of the European Union, L-2995 Luxembourg, Luxembourg; 2013. p. 176. ISSN: 1018-5593

[28] Delamuta JRM, Ribeiro RA, Ormeño-Orrillo E, Melo IS, Martínez-Romero E, Hungria M. Polyphasic evidence supporting the reclassification of *Bradyrhizobium japonicum* group Ia strains as *Bradyrhizobium diazoefficiens* sp. nov. International Journal of Systematic and Evolutionary Microbiology. 2013;**63**(Pt 9):3342-3351. DOI: 10.1099/ijs.0.049130-0

[29] Unkovich M, Herridge D, Peoples M, Cadisch G, Boddey B, Giller K, et al. Measuring Plant-Associated Nitrogen Fixation in Agricultural Systems. Canberra, Australia: Australian Centre for International Agricultural Research (ACIAR); 2008. p. 258

[30] Shearer G, Kohl DH. Natural ^{15}N abundance as a method of estimating the contribution of biologically fixed nitrogen to N_2-fixing systems: Potential for non-legumes. Plant and Soil. 1988; **110**(2):317-327. Available from: http://www.jstor.org/stable/42937618

[31] Nebiyu A, Huygens D, Upadhayay HR, Diels J, Boeckx P. Importance of correct B value determination to quantify biological N_2 fixation and N balances of faba beans (*Vicia faba* L.) via ^{15}N natural abundance. Biology and Fertility of Soils. 2014;**50**(3):517-525. DOI: 10.1007/s00374-013-0874-7

[32] López-Bellido FJ, López-Bellido RJ, Redondo R, López-Bellido L. B value and isotopic fractionation in N_2 fixation by chickpea (*Cicer arietinum* L.) and faba bean (*Vicia faba* L.). Plant and Soil. 2010;**337**(1):425-434. DOI: 10.1007/s11104-010-0538-4

[33] Maskey SL, Bhattarai S, Peoples MB, Herridge DF. On-farm measurements of nitrogen fixation by winter and summer legumes in the hill and Terai regions of Nepal. Field Crop Research. 2001;**70**(3):209-221. DOI: 10.1016/S0378-4290(01)00140-X

[34] Farquhar GD, O'Leary MH, Berry JA. On the relationship between carbon isotope discrimination and the intercellular carbon dioxide concentration in leaves. Australian Journal of Plant Physiology. 1982;**9**(2):121-137

[35] Ehleringer JR, Cerling TE. Atmospheric CO_2 and the ratio of intercellular to ambient CO2 concentrations in plants. Tree Physiology. (Victoria, Canada: Heron Publishing). 1995;**15**:105-111. DOI: 10.1093/treephys/15.2.105

[36] Institute SAS. SAS/STAT 15.1® User's Guide: High-Performance Procedures. Cary, NC: SAS Institute Inc; 2018. p. 805

[37] Moore KJ, Dixon PM. Analysis of combined experiments revisited. Agronomy Journal. 2015;**107**(2):763-771. DOI: 10.2134/agronj13.0485

[38] Moretti LG, Lazarini E, Bossolani JW, Parente TL, Caioni S, Araujo RS, et al. Can additional inoculations increase soybean nodulation and grain yield? Agronomy Journal. 2018;**110**(2):715-721. DOI: 10.2134/agronj2017.09.0540

[39] Martyniuk S, Kozieł M, Gałzzka A. Response of pulses to seed or soil application of rhizobial inoculants. Ecological Chemistry and Engineering. 2018;**25**(2):323-329. DOI: 10.1515/eces-2018-0022

[40] Cerezini P, Kuwano BH, dos Santos MB, Terassi F, Hungria M, Nogueira MA. Strategies to promote early nodulation in soybean under drought. Field Crop Research. 2016;**196**: 160-167. DOI: 10.1016/j.fcr.2016.06.017

[41] Mathenge C, Thuita M, Masso C, Gweyi-onyango J. Variability of soybean response to rhizobia inoculant, vermicompost, and a legume-specific fertilizer blend in Siaya County of Kenya. Soil & Tillage Research. 2019; **194**:104290. DOI: 10.1016/j.still.2019.06.007

[42] Thao T, Singleton P, Herridge D. Inoculation responses of soybean and liquid inoculants as an alternative to peat-based inoculants. In: Herridge D, editor. Inoculants and Nitrogens Fixation of Legumes in Vietnam. ACIAR Proceedings 109e, 17-18 February 2001. Hanoi, Vietnam: Australian Centre for International Agricultural Research (ACIAR); 2001. pp. 67-74

[43] Cardoso JD, Gomes DF, Goes KCGP, Fonseca NS Jr, Dorigo OF, Hungria M, et al. Relationship between total nodulation and nodulation at the root crown of peanut, soybean and common bean plants. Soil Biology and Biochemistry. 2009;**41**(8):1760-1763. DOI: 10.1016/j.soilbio.2009.05.008

[44] Divito GA, Sadras VO. How do phosphorus, potassium and sulphur affect plant growth and biological nitrogen fixation in crop and pasture legumes? A meta-analysis. Field Crops Research. 2014;**156**:161-171. DOI: 10.1016/j.fcr.2013.11.004

[45] Radzka E, Rymuza K, Wysokinski A. Nitrogen uptake from different sources by soybean grown at different sowing densities. Agronomy. 2021;**11**(4):720. DOI: 10.3390/agronomy11040720

[46] Herridge DF, Peoples MB, Boddey RM. Global inputs of biological nitrogen fixation in agricultural systems. Plant and Soil. 2008;**311**(1):1-18. DOI: 10.1007/s11104-008-9668-3

[47] Fermont AM, van Asten PJA, Tittonell P, van Wijk MT, Giller KE. Closing the cassava yield gap: An analysis from smallholder farms in East Africa. Field Crop Research. 2009;**112**(1):24-36. DOI: 10.1016/j.fcr.2009.01.009

[48] van Vugt D, Franke AC, Giller KE. Understanding variability in the benefits of N2-fixation in soybean-maize rotations on smallholder farmers' fields in Malawi. Agriculture Ecosystems & Environment. 2018;**261**(36):241-250. DOI: 10.1016/j.agee.2017.05.008

[49] Sanginga N, Dashiell K, Okogun JA, Thottappilly G. Nitrogen fixation and N contribution by promiscuous nodulating soybeans in the southern Guinea savanna of Nigeria. Plant and Soil. 1997;**195**(2): 257-266. DOI: 10.1023/A:1004207530131

[50] Zhao B, Kondo M, Maeda M, Ozaki Y, Zhang J. Water-use efficiency and carbon isotope discrimination in two cultivars of upland rice during different developmental stages under three water regimes. Plant and Soil. 2004;**261**(1–2): 61-75. DOI: 10.1023/B:PLSO.0000035562.79099.55

[51] Yang Y, Gou R, Zhao J, Qi N, Wen Z, Kassout J, et al. Variation in carbon isotope composition of plants across an aridity gradient on the loess plateau, China. Global Ecology and Conservation. 2021. DOI: 10.1016/j.gecco.2021.e01948

[52] Farquhar GD, Ehleringer JR, Hubick KT. Carbon isotope discrimination and photosynthesis. Annual Review of Plant Physiology and Plant Molecular Biology. 1989;**40**: 503-537. DOI: 10.1146/annurev.pp.40.060189.002443

[53] Arens NC, Jahren AH, Amundson R. Can C3 plants faithfully record the carbon isotopic composition of atmospheric carbon dioxide? Paleobiology. 2000;**26**(1):137-164. DOI: 10.1666/0094-8373(2000)026 <0137:CCPFRT>2.0.CO;2

[54] Bunce J. Consistent differences in field leaf water-use efficiency among soybean cultivars. Plants. 2019;**8**(123): 1-8. DOI: 10.3390/plants8050123

[55] MASA. Anuário de Estatísticas Agrárias 2015. Maputo, Mozambique; 2016. Available from: https://www.masa.gov.mz

[56] Gyogluu C, Boahen SK, Dakora FD. Response of promiscuous-nodulating soybean (*Glycine max* L. Merr.) genotypes to Bradyrhizobium inoculation at three field sites in Mozambique. Symbiosis. 2016;**69**:81-88. DOI: 10.1007/s13199-015-0376-5

[57] Hungria M, Nogueira MA, Campos LJM, Menna P, Brandi F, Ramos YG. Seed pre-inoculation with Bradyrhizobium as time-optimizing option for large-scale soybean cropping systems. Agronomy Journal. 2020; **112**(6):5222-5236. DOI: 10.1002/agj2.20392

Chapter 6

Optimization of Application Technology for Plant Protection Products in Soybean Crops in Brazil

Carlos Gilberto Raetano, Matheus Mereb Negrisoli, Flávio Nunes da Silva, Danilo Morilha Rodrigues and Evandro Pereira Prado

Abstract

Soybean is the main commodity of Brazilian agribusiness, and the country stands out for the largest world production of this oilseed. The culture is carried out under two main forms of cultivation, conventional and in the form of no tillage. The possibility of two to three agricultural crops per year contributes to the emergence of various plant protection problems, including soybean rust, the stinkbug complex, defoliating caterpillars, nematodes, in addition to competition with weeds. Thus, the purpose of this chapter is to describe the main application techniques of chemical or biological products in the control of agents that are harmful to the soybean crop, as well as to bring technological innovations involving remote sensing, unmanned aerial vehicle, and other techniques of application in the control of these harmful agents to the crop. Also comment on the benefits of spray adjuvants and the limitations of tank-mixes with plant protection products intended for soybean cultivation.

Keywords: pesticide application methods, remote sensing, unmanned aerial vehicle, chemical and biological control, *Glycine max*

1. Introduction

Soybean [*G. max* (L.) Merrill] is one of the most important crops in the world, principally when it comes to oilseeds. The main product of this commodity after processing is the oil and soybean meal as a protein supplement. The world production of this crop in 2020/2021 season is about 366.23 million tons over 127.88 million ha resulting in a yield of 2.86 kg ha^{-1} [1].

The countries that lead as soybean producers are Brazil and the United States of America, which account for up to 69% of the total world production of the commodity, for both countries soybean is the most exported commodity. The amount (US$) exported of soybean considering the top 10 exporters in 2020 was 54.4 billion US dollars, most of the production going to China [2].

As the biggest producer of soya, Brazil had in the 2020/2021 season a crop production of 135.9 million tons over 38.5 million ha resulting in a productivity of 3.52 kg ha^{-1}. The productivity above the global average indicates high technification of this crop production. In relation to the previous crop season, Brazil has increased the production by 8.9% and productivity by 4.4%. The income to the Brazilian's commercial balance considering all the products from soybean was US$35.2 billion in 2020 [3].

In Brazil, the soybean production depends on biological nitrogen fixation, on the no-tilled area that represents more than 70% of the cultivated area, and on integrated pest management. However, the biggest challenge is the monitoring of crops due to the large extension of cultivated areas.

2. The plant protection problems in soybean crop

In soybean crop plant protection, problems begin at sowing, especially in areas of crop succession with insect pests in the cultural remains of the previous crop, such as stink bugs and pathogens in the soil whose main target is the seeds. Currently, in Brazil, seed treatment is over 95% of all soybeans planted in the country, carried out by manufacturers of plant protection products or companies specialized in this activity.

Initially, soybean seed treatment is aimed to control soil-dwelling fungi as a seed protection measure, but with little curative effect. In Brazil, currently, mixtures containing three or four different fungicides are available to the farmer for the treatment of seeds. The aim is to protect the seed and seedling in the early stages of development. For the control of insect pests that damage seeds for a long period, carbofuran was used, a very toxic product and sometimes toxic to seedlings. Currently, diamide-based products such as cyantraniliprole are used in seed treatment for the control of caterpillars and neonicotinoids such as imidacloprid and thiamethoxam for the control of sucking insects. More recently, chemical or biological products are available for the control of nematodes in seeds, such as abamectin and *Pasteuria nishizawae*, respectively. Other natural products have been associated with biological products for nematode control, such as root growth promoters associated with bacterial spores.

The treatment of the seed with nitrogen-fixing bacteria can be carried out together with the treatment with chemical products or later before sowing, directly in the seed or in the planting furrow. The seed protection period does not exceed 25 days, and depending on the occurrence of caterpillars and other insects in the area, it may be necessary to apply an insecticide to protect the seedlings, which can be associated with the application of non-selective herbicide for weeds control.

2.1 Pathogens

Plants in the initial growth stages can be attacked by soil-dwelling fungi of the *Pythium*, *Rhizoctonia,* and *Fusarium* genera, and depending on the inoculum potential of the fungus *Phakopsora pachyrhizi*, soybean rust disease can be the main plant protection problem of the crop. This fungus is an obligatory or biotrophic parasite as it only survives in living hosts. When environmental conditions are favorable for the development of the fungus, the rust disease is highly destructive and can reduce production by up to 90% [4–6]. It is difficult to control, due to its epidemiology, which starts in the lower parts of the canopy, thus making it difficult for spray droplets to penetrate the lower parts of the plant and to the rapid breakdown of resistance genes by some cultivars.

Disease control is essentially made by use of fungicides with a specific site of action and/or protectors with multiple sites of action. Due to the frequent use of fungicides in the crop, the pathogen's resistance to chemical molecules is of increasing concern. There are reports of selection of populations of *P. pachyrhizi* resistant to triazoles [7], strobilurins [8] and carboxamides [9], three site-specific chemical groups most used in this crop to control soybean rust disease.

In addition to chemical control with fungicides, methods such as sanitary vacuum are also adopted, which aim to reduce the inoculum of the fungus during the off-season due to the absence of a host, some tolerant cultivars, and crop rotation with grass instead of soybean, favoring the management of plants from the previous culture that spontaneously germinated in the area of the current culture (voluntary plant). The sanitary vacuum is a law of the Brazilian Ministry of Agriculture and defined as the period of at least 90 days without the culture and voluntary plants in the field. The purpose of the sanitary vacuum is to reduce the survival of the fungus that causes Asian rust during the off-season and thus delay the occurrence of the disease in the crop.

In addition to the traditional diseases at the beginning and full development of soybeans, a group of diseases is characterized by affecting soybeans mainly at the end of the crop development cycle. This group of diseases are called late season diseases (DFC). The main diseases and pathogens of this group are cercospora leaf blight (*Cercospora kikuchii),* brown spot (*Septoria glycines*), Anthracnose (*Colletotrichum* spp.), and Frogeye leaf spot (*Cercospora sojina*). Cercospora leaf blight and brown spot are commonly found at the same time on soybeans from the R6 development stage and can cause over 30% yield loss. Anthracnose can be found during all soybean cycles, but as a late season disease can be present on the soybean pods, causing them to fall, the soybean seeds can also be infected. Frogeye leaf spot can be observed on the leaves, soybean pods, and seeds [10].

2.2 Insects

After sowing, young forms of beetles, aphids, bugs, and caterpillars that cut the seedlings causing failures in the stand constitute the biggest problems for the establishment of the soybeans crop. Thus, chemical products based on diamides and neonicotinoids have been frequently used both in seed treatment and in spraying up to 30 days after crop emergence. In the initial stages of development, the stink bugs existing in the area may come from the previous culture that remained in the straw and are harmful to plants since the beginning of their development.

In the vegetative development stages, defoliating caterpillars and, depending on the growing region, whitefly can cause direct and indirect damage to the crop with the transmission of viruses. Diamide-based products can be interspersed with active ingredients from other chemical groups or biological products based on *Bacillus* sp. and *Baculovirus anticarsia* for the control of lepidopteran pests. Growth-regulating products, such as pyriproxyfen, have become an important tool in the management of whitefly in soybeans.

In the reproductive development stages, the stinkbug complex becomes the biggest problem, as in addition to causing direct damage to the grains, it can cause physiological disorders in the plant, such as leaf retention. Like defoliating caterpillars, bed bugs in the hottest hours of the day have the habit of staying closer to the ground, and in times with milder conditions, they become more exposed to spraying, which can make a difference for the control of these insects.

2.3 Weeds

Weed control depends on the management adopted in the area, especially in crop succession, and two herbicide applications may be necessary in the initial post emergence of the crop, spaced 15–20 days apart. On weed management in soybeans, there is an initial period that the presence of weeds does not affect soybean yield; this interval is between 11 and 24 days after plant emergence. This interval can vary depending on the cultivar, weed infestation, and weather conditions [11]. The advance of GMO technology in cultivars that are resistant to different types of herbicides (i.e., Roundup Ready® soybean and Intacta 2 Xtend® soybean) allowed the use of post-emergence herbicides. The use of herbicides is the most economical and used technique for weed control. However, the inadequate use of the herbicides can select resistant or less tolerant weeds. In soybean crops, 335 cases of resistance have been reported worldwide since 1996 regarding different modes of action. In Brazil, *Bidens pilosa, Euphorbia heterophylla, Bidens subalternans, Brachiaria plantaginea, Digitaria ciliaris, Eleusine indica, Lolium perenne ssp. multiflorum, Parthenium hysterophorus, Conyza bonariensis, Conyza canadensis, Digitaria insularis, Conyza sumatrensis, Ageratum conyzoides, Amaranthus retroflexus, Chloris elata, Amaranthus palmeri, Amaranthus hybridus, Echinochloa crus-galli var. crus-galli* already had confirmed cases of resistance [12]. With the emergence of transgenic soybeans in Brazil, resistance to glyphosate herbicide applications has increased dramatically. This promoted the selection of weeds resistant to the herbicide, making it difficult to manage and harvest the crop. *D. insularis*, *Conyza* sp., and *A. palmeri* are weeds that are difficult to control in soybeans.

Weeds resistant to the 5-enolpyruvylshikimate-3-phosphate synthase (EPSPs) enzyme inhibitor herbicide started to be controlled with acetolactate synthase (ALS) enzyme inhibitor herbicides and are currently resistant to these two chemical groups.

3. Application techniques of plant protection products on soybean crops

In no-tillage cultivation areas, the application of plant protection products begins with the desiccation of the crop that precedes the soybean crop. Self-propelled sprayers, tractor-pulled or mounted on the three points of the hydraulic system are quite common in soybeans.

In the more advanced vegetative growth stages and in the reproductive stages, the high leaf area index, in some cultivars, reaches its maximum at the R2 development stage, making it difficult to control pests and diseases by the leaves forming a barrier to the penetration of spray droplets inside the crop canopy. The use of air assistance and transfer of electrical charge to the droplets at the spray boom can contribute to the reduction of these harmful agents to culture.

The adoption of technologies in a combined way can allow greater penetration of the droplets inside the canopy of the culture, favoring the contract of the chemical or biological product with the target to be reached. During this period, depending on the climatic conditions in the field, the appearance of soybean rust is common and, in Brazil, the existence of cultivars with an indeterminate growth habit makes it difficult to place the product in the lower parts of the canopy.

In some countries due to the use of early cultivars with determinate growth habits and greater spacing between planting rows, single or double, it is common to use hose drops. The hose drops are pendulum artifacts in the spray boom, and commonly with two spray tips at the distal end of the structure, spraying laterally, and providing better distribution of fungicides in the lower portion of the canopy.

The applications of plant protection products with drones are complementary to those carried out with self-propelled and aircraft on borders or specific points, depending on the harmful agent to be controlled. In the Midwest, Northeast, and North regions of Brazil, the predominance of large properties and the short interval between applications favor the use of agricultural aircraft and self-propelled for plant protection treatments in soybean crops. In areas of crop succession, anticipating the harvest through desiccation with herbicides has been a common practice.

4. Advances in ground boom sprayers

In Brazil, sprayers coupled to the tractor's hydraulic system, dragging and self-propelled, are still the most used equipment in the protection of the soybean crop. Self-propelled equipment appeared in Brazil in 1987 [13] and is currently the most widespread equipment in large agricultural properties. In recent years, boom sprayers have increased boom sizes from 9 m to up to 58 m. On the national market, boom supports or parts of the boom support made of metal, aluminum, or carbon fiber are available, thus making the structure lighter. The boom supports started to be developed with coil spring and shock absorber and the frame for fixing the spray boom in the system in a pendulum or trapezoidal shape, providing greater stability to the boom and greater uniformity in the application of phytosanitary products.

With the increase in the size of the spray boom and in an attempt to improve the stability of the entire system, some sprayer manufacturers opted for placing the spray boom in the front or middle part of the self-propelled. The new system for spray boom opening and closing is no longer mechanical and has become electronic. However, the greatest difficulty is the opening of the crop canopy to allow greater penetration of spray droplets in the lower parts of the canopy, especially in cultivars with indeterminate growth habits.

Soybean rust, due to the epidemiology of the disease, starts in the lower parts of the canopy, and the placement of the chemical in this region of the crop canopy becomes essential to control the disease. Different devices were developed for canopy opening, such as hose systems dragged over the crop and roller boom attached to the spray boom "canopy opener" [14]. These devices can potentially spread the pathogen's inoculum faster in the growing area, in addition to causing the flower to fall in the crop's reproductive development stage. Although the use of air assistance at the spray boom appeared in the 1980 [15], only in 1996, in Germany, air-assisted sprayers were shown at the spray boom manufactured by the Hardi company [16].

In 1997, the biggest Brazilian manufacturer of sprayers started the production of sprayers mounted on the tractor's hydraulic system and pulled by an air-assisted tractor at the spray boom. Air assistance at the spray boom gives additional kinetic energy to those smaller droplets, making it possible to reach into the crop canopy by plant movement, provided there is sufficient leaf area index in the crop. The characterization and benefits of this technology associated with the spray boom are reported in research carried out by many researchers [17–23]. With the use of technology in different crops and plant protection problems, the limitation of the use of air assistance at the spray boom on bare soil or in crops at early stages of development was evident, as drift by air deflection through the soil can be incremented. Currently, for the acquisition of equipment with this technology, it is essential for the farmer to analyze the cost-benefit ratio, considering that the use of the technology is limited.

In recent decades, another technology that was definitely adopted in the spraying of agricultural crops with the possibility of reducing application rates, drift, and environmental impact was the transfer of electrical charge to the drops (by indirect induction). This technology is dependent on the load-mass ratio and the distance from the target for good functioning, when not associated with air assistance. However, using only the transfer of electrical charge to the droplets is not enough to improve the penetration of spray droplets inside the crop canopy. The attraction of droplets by the plants promotes better spray coverage only at the top of the crop. Thus, in 2019, at Show Rural in the city of Cascavel, PR, Brazil, the self-propelled device was shown with air assistance at the spray boom and transfer of electrical charge to the drops combined.

In China and the United States, other companies also develop sprayers with the combination of both technologies (air assistance combined with the transfer of electrical charge to the drops). The first self-propelled prototype with a combination of both technologies compared with air assistance and conventional spraying on spray deposits on soybean crop was evaluated by our team in the Midwest region of Brazil, and the results are reported in [23].

5. Spray adjuvants in soybean crop

Adjuvants can be defined by "a product added to the formulation or the spray application mixture that helps or modifies the pesticide action aiming to guarantee efficacy and safety of the application process." The real role of the adjuvants in the spray process needs yet to be more discussed. However, the main characteristics of the adjuvants can be divided in modifiers of the compatibility, solubility, stability, foam formation and pH of the spray mix, reducer of drift and evaporation, also can interfere in the process of retention, adhesion, wetting, scattering, and mobility (absorption, penetration, and translocation) in the target (**Table 1**) [24].

A study [25] using different adjuvants mixed with the fungicide mancozeb conducted to evaluate the retention of the spray mix in soybean leaves showed that each adjuvant has a different interaction with the foliar surface. The adjuvants that promoted a greater retention, compared with the others tested, were a mineral oil and a polymer and the lower retention a surfactant. Also, this study concluded that the retention was associated with the surface tension of the spray mix, where both variables had a positive linear relationship, when both variables increase or decrease concurrently and at a constant rate.

Classification	Recommendation	Examples
Spreader	Increase leaf surface covered by spraying	Surfactants
Adhesives and penetrating agents	Increase penetration, absorption or adhesion of the	Mineral oils, vegetable oils, latex derivatives
Drift reducers	Decrease of very fine droplet size formation or increase the size of the droplet	Polymers, polysaccharides, oils, phospholipids

Table 1.
Classification, recommendation of use and examples of adjuvants used in soybean crops.

Another variable frequently added to the use of adjuvants is the reduction of the spray volume, a technique often used to increase the operational capacity of sprayers, which may impact on the leaf coverage. The leaf area covered can increase exponentially as droplet diameter increases. However, doubling droplet diameter requires an eightfold of spray volume. The use of a soybean methylated oil, nonylphenol ethoxylate blend, for example, can increase the average wetted area in plant surfaces from 0.055 mm^2 (water only) to 0.229 mm^2. The addition of the adjuvant results in 4.16 times reduction in spray volume with equal spray coverage [26].

The control of soybean rust, one of the most important diseases for soybean crop, showed greater dependence of the surfactant at low spray volumes applications, which provided increases in the leaf surface coverage [27]. A big concern regarding spray application, even more important for nonselective herbicides, is spray drift and volatility. A study conducted to evaluate dicamba volatilization and drift showed that the addition of adjuvants (lecithin + propionic acid, lecithin + soybean methylated ester + ethoxylated alcohol or soybean methylated oil) can decrease droplet size and increase driftable droplet percentage [28]. However, when also considering surface tension and contact angle results also measured, the dicamba-only treatment has low droplet spread potential, which may negatively affect herbicide efficacy. These results demonstrate that it is not adequate to consider only one variable, but the interaction among them when choosing an adjuvant to mix. Therefore, the choice of which adjuvant is best in each situation should take in consideration the whole scenario.

6. Application techniques of biological control agents

When it comes to biopesticides (i.e., natural organisms, including their genes or metabolites or substances derived from natural materials, for controlling pests), the application technique is one of the most challenging steps because originally the concepts involving spraying were designed to synthetic molecules and not live organisms such as the biocontrol agents. Therefore, the correct method of delivery is as crucial as it is for synthetic molecules for an optimal performance, once its control efficacy against many pathogens and pests has been already proven throughout many years of research, not only on soybean but many other crops. Therefore, the concepts of spray application technology must be improved or modified to attend to the needs of the biocontrol agent.

There are three major types of augmented biological control: classical, inoculative and inundative. In this section of the chapter, the application techniques are going to regard mostly the inundative method, which is the massive production and release of biocontrol agents or natural enemies to control the pest or pathogen quickly. In this method, no significant reproduction by the natural enemy is expected in order to control the pest population. The disease/pest control is only a result of the released individuals. This strategy can be compared with those used for synthetic chemical pesticides.

The first step when spraying biocontrol agents is to understand the life cycle of the pathogen/pest and its dynamic in the environment to decide which is the better application technique for that specific target and is crucial to understand if the methodology chosen is adequate for the biocontrol agent, beyond that, the knowledge of the biocontrol agent mode of action is also a factor that should be accounted for.

Among the application techniques, there are three that stand out for biopesticides: spraying over the crop, seed treatment, and in furrow application, where each of them has its particularities.

The spray application over the crop is recommended mostly for most of the bioinsecticides and the biofungicides recommended to control foliar diseases and white mold. As an example of the importance of understanding the biocontrol agent and its target, white mold (*Sclerotinia sclerotiorum*) a soilborne disease, the recommended technique is the spray application once the mode of action of the *Trichoderma sp.* a biocontrol agent recommended to control this plant disease, attacks the sclerotia, a survival structure of the fungus that stays above the soil surface. This technique's efficacy is very influenced by the application conditions, once the biocontrol agents are subject to the environmental conditions such as UV light, temperature, and relative humidity.

The seed treatment application recommendation for biocontrol agents follows almost the same principles of disease control as the synthetic fungicides. Biocontrol agents can be precoated or encapsulated onto the seed, the formulation of the biopesticide should guarantee a surviving period of desiccation. The seed encapsulation involves enveloping the seed, the biocontrol agent, and possibly other components such as pesticides or micronutrients, in a gelatinous or polymer gel matrix, promoting an enhancement of survival of the biocontrol agents on seed [29]. The products used in the seed treatment must be compatible with the biocontrol agent. Moreover, the seed has a maximum amount of product that can be added onto it, usually 10 mL per soybean seed.

In furrow application, targets are commonly nematodes and soilborne pathogens, this methodology applies the biocontrol agent during the soybean sowing directly on the seed. The seeder is adapted with a tank specific for this purpose, and in each row a solid stream nozzle is placed and the biopesticide is applied in a furrow. This technique when compared with seed treatment delivers the biopesticide in a more precise dose and at the correct moment.

7. Remote sensing on diseases and weed control in soybean

Remote sensing is the process to retrieve important information from an object without physical contact between the sensor and the target [30, 31]. Along with the global positioning system (GPS) and geographic information system (GIS), remote sensing is part of the precision farming techniques. Remote sensing works by capturing information from a specific area (data or imaging), while GPS is responsible for georeferencing it to locate field variables as precisely as possible, and GIS is used to interpret the data obtained to produce the final outcomes, such as application maps. Therefore, through the integration of these techniques, it is possible to obtain sufficient data to interpret and decide on the most appropriate management for a field [31, 32].

From an electromagnetic flux emitted in different wavebands, there is an interaction with the object that can be manifested in reflectance, transmittance, absorbance, fluorescence, and phosphorescence [33]. Sensors have the ability to detect differences in these values, generating indicative data regarding its physical characteristics [31, 32, 34]. These characteristics are used to identify morpho-physiological changes in plants or vegetation, such as the incidence of pathogens or insects [35] and to distinguish between plant species, biomass level, and soil, helping to identify weeds within the crop [36]. The processes of acquiring spectral data, data modeling, and detection model construction are usually extensive and require a great amount of time and research.

For pesticides technology application, remote sensing is a powerful tool for a specific branch of precision agriculture known as spot spraying or site-specific spraying [37]. Firstly, it is necessary to acquire spectral data in the field, which is usually conducted through a passive sensor that does not have its own source of energy, such as satellites and spectroradiometers, but instead is capable of retrieving information through spectral analysis (spectral signature) or imaging [31]. All information from a field is gathered with the geographical positions of certain points of interest and finally supplied to the spraying system to spray locally only where it is necessary. These places can be defined based on the incidence of disease, pests, or weeds in the field. One of the greatest benefits is the savings of product waste where application is not necessary, environmental preservation, and a more conscious application toward a sustainable agriculture [36]. The capacity of identifying diseases even prior to visual symptoms is also a great opportunity to improve control efficacy as well as the spraying technology [33].

Another example for fungicides and insecticides application is to use remote sensing to identify different biomass levels of the crop canopy during the application and using this information to adjust the water volume according to its foliage density [38]. In the study [38], a mechanical sensor was used in front of a sprayer to evaluate the crop canopy density and to vary the spraying volumes based on the canopy foliar density. This operation allowed them to reduce 13% of insecticide use, while maintaining pest control efficacy and improving natural enemies' preservation.

Nowadays, a more common monitoring by remote sensors is the one applied to herbicide treatments in preplant or over-the-top applications. An active sensor is placed at different sections of the sprayer boom and is capable of detecting weeds incidence as it moves through the field (real time monitoring), transmitting the information to a data center analyzer that uses it to make the application at the exact spot where the weed is located (**Figure 1**). These systems are equipped with dedicated valves at each spray nozzle in order to allow a unique nozzle control, such as actuated ball valves, solenoids, or PWM (pulse width modulation) valves.

Few innovations have been applied to these real-time sensors for herbicide management in soybean. At first, most of the systems were dependent on the weed detection in contrast to the soil (Green-on-Brown), therefore could only be applied to fallow lands [37]. Due to similarities in morphologies, physical characteristics, and

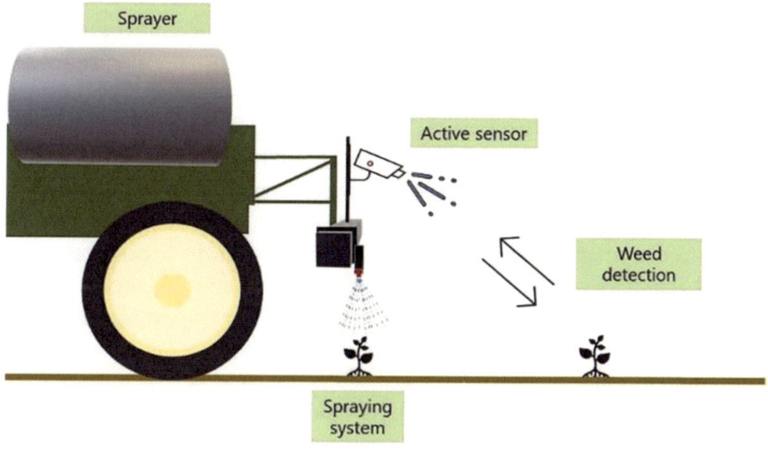

Figure 1.
Schematic demonstration of the real-time weed detection in the field by an active sensor, sending the signal to the spraying system that opens the spray nozzle right onto the weed.

colors, differentiation between plant species required advanced technology such as hyperspectral imaging sensors and deep learning algorithms [39]. This new system enables the differentiation of weeds from the main crop and is known as Green-on-Green technology.

Moreover, spot spraying application is promising to promote a proper pesticide resistance management of weeds and pathogens populations due to minimal exposure to the herbicides [40]. These technologies have potential to significantly improve spraying techniques toward a more sustainable application, acting heavily on chemicals waste reduction, resistance management, and a more precise application for a better control efficacy.

8. Tank mix of pesticides for soybean crop applications

Tank mixing seeks to improve spraying operational capacity and reduce application costs by associating different plant protection products, fertilizers, as well as other products in the spray tank moments before application [41]. This practice was regulated in Brazil only in 2018 (Normative Instruction N°. 40, October 11, 2018) [42], although it was widely performed a long time before this period, often with complex mixtures of products [43]. According to [43], 97% of applications in Brazil contained at least two products mixed in the spray tank.

Many applications of herbicides, insecticides, and fungicides in soybean crop are conducted at the same period, making it more convenient to combine the products in a single application. Therefore, it reduces the cost and time spent in operations, the movement of machines in the fields and, thereby, saves fuel, water and even reduces the number of machines needed [44]. Another great advantage of tank mixing different products is to help reduce pesticide resistance of pathogens and weeds. In Ref. [45], it is reported that applications of more than two modes of action of herbicides in tank mix reduce 83 times the likelihood of herbicide resistance selection.

Although very beneficial in terms of practicality, tank mixing requires guidelines to be correctly conducted, especially due to chemicals compatibility in the tank and positioning of technology application parameters, such as spray volume and droplet size. Overall, chemical compatibility is the main factor defining whether certain products can be mixed in the spray tank, regarding its physical and chemical compatibility [46].

Physical incompatibility may cause loss of spray solution stability, leading to the formation of precipitates, complexation, and phase separation. These changes are influenced by the pH, electrical conductivity, and surface tension of the spray liquid [46]. Furthermore, degradation of active ingredients is possible due to oxidation, hydrolysis, and encapsulation reactions [41]. On the other hand, chemical compatibility is related to the product effectiveness, in which certain products, when mixed, may vary the control efficacy it was expected to have when used alone [47]. Among several factors, incompatibilities can also occur due to improper agitation systems, water quality, application rate, and the solution pH [41].

Physical compatibility can be quickly tested in the field or laboratory through a common test known as "the jar test" [41]. The compatibility can be evaluated in a small proportion by mixing all products into a jar at the same rate or dose that it would be used in the tank, as well as the proportion of water and order added in the tank. Although it does not represent entirely how this mixing process would happen in the sprayer due to additional effects of pumps and accessories, it can be a great advantage to foresee any formation of precipitates or phase separation. Chemical compatibility can be evaluated through an efficacy test in laboratory or in the field.

One of the most important factors is the order that each product is added to the tank. In general, oily solutions are added lastly to the tank using water as a carrier. Among several methodologies, one of the most common is the addition of compatibility agents firstly, then water-dispersible granules, wettable powder, concentrated suspension, concentrated solution, and emulsifiable concentrate [41]. The product label should always be followed in this mixing process.

Tank mixes of different products are a common activity in soybean crops and have the potential to develop even more. More studies need to be conducted to assess different chemicals compatibility as well as to instruct farmers and operators on how to do it properly. However, its benefits are evident, and the need to gain operational conditions will still increase, which makes it even more important and applicable nowadays.

9. Insecticide application on planting furrow

Several species of pests can attack during the initial development of the soybean/ reduction in the crop stand and/or harming plant development, which consequently negatively affects the production of grains and/or seeds [48]. Among the soil pests, insects and nematodes may occur throughout the entire crop cycle, causing severe damage in the early stages of development of plants.

The preventive insecticide application on sowing furrowing is a promising alternative to avoid or minimize the damages caused by some pests on soybean plants [49–51]. In crop areas with high soil pest infestations, the application on planting furrows can be an option to ensure better distribution of pesticides [52]. Spraying pesticides in a planting furrow, especially when the pests are located nearby of the soil surface, can provide a good pest control [53].

In general, the insecticide application on sowing furrows is performed at the time of sowing and before covering the furrow with soil by the use of a sowing—fertilizer machine with a tank to store the pesticide solution or pesticide granules. This equipment may at the same operations fertilizer, planting and apply pesticide.

In Brazil, the application of insecticides in the seed furrow is still little used by soybean farmers with few studies in this area. However, it is noted that the insecticides application in the seed furrow has a great potential in soil pests control, promoting an initial protection to the plants with an adequate initial stand [54].

The use of insecticides applied in the seed furrow is limited, due to the need to use special machines for this operation, in addition to the higher costs of granules formulations available on the pesticide market [54]. Although this application technique is limited, some studies reported that this method of applying insecticides at label doses is effective in controlling pests that attack seeds, plants, or other subterranean plants parts for a period of 35–49 days [55].

As mentioned above, the use of pesticides on furrows to control soil pests and nematodes is not usually practiced by farmers. Most of the studies aimed to verify the effectiveness of insecticides. The effects of insecticides spraying on sowing furrow and seed treatment in order to control the insect *Phyllophaga cuyabana*, an important soil pest that consumes the soybean roots reducing the absorption of water and nutrients, were studied [49]. The authors concluded that insecticide applications are important tactics, and it may be employed in the management of *P. cuyabana* larvae in soybean crops. In this context, the insecticides fipronil, thiamethoxam, imidacloprid, and clothianidin, applied to soybean seeds, and chlorpyrifos, applied on the seed furrow, can ensure satisfactory soybean yields in areas with high pest infestation.

The same authors mentioned that the compatibility of insecticides with nitrogen fixing bacteria (*Bradyrhizobium*) should be considered and investigated, since the largest amount of nitrogen required by the crop is obtained through biological fixation.

Greater grain yield values have been reported when the insecticides fipronil, clothianidin, endosulfan, and chlorpyrifos were applied in the sowing furrow of soybean in order to control larvae of *P. cuyabana* and *Liogenys fuscus* [56]. All insecticides treatments mentioned exceeded statistically the yields of the treatment control (no insecticide application).

Increased soybean yield (approximately 20%) by application of disulfoton to the sowing furrow against two-striped leaf beetles (*Medythia nigrobilineata*) was reported [57]. According to the authors, this method can be easily applied to the use of fertilizing equipment mounted on the seed sowing machine with lower costs. The authors also believe that this pest control method is a key method for maximizing soybean yield with the labor-saving feature demanded by large-scale farming.

The performance of insecticide efficiency applied in the sowing furrow against *Sternechus subsignatus* in soybean crop was studied [54]. The author verified control up to 80% 31 days after plants emergence although this insecticide had affected negatively the final stand. The application of thiamethoxam (granule formulation) increased the soybean yield. Reduction of nematode population *Heterodera glycines* both in soil and soybean roots by nematicides/insecticides application in sowing furrow with greater soybean yield values in the treatments with nematicides was reported [50].

The applications of insecticides in sowing furrows in soybean crops can be a viable and economic alternative to be used as one more tool in Integrated Pest Management (IPM). Studies with new insecticides molecules, recently available on the pesticide market, should be encouraged in order to elucidate the control effectiveness and compatibility with nitrogen-fixing bacteria, which is widely used by Brazilian farmers. The appropriate method of applications also should be considered to maximize the insecticide performance and reduce environmental contamination.

10. Unmanned aerial vehicles (UAVs): sprayer drone in soybean crop

nlike aircraft with the presence of an on-board pilot, UAVs are used as a tool for applying pesticides in more complex terrain, with the surface with greater undulation, presence of obstacles, and in smaller areas [58]. In addition to spraying plant protection products, UAVs can also be used for image capture and remote sensing [59–62].

The use of UAVs in spraying, already well established in some countries such as China and Japan, is due to the compatibility with the farms that predominate in the region [60]. In the year 2018, there were about 30,000 spraying UAVs in China, covering an area of approximately 17.8 million hectares [63].

It is possible that, even in large areas, as in the case of soybean crops that require a large number of applications, the use of UAVs can optimize the operational capacity of the application, reduce costs, and ensure the effectiveness of phytosanitary management, which can be used in localized applications for the control of weeds, pests, and diseases.

UAVs can be classified into two segments, fixed wing and rotary wing. Fixed-wing UAVs follow the same operating principle as an airplane and are primarily used in remote sensing to capture images and map creation used in precision agriculture. The rotary wing UAVs are divided into helicopters and multirotors, which have shown a high growth in use, presenting models that are widely used to obtain images and also in the application/spraying of agricultural pesticides. Rotary-wing UAVs, especially multirotor UAVs, have the advantage of remaining stable in the air even with fluctuating winds and perform maneuvers with greater precision [58, 62, 64, 65].

The majority of spray drones available are capable of carrying 10–15 kg of liquids corresponding to 10–15 L of spray liquid. Ability to spray from 1 to 4 ha h^{-1} (10–50 ha day^{-1}), which can be 40–60 times faster than a manual operation. They have 4–6 spray nozzles, where each one can be positioned directly below a rotor or arranged on a spray boom. The application range can vary from 3 to 7 m in width [61–63]. Nozzles can be hydraulic (same as those used on land sprayers and airplanes) and also rotatory. As in airplanes, in UAVs, the electrostatic spraying system can also be used [66].

The drone's flight time is highly dependent on its weight and the presence and/or intensity of winds, but in general, it varies from 10 to 30 minutes with a fully charged battery [63, 64].

The main limitation in the use of UAVs for spraying agricultural pesticides is related to the flight autonomy provided by the batteries. One solution would be to increase the capacity and size of the batteries; however, the greater the capacity and size, the greater the weight and consequently reducing autonomy [62, 64].

In 2021, the Ministry of Agriculture, Livestock and Supply—MAPA, Brazil, through Ordinance No. 298, of September 22, 2021, established rules for the operation of remotely piloted aircraft intended for the application of pesticides and the like, adjuvants, fertilizers, inoculants, correctives, and seeds [67].

In the soybean crop, research has been conducted comparing weed control, soybean rust and stink bug control by conventional spraying methods (tractor-mounted sprayer) and with drones varying the application rate. Although research has been conducted on soybean crop with spraying, its aim is to obtain information to support applications in areas with difficult access with land sprayers, greater assertiveness in localized applications and in borders in soybean crop.

Some studies show the potential of using UAVs in soybean crops, evidencing that the deposition of the marker in the upper and middle thirds of soybean plants in applications carried out with UAV was similar to the deposition obtained with a CO2-propelled knapsack boom sprayer [68].

Recently, a field trial was installed on soybean crops at V4 stage crop to compare weed control between spraying performed with Sky Drones Pelicano 2020 drone (application rate 15 L ha^{-1}, flight height 3.0 m, displacement velocity of 18 km h^{-1}) and with the boom sprayer (application rate 120 L ha^{-1} and displacement velocity of 8.5 km h^{-1}). At 7 days after application (DAA), weed control with both application techniques was very similar after application of the herbicide mixture clethodim plus glyphosate. In the next evaluations, the differences in the control can be better evidenced.

In Brazil, the potential for use is greater for small farmers and agricultural crops with greater added value. The biggest difficulty in using this technology is in the adjustment and calibration of the equipment in view of the variation in the application rate, flight height, droplet size, and spray tips more suited to the target to be reached.

Other challenges are the size of the plots and the experimental design, as the dimensions of the plots are larger when compared with spraying with terrestrial equipment. The potential risk of drift is little known with remotely piloted aircraft. In this sense, research has been conducted with adjuvants to minimize problems with drift in drone spraying. The use of rotating nozzles has also been adopted in this technology for providing a more homogeneous droplet spectrum and, depending on the droplet size, with greater control of spray drift.

Considerable advances have been made with another one tactic within the IPM using remotely piloted aircraft, which is the distribution of parasitoids in capsules in different crops for the control of insect pests. The use of this technique presupposes knowledge of the parasitoid's ability to disperse, and through georeferenced points in the sites, the release of these insects is established.

11. Final considerations

In order to optimize the pesticide application technology in plant protection, it is necessary to overcome challenges such as connectivity in the field to better use the available resources of precision agriculture, as well as the training of people with specialized labor and the joint use of techniques of integrated management in the control of harmful agents to plants. In Brazil, especially in soybean, due to the existence of large areas of cultivation, the implementation of integrated management in plant protection has been a constant challenge, from the monitoring of pests, diseases, and weeds to decision-making and selection of techniques of control. From the south to the north of the country, the climatic conditions are very different, requiring different procedures for specific regions. In this sense, the Brazilian agricultural research company Embrapa has contributed significantly to the diffusion of technology to different regions. Regulatory activities in the commercialization of adjuvants with the inspection of the actual functionality of the adjuvants, the implementation of periodic inspection of agricultural machines, and the establishment of spray drift limits for manual, land (mounted, trailed, or self-propelled), and aerial (unmanned or with on-board pilot) applicator equipment could significantly contribute to the optimization of the entire national agricultural production process.

Author details

Carlos Gilberto Raetano[1*], Matheus Mereb Negrisoli[1], Flávio Nunes da Silva[1], Danilo Morilha Rodrigues[1] and Evandro Pereira Prado[2]

1 Department of Plant Protection, School of Agriculture, Sao Paulo State University, Botucatu, SP, Brazil

2 Department of Plant Production, College of Agricultural and Technological Sciences, Sao Paulo State University, Dracena, SP, Brazil

*Address all correspondence to: carlos.raetano@unesp.br

References

[1] USDA. World Agricultural Production [Internet]. 2021. Available from: https://downloads.usda.library.cornell.edu/usda-esmis/files/5q47rn72z/wp989m25f/9z904211s/production.pdf [Accessed: December 12, 2021]

[2] FAOSTAT. Countries by Commodity [Internet]. 2021. Available from: https://www.fao.org/faostat/en/#rankings/countries_by_commodity_exports [Accessed: December 26, 2021]

[3] CONAB. Boletim da safra de grãos [Internet]. 2021. Available from: https://www.conab.gov.br/info-agro/safras/graos/boletim-da-safra-de-graos [Accessed: December 26, 2021]

[4] Yorinori JT, Paiva WM, Frederick RD, Costamilan LM, Bertagnolli PF, Hartman GL, et al. Epidemics of Soybean Rust (*Phakopsora pachyrhizi*) in Brazil and Paraguay from 2001 to 2003. Plant Disease. 2005;**89**:675-677

[5] Henning AA, Almeida AMR, Godoy CV, Seixas CDS, Yorinori JT, Costamilan LM, Ferreira LP, Meyer MC, Soares RM, Dias WP. Manual de identificação de doenças de soja. 5.ed. Londrina: Londrina: Embrapa Soja; 2014. 76p (Embrapa Soja Documentos 256). Available from: https://ainfo.cnptia.embrapa.br/digital/bitstream/item/105942/1/Doc256-OL.pdf [Accessed: August 08, 2022]

[6] Hartman GL, Sikora EJ, Rupe JC. Rust. In: Hartman GL, Rupe JC, Sikora EJ, Domier LL, Davis JA, Steffey KL, editors. Compendium of Soybean Diseases and Pests. 5th ed. Saint Paul: APS Press; 2015. pp. 56-59

[7] Godoy CV. Risk and management of fungicide resistance in the Asian soybean rust fungus *Phakopsora pachyrhizi*. In: Thind ST, editor. Fungicide Resistance in Crop Protection. India: Punjab Agricultural University; 2012. pp. 87-95. DOI: 10.1079/9781845939052.0087

[8] Klosowski AC, May de Mio LL, Miessner S, Rodrigues R, Stammler G. Detection of the F129L mutation in the cytochrome b gene in *Phakopsora pachyrhizi*. Pest Management Science. 2016;**72**:1211-1215

[9] Simões K, Hawlik A, Rehfus A, Gava F, Stammler G. First detection of a SDH variant with reduced SDHI sensitivity in *Phakopsora pachyrhizi*. Journal of Plant Diseases and Protection. 2018;**125**(1):21-26. DOI: 10.1007/s41348-017-0117-5

[10] Godoy CV, Almeida AMR, Costamilan LM, Meyer MC, Dias WP, Seixas CDS, et al. Doenças da soja. In: Amorim L, Rezende JAM, Bergamin Filho A, Camargo LEA, editors. Manual de Fitopatologia, v.2. Doenças de plantas cultivadas. 5th ed. São Paulo: Ceres; 2016. pp. 657-676

[11] Silva AF, Concenço G, Aspiazú I, Ferreira EA, Galon L, Freitas MAM, et al. Period before interference in soybean-RR crop under low, medium and high infestation level conditions. Planta Daninha. 2009;**27**(1):57-66. DOI: 10.1590/S0100-83582009000100009

[12] Heap I. The International Herbicide-Resistant Weed Database. 2022. Available from: www.weedscience.org [Accessed: August 02, 2022]

[13] Boller W. Tecnologia de Aplicação de Defensivos II. Passo Fundo: RS; 2007. pp. 23-35

[14] Zhu H, Derksen RC, Ozkan HE, Reding ME, Krause CR. Development

of a canopy opener to improve spray deposition and coverage inside soybean canopies: Part 2. Opener design with field experiments. Transactions of the ASABE. 2008;**51**(6):1913-1921

[15] Robinson TH. Large-scale ground-based application techniques. In: Matthews GA, Hislop EC, editors. Application Technology for Crop Protection. Wallingford, Oxon, UK: CAB International; 1993. pp. 163-186

[16] Koch H. The evolution of application techniques in Europe. In: Proceedings of 1rst Simpósio Internacional de Tecnologia de Aplicação de Produtos Fitossanitários, Águas de Lindóia, São Paulo, Brazil. 26-29 March 1996. pp. 30-38

[17] Cooke BK et al. Air-assisted spraying of arable crops in relation to deposition, drift and pesticide performance. Crop Protection. 1990;**9**(4):303-311

[18] Taylor WA, Andersen PG. A review of benefits of air assisted spraying trials in arable crops. Aspects of Applied Biology. 1997;**48**:163-174

[19] Bauer FC, Raetano CG. Assistência de ar na deposição e perdas de produtos fitossanitários em pulverizações na cultura da soja. Scientia Agricola. 2000;**57**(2):271-276

[20] Matthews GA. Pesticide Application Methods. 3rd. ed. Oxford, UK: Blackwell Science; 2000

[21] Christovam RS, Raetano CG, Aguiar-Júnior HO, Dal Pogetto MHFA, Prado EP, Júnior HOA, et al. Assistência de ar em barra de pulverização no controle da ferrugem asiática da soja. Bragantia. 2010;**69**(1):231-238

[22] Prado EP, Raetano CG, Aguiar-Júnior HO, Dal Pogetto MHFA, Christovam RS, Gimenes MJ, et al. Velocidade do ar em barra de pulverização na deposição da calda fungicida, severidade da ferrugem asiática e produtividade da soja. Summa Phytopathologica. 2010;**36**(1):45-50

[23] Raetano CG, Rezende DT, Prado EP. Application technologies for Asian Soybean rust management. In: El-Shemy HA, editor. Soybean Physiology and Biochemistry. Rijeka, Croatia: InTech; 2011. pp. 117-138

[24] Raetano CG, Chechetto RG. Adjuvantes e formulações. In: Antuniassi UR, Boller W, editors. Tecnologia de aplicação para culturas anuais. 2nd ed. Passo Fundo: Aldeia Norte; Botucatu: FEPAF; 2019. pp. 29-40

[25] Silva FN. Spray Retention of Mancozebe with Adjuvants and Application Rate on Soybean Rust Control [thesis]. Botucatu: State University of Sao Paulo; 2020

[26] Gimenes MJ, Zhu H, Raetano CG, Oliveira RB. Dispersion and evaporation of droplets amended with adjuvants on soybeans. Plant Protection. 2013;**44**: 84-90. DOI: 10.1016/j.cropro.2012.10.022

[27] Roehrig R, Boller W, Forcelini CA, Chechi A. Use of surfactant with different volumes of fungicide application in soybean culture. Engineering Agriculture. 2018;**38**(4): 577-589. DOI: 10.1590/1809-4430-Eng. Agric.v38n4p577-589/2018

[28] Ferreira PHU, Thiesen LV, Pelegrini G, Ramos MFT, Pinto MMD, Ferreira MC. Physicochemical properties, droplet size and volatility of dicamba with herbicides and adjuvants on tank-mixture. Scientific Reports. 2020;**10**:18833. DOI: 10.1038/ s41598-020-75996-5

[29] Bharti V, Ibrahim S. Biopesticides: Production, formulation and application

systems. International Journal of Current Microbiology and Applied Science. 2020;**9**(10):3931-3946. DOI: 10.20546/ijcmas.2020.910.453

[30] Nilsson HE. Remote sensing and image analysis in plant pathology. Canadian Journal of Plant Pathology. 1995;**17**:154-166

[31] Shiratsuchi LS, Brandão ZN, Vicente LN, Victoria DC, Ducati JR, Oliveira RP, et al. Sensoriamento Remoto: conceitos básicos e aplicações na Agricultura de Precisão. In: Bernardi ACC, Naime JM, Resende AV, Bassoi LH, Inamasu RY, editors. Agricultura de precisão: Resultados de um novo olhar. 2nd ed. Brasília: Embrapa; 2014. pp. 58-73

[32] Kelly M, Guo Q. Integrated agricultural pest management through remote sensing and spatial analyses. In: Ciancio A, Mukerji KG. editors. General Concepts in integrated Pest and Disease Management. Dordrecht: Springer; 2007. pp. 191-207

[33] Ahmed MR, Yasmin J, Mo C, Lee H, Kim MS, Hong SJ, et al. Outdoor applications of hyperspectral imaging technology for monitoring agricultural crops: A review. Journal of Biosystems Engineering. 2016;**41**(4):396-406

[34] Mahlein AK. Plant disease detection by imaging sensors: Parallels and specific demands for precision agriculture and plan phenotyping. Plant Disease. 2016;**100**(2):241-251. DOI: 10.1094/PDIS-03-15-0340-FE

[35] Ali MM, Bachik NA, Muhadi NA, Yusof TNT, Gomes C. Non-destructive techniques of detecting plant diseases: A review. Physiological and Molecular Plant Pathology. 2019;**108**:101426

[36] Genna NG, Gourlie JA, Barroso J. Herbicide efficacy of spot spraying systems in fallow and postharvest in the pacific northwest dryland wheat production region. Plants. 2021;**10**(12):2725

[37] Calvert B, Olsen A, Whinney J, Rahimi Azghadi M. Robotic spot spraying of Harrisia Cactus (*Harrisia martinii*) in grazing pastures of the Australian Rangelands. Plants. 2021;**10**(10):2054

[38] Dammer KH, Adamek R. Sensor-based insecticide spraying to control cereal aphids and preserve lady beetles. Agronomy Journal. 2012;**104**(6):1694-1701

[39] Hasan AM, Sohel F, Diepeveen D, Laga H, Jones MG. A survey of deep learning techniques for weed detection from images. Computers and Electronics in Agriculture. 2021;**184**:106067

[40] Ghanizadeh H, Harrington KC. Herbicide resistant weeds in New Zealand: State of knowledge. New Zealand Journal of Agricultural Research. 2021;**64**(4):471-482

[41] Raetano CG, Chechetto RG. Misturas em tanque. In: Antuniassi UR, Boller W, editors. Tecnologia de Aplicação para Culturas Anuais. 2nd ed. Passo fundo: Aldeia Norte; 2019. pp. 49-86

[42] Brasil. Instrução Normativa n° 40, de 11 de outubro de 2018. Instrução Normativa N° 40. 2018

[43] Gazziero DLP. Misturas de agrotóxicos em tanque nas propriedades agrícolas do Brasil. Planta Daninha. 2015;**33**(1):83-92

[44] Santra G, Mohant A, Rout S. A survey on chemical sprayer influenced by various parameters of AAS. International Journal of Modern Agriculture. 2020;**9**(3):544-549

[45] Evans JA, Tranel PJ, Hager AG, Schutte B, Wu C, Chatham LA, et al. Managing the evolution of herbicide resistance. Pest Management Science. 2016;**72**(1):74-80

[46] Petter FA, Segate D, Almeida FA, Alcântara Neto F, Pacheco LP. Incompatibilidade física de misturas entre inseticidas e fungicidas. Comunicata Scientiae. 2013;**4**:129-138

[47] Petter FA, Segate D, Pacheco LP, Almeida FA, Alcântara NF. Incompatibilidade física de misturas entre herbicidas e inseticidas. Planta Daninha. 2012;**30**:449-457

[48] Hoffmann-Campo CB. Pragas iniciais da soja: tamanduá-da-soja, piolho-de-cobra e torrãozinho. In: Seminário de Manejo de Pragas e Doenças Iniciais das culturas de soja e milho em mato grosso do sul, 1., 2002, Dourados. Anais... Dourados: Embrapa Agropecuária Oeste; 2002. pp. 45-54 (Embrapa Agropecuária Oeste. Documentos, 48)

[49] Ávila CJ, Gomez SA. Efeito de inseticidas aplicados nas sementes e no sulco de semeadura, na presença do coró-da-soja, *Phyllophaga cuyabana*. Dourados: Embrapa Agropecuária Oeste; 2003. (Embrapa Agropecuária Oeste. Documentos, 55)

[50] Santos PS. Aplicação em sulco de nematicidas em soja [masters dissertation]. , Santa Maria/RS: Universidade Federal de Santa Maria; 2015. Available from: https://repositorio.ufsm.br/handle/1/7622 [Accessed: December 28, 2021]

[51] Guarnieri CCO et al. Eficácia de tiodicarbe, cadusafós e condicionador de solo via tratamento de sementes e/ou sulco de plantio no controle de nematoides na cultura da soja. In: Brazilian Congress of Nematology. Brazil: Bento Gonçalves, RS; 2018

[52] Corte GD. Tecnologia de aplicação de agrotóxicos no controle de fitonematoides em soja [masters dissertation]. Santa Maria: Universidade Federal de Santa Maria; 2013. p. 61. Available from: https://repositorio.ufsm.br/bitstream/handle/1/7572/DALLA%20CORTE%2C%20GERSON.pdf

[53] Ávila CJ, Grigolli JFJ. Pragas de soja e seu controle. Embrapa Agropecuária Oeste-Chapter in Scientific Book. 2014 (ALICE)

[54] Possebon SB. Desempenho de uma semeadora-adubadora e métodos de aplicação de inseticidas no sulco em plantio direto [thesis]. Universidade Federal de Santa Maria; 2011. 113p. Available from: http://repositorio.ufsm.br/handle/1/3592

[55] Gassen D. Aula prática: pequenas e rápidas dicas sobre o uso de inseticidas em lavouras sob plantio direto podem economizar dinheiro e evitar dores de cabeça para o produtor. Revista Cultivar Grandes Culturas. [Internet]. 2015. Available from: https://revistacultivar.com.br/artigos/aula-pratica [Accessed: December 26, 2021]

[56] Vivan LM et al. Eficácia de inseticidas aplicados no sulco de semeadura no controle dos corós *Phyllophaga cuyabana* e *Liogenys fuscus* na cultura da soja. In: Reunião Sul-Brasileira sobre pragas de solo, 10., 2007, Dourados. Anais... Dourados: Embrapa Agropecuária Oeste; 2007. pp. 199-202 (Embrapa Agropecuária Oeste. Documentos, 88)

[57] Takei M et al. Assessment of damage caused by two-striped leaf beetle (*Medythia nigrobilineata* Motschulsky) larval feeding of root nodules in soybean

and its control during furrow cultivation at seeding time. Plant Production Science. 2014;**17**(3):276-283. DOI: 10.1626/pps.17.276

[58] Chen H, Lan Y, Fritz BK, Hoffmann WC, Liu S. Review of agricultural spraying technologies for plant protection using unmanned aerial vehicle (UAV). International Journal of Agricultural and Biological Engineering, Beijing. 2021;**14**(1):38-49. DOI: 10.25165/j.ijabe.20211401.5714

[59] Inamasu RY, Bernardi AC de C. Agricultura de precisão. In: de Bernardi AC, de Naime JM, de Resende AV, Bassoi LH, Inamasu RY. (Ed.). Agricultura de precisão: resultados de um novo olhar. Brasília, Embrapa, 2014: p. 21-33

[60] Xiongkui H, Bonds J, Herbst A, Langenakens J. Recent development of unmanned aerial vehicle for plant protection in East Asia. International Journal of Agricultural and Biological Engineering. 2017;**10**(3):18-30. DOI: 10.3965/j.ijabe.20171003.3248

[61] Xin F, Zhao J, Zhou Y, Wang G, Han X, Fu W, et al. Effects of dosage and spraying volume on cotton defoliants efficacy: A case study based on application of unmanned aerial vehicles. Agronomy. 2018;**8**(6):15p. DOI: 10.3390/agronomy8060085

[62] Grammatikis PR, Psarigiannidis P, Lagkas T, Moscholios I. A compilation of UAV applications for precision agriculture. Computer Networks. 2020;**172**:107148. DOI: 10.1016/j.comnet.2020.107148

[63] Xiao O, Xin F, Lou Z, Zhou T, Wang G, Han X, et al. Effect of aviation spray adjuvants on defoliant droplet deposition and cotton defoliation efficacy sprayed by unmanned aerial vehicles. Agronomy. 2019;**9**:217. DOI: 10.3390/agronomy9050217

[64] Kim J, Kim SC, Ju C, Son HI. Unmanned aerial vehicles in agriculture: A review of perspective of platform, control, and applications. IEEE Access. 2019;**7**:105100-105115. DOI: 10.1109/ACCESS.2019.2932119

[65] Chen HB, Fritz KB, Lan YB, Zhou ZY, Zheng JF. An overview of spray nozzles for plant protection from manned aircrafts: Present research and prospective. International Journal of Precision Agricultural Aviation. 2020;**3**(2):1-12

[66] Zhang Y, Huang X, Lan Y, Wang L, Lu X, Yan K, et al. Review: Development and prospect of UAV-based aerial electrostatic spray technology in China. Applied Science. 2021;**11**(4071):15p. DOI: 10.3390/app11094071

[67] MAPA – Ministério da Agricultura, Pecuária e Abastecimento. Portaria N° 298, 09-22-2021. D.O.U. Ed. 182, Sec. 1, p. 14. Available from: https://www.in.gov.br/en/web/dou/-/portaria-mapa-n-298-de-22-de-setembro-de-2021-347039095 [Accessed: December 28, 2021]

[68] Cunha JPAR, Silva MRA. Deposition of spray applied to a soybean crop using an unmanned aerial vehicle. International Journal of Precision Agricultural Aviation. 2021;**4**(2):8-13. DOI: 10.33440/j.ijpaa.20210402.167

Chapter 7

Climate-Resilient Technologies for Enhancing Productivity of Soybean in India

Bhagwan Asewar, Megha Jagtap, Gopal Shinde, Shivaji Mehetre and Madan Pendke

Abstract

Soybean (*Glycine max* [L.]) contributes 25% of the global edible oil. Globally, soybean area and production in 2020 were 127.9 million ha and 379.8 million tons, respectively. Soybean has got early acceptance as an important oilseed crop in India in with approximately 10–11 million ha area. Now, it has become a major crop by replacing the traditional and contemporary popular crops in nearly all parts of India. Climate change effects, like change in *Monsoon* pattern, increase in dry spell frequencies, heavy rainfall event during crop growth period, extended monsoon at harvesting stage, has drastically influenced the productivity of soybean, which needs attention to identify the constraints and accordingly adapt the climate-resilient technologies. The recent research conclusions revealed that the climate-resilient technologies like selecting suitable varieties, sowing within proper sowing window, broad bed furrow (BBF) method of sowing, dry spell management practices to reduce moisture stress, rainwater harvesting, and soil conservation through BBF method of planting, reuse of harvested and stored rainwater during moisture stress conditions, the adaptation of micro-irrigations systems for protective irrigation, intercropping in soybean, resource conservation techniques can mitigate the effects of climate change and enhance the productivity of soybean in a sustainable manner.

Keywords: broad bed and furrow, dry spell management, resilient technologies, soybean yield

1. Introduction

Soybean cultivation and its use was started from the beginning of Chinas agriculture age. Its utilization for human consumption was mentioned [1] long ago 6000 years back in Chinese medical compilations. Soybean has meant for milk, meat, bread, cheese and oil from centuries ago to common people of China, Korea, Manchuria, Japan, Philippines and Indonesia. The center for origin of soybean is China while some reports second center as India [2–4]. In India initially, soybean has been traditionally grown in patches of the Kumaon Hills of Uttarakhand (earlier Uttar Pradesh), the Khasi Hills, the Naga Hills, Manipur, Himachal Pradesh, and parts of central India encompassing Madhya Pradesh.

IntechOpen

During 2020 the global area and production of soybean was 127.9 million ha and 379.8 million tons respectively. India stands fifth rank in the area and production in the world after countries USA, Brazil, Argentina and China. The productivity of soybean in India is low (1.1 tons/ha) as compared to the world (2.3 tons/ha).

Rainfed agriculture produces much of the food consumed globally. It accounts for more than 95% of farmed land in Sub-Saharan Africa, 90% in Latin America, 65% in East Asia and 60% in South Asia. The rainfed agriculture across the world is being affected by climate change.

1.1 Indian situation

On global map India lies between 8° N and 37°. In all climate over Indian subcontinent is dominantly tropical except Northern parts where the climate is temperate. Climate over India is characterized with distinct hot and dry climate altered with wet and humid climate in cyclic pattern. This is due to monsoon winds. Typical monsoon over India has two branches, i.e., Arabian sea branch and Bay of Bengal branch.

In Indian context soybean has occupied the position as major crop of India by replacing the traditionally grown major region-specific contemporary crops like cotton, paddy, and wheat. This has derived the agriculture completely new cropping systems also. Reasons of early popularity of the soybean crop in India are good demand of crop in international market, assured economic value of the crop along with the nutrient rich in both oil content (16–21%) and protein content (36–42%). Climatic requirements of soybean crop in more or less variations matches with existing climatic conditions over India which may be one of the reasons for steady performance of soybean in terms of production and productivity.

Based on productivity levels of soybean, total Indian geographical area is categorized in six agroclimatic zones which are Northern Hill Zone, Northern Plain Zone, Eastern Zone, North Eastern Hill Zone, Central Zone, and Southern Zone. The breeding for improving varieties with suitability and management practices specific to the particular zone are identified to achieve higher levels of productivity. In last years the area under soybean in India is constantly increasing with increase in productivity. In year 2020–2021 among various soybean growing states of India, Madhya Pradesh followed by Maharashtra were the first two states to grow soybean on 5.85 million ha and 4.32 million ha area respectively. In other states like Karnataka, Telangana, Gujarat, and Chhattisgarh there is considerable scope for the area enlargement (**Table 1**).

The average annual rainfall of India is 1192 mm, with spatial and temporal variability which delineate the Indian agriculture in to dry farming, dryland farming and rainfed farming. The major soil types of India are Vertisols, Inceptisols and Entisols. Vertisols are deep black cotton soils having swelling and shrinkage properties, good water holding capacity. In India soybean is promising and prominent rainfed crop where seasonal rainfall is 900 mm and soil type is vertisols. In some parts of India traditionally adopted cropping system was *Kharif* fallow followed by post-rainy wheat or chick pea, but soybean has substituted this cropping system as soybean followed by wheat or chick pea. Gain in the *kharif* season crop has bring out improvement of small and marginal farmers profitability, socio-economic condition. The small and marginal farmers are subjected to subsistence farming due to limited production inputs. Even under aberrant weather conditions soybean has maintained its steady performance in terms of profitable returns to farmers [5].

Indian Council of Agricultural Research—Indian Institute of Soybean Research, at Indore, Madhya Pradesh was established to provide centralized research to support

Sr. no.	State	Sowing area (ha)	Expected yield (kg/ha)	Estimated production (100,000 million tons)
1	Rajasthan	9.25	761	7.05
2	Madhya Pradesh	55.69	939	52.29
3	Maharashtra	43.85	1102	48.32
4	Chattisgarh	051	910	0.46
5	Gujrat	2.237	1015	2271
6	Karnataka	3.827	1005	3846
7	Others	1.129	975	1101
8	Telangana	3.488	1015	3540
Total		119.982	991	118.888

Source: From [5].

Table 1.
State-wise expected area, production, and productivity of soybean 2021.

soybean production system research with basic technology and breeding material. Under AICRPS (All India Coordinated Research Project on Soybean) system 102 improved soybean varieties have been developed which were suitable to different agroclimatic zones, and promising with high yields due to tolerance to various biotic and abiotic stresses.

After independence through green revolution, high yielding varieties, India achieved record breaking/higher food grain production and achieved self-sufficiency in cereals production. This achievement of the higher production was associated with several extensive problems which encounter to degradation of the natural resources and challenged sustainability. The concepts of conservation have gained importance due to the threat of resource degradation problems, the necessity to reduce production costs, increase productivity and profitability and make agriculture more inexpensive. Over exploitation of natural resources, conducive to unsustainable growth; these strategies need to change in coming years for increasing the sustainability. For Indian agriculture it is the time where it must search for new guidelines—may be through strategies, policies, and actions which must be accepted to move toward sustainability.

In recent past decades, at global level, rapid moves have been made in order to develop and spread the technologies which will conserve resources. Conservation of soil and rainwater can be achieved with change in land configuration, and method of planting like broad bed method of sowing, minimizing soil disturbance through restricted tillage. Adoption of conservation agriculture (CA) include the basic three pillars or principles which are

1. Minimum soil disturbance

2. Crop rotations

3. Permanent soil cover (through cover crops, crop residues)

These principles can be effective possible ways towards the realization of the sustainability in agriculture and will help in achieving the goals of increasing productivity with prevention of further degradation of natural resources. To sustain soil fertility, improving fertilizer/water use efficiency, physical properties of soil, and enhanced crop productivity, techniques of crop residues retention on the soil surface, organic and inorganic combined fertilization and involvement of legumes in crop rotation are essential. Above techniques coupled with minimum or no-tillage practices play a major role in conservation of natural resources. Thus, resource- conserving technologies can be used to enhance soil health, water, and nutrient use efficiency.

2. Constraints responsible for low productivity of soybean in India

The national and regional yield or productivity levels are low as compared to global level. There are many limiting factors responsible for low productivity which are the gap in genetic and breeding for variety improvement, gap in management practices, post-harvest handling, and socio-economic factors. One or more than one limiting factor will aggravate the adverse effect on productivity, which is commonly observed in dryland farming areas.

A. Genetic and breeding for variety improvement,

i. Inherent poor seed longevity,

ii. cultivation of single variety on large area mono-variety

iii. shattering losses due to delayed harvesting,

B. Inappropriate management practices

i. Cultivation of soybean crop under rainfed condition.

ii. sowing under delayed conditions (after sowing window).

iii. Non application of seed treatment and inoculation at the time of sowing.

iv. No/little use of organic manures

v. mixed sowing of seed with fertilizers

vi. Low or excess population of plant per hectare

vii. Imbalanced and improper fertilizer application

viii. Inefficient water management practices,

ix. Lack of proper insects and pests control measures

x. Disproportionate use of water in spraying of pesticides,

C. Post-harvest handling

i. Proneness of soybean to field weathering

ii. Storage of soybean at higher seed moisture content can reduce the quality of seed for its use as seed material for sowing and may result in low or no germination of seeds

iii. Mechanical harvesting of soybean more than 450 RPM speed of thresher will adversely affect the seed germination.

D. Socio-economic factors

i. Timely unavailability of quality inputs,

ii. Poor coping techniques,

iii. Unavailability of credit loans to farmers at the time of sowing.

3. Climate resilient soybean production technology

3.1 Selection of field

Soybean production is remunerative when it is cultivated on a well-drained, sandy loam to clay soils. Soils with medium available water holding capacity and reasonable depth, can also be suitable for soybean. It requires soil with comparatively rich in organic carbon content. Neutral pH is ideal for harnessing maximum yield of soybean. The soil with excessive salts/sodium and poor drainage conditions are not appropriate for cultivation of soybean due to inhibition of seed germination.

3.2 Requirement of tillage

One summer ploughing followed by two cross cultivation or harrowing for breaking of soil clods or one deep ploughing once in 2–3 years can prepare perfect seedbed for a good growth and yield of soybean. It is the need of time to adopt minimum and conservation tillage (CT) to attain sustainability in soybean production.

3.3 Moisture conservation through land management

Soybean is cultivated majorly on medium to heavy soils and in Indian context, it is cultivated as major crop in rainfed areas which are associated with the dominance of vertisol soils. The rainfed areas of India are having the variation in rainfall distribution and its frequency. Extreme events like incidence of nearly 50% of seasonal rainfall in 2–3 rainy days are also recorded in many places in rainfed areas. As the Vertisol soils are high in clay content and their characteristic feature of swelling and shrinkage associated with slow infiltration rate, in such extreme events, they are prone to heavy runoff losses. The loss of rainwater not only removes soil nutrients but also its availability for utilization by crop toward its growth over season is also lost. This situation demands the essentiality for *in situ* soil and water conservation. Also, safe removal of excess water from the field is essential under water stagnation conditions. For attaining good drainage and *in situ* soil and water conservation land lay outs can be altered to achieve these objectives, which are as below.

1. Conservation furrows

Under heavy rainfall events, to conserve moisture and draining out excess water from the field, can be achieved by a very simple technique of opening conservation furrows after every 4/6 rows of soybean.

2. Broad bed and furrow

Broad bed and furrow (BBF) systems have been found to agreeably attain both the objectives specially in deep vertisols.

3.4 Selection of suitable variety

A good soybean variety should be high-yielding and should exhibit stable performance across abiotic and biotic stresses during crop growth. Growing more than one (preferably three to four) variety with different maturity durations is most advisable particularly when the planting area is on quite a large area to achieve sustainability in soybean production and to make efficient use of harvesting equipment and laborer.

3.5 Sowing window

In the northern, northeast, northwest, and central region of India, soybean is mostly cultivated as the sole crop in *Kharif* (June–October). But, in the peninsular and the North– Eastern region of the country, successful soybean cultivation is also feasible during *rabi*/spring and summer seasons (November–April). In Maharashtra also the area under cultivation of spring and summer soybean is increasing. Two successive seasons of soybean cultivation is not scientifically recommended due to the recent outbreak of pest and diseases.

During *Kharif* season, under irrigated conditions, one presoaking irrigation in the second fortnight of June is given for pre-monsoon sowing of soybean which has been found to give optimum plant population, good growth and high yields. When irrigation is not available, sowing should be done after the commencement of monsoon rains. Sowing should be started only after receipt of minimum 100–150 mm rainfall (depending on soil type) to avoid initial crop failure due to moisture stress or dry spell immediately after sowing. The date of onset of Monsoon varies in various parts of country. Appropriate sowing window for *Kharif* soybean is in between the third week of June to the second week of July (as per the commencement of monsoon). Timely sowing is very important non monitory input. Late planting leads to manifold problems like a poor plant population, reduced vegetative growth, reproductive efficiency, pod development, and yield. Sowing of *Kharif* soybean after 20th July is not recommended.

3.6 Optimum plant population

Optimum plant population is prerequisite for higher yields. Per plant yield contributing characteristics are in good quantity when the optimum plant population is maintained. One of the reasons for poor soybean yield in India is the sub-or super-optimal plant population. A plant population of about 450 thousand is optimum with a range of 400–600 thousand plants per hectare. In general, a comparatively higher plant population is desirable for determinate varieties. Use of quality seed and good seedbed condition are pre- requisite to maintain the desired field emergence.

3.7 Planting geometry

The ideal planting geometry with row spacing of 45 and 30 cm is recommended for northern, central and southern zones respectively while plant to plant spacing is at 4–5 cm and 7.5–10 cm respectively. For delayed sowing due to late-onset of monsoon, to tcompensate yield reductions owing to reduced vegetative growth along with early flowering can be achieved to some extent by reducing the row to row spacing (30 cm) and increasing the seed rate by 25%. To ensure good germination and subsequent plant stand the seed should be placed at the depth of 3–5 cm.

3.8 Seed rate

Seed rate is dependent on seed index and germination ability, and percentage.
Small seeded variety—60–65 kg/ha
Medium seeded variety—65–70 kg/ha
Bold seeded variety—70–75 kg/ha
The required seed rate is generally 65 kg/ha.

3.9 Seed treatment, seed inoculation

Seed treatment before sowing achieves the objectives of good seed germination and uniform plant stand, reduction in seed born inoculum, protection of seed and seedling from soil born pathogenic infestations. More than 20 species of fungi are known to affect soybean plant as seed-borne. Therefore, seed treatment becomes most important preventive measure which include

1. Fungicidal seed treatment of Thiram + Carbendazim (2:1) to the seeds at the rate of 3 g/kg seed of soybean.

   OR

   Treating soybean seed with microbial inoculum of *Trichoderma Viridi* at the rate 5 g/kg seed will effectively prevent collar rot disease and assure good plant establishment.

2. For control of yellow mosaic virus (YMV) infestation to soybean, recommended fungicides are used for seed treatment. Then immediately after the treatment with insecticides like Thiamethoxam 30 FS (flowable concentrates for seed treatment) at the rate of 10 ml/kg seed or Imidacloprid 48 FS (flowable concentrates for seed treatment) at the rate of 1.25 ml/kg seed is recommended.

3. Seed must be treated with *Bradyrhizobium japonicum* culture at the rate 500 g/65 kg seed, for increasing the number of effective nodules and fixation of atmospheric nitrogen in root nodules of soybean plants.

4. For increasing the solubility of the unavailable phosphorous in soil, it is recommended to treat the seed with phosphorous solubilizing bacteria (PSB) at the rate 500 g/65 kg seed.

3.10 Nutrient management

Soybean is considered to be a moderately exhaustive crop. Application of balanced fertilizer dose ensures better performance of soybean in terms of yield. The integration of organic manures like 5–10-ton farm yard manure (FYM) or 2.5-ton poultry manure/ha before last harrowing with inorganic fertilizers as the basal dose of 25:60–80:40–50:20 N: P2O5:K2O:S kg/ha (at the time of sowing) generally provides nutrition in balanced which will help in harnessing the yield potential of soybean.

As per the soil test report, the recommended micronutrients should be applied by soil or foliar method of application

3.11 Weed management and intercultural operations

Weeds poses major threat to soybean production by the way of yield losses to the extent of 30.5–53%. The critical crop weed coemption period for soybean crop is 20–45 days after sowing (DAS) In general, the mechanical weed control measures in soybean are hand weddings twice, at 20 and 40 DAS are recommended. In *kharif* season during incessant rains or in case of labor unavailability, chemical weed control promises effective weed control. Application of pre-plant incorporation (PPI) or pre-emergence (PE) or post-emergence (PoE) herbicides and two hands weeding were found equally effective to reduce the weed load in soybean.

1. PE application of Alachlor at the rate 2 kg/ha or Pendimethalin 30% EC at the rate 0.50–0.75 kg ha or Butachlor 50% EC at the rate 1 kg/ha dissolved in 500 l of water is sprayed after sowing but before emergence of soybean, i.e., in 2–3 DAS results in better weed suppression.

2. PoE application of Imazethapyr 10% SL at the rate of 0.25 kg ha dissolved in 500 l of water can sprayed within 7–21 days after emergence of soybean will effective for good weed control.

3.12 Water management

Proper water management at critical growth stages is essential to optimize yield. Stress on account of surplus or shortage soil moisture would be unfavorable growth and yield of soybean. Water requirement of soybean for planting and germination is approximately 100 mm, the total water requirement of soybean is around 500 mm.

Generally, soybean is cultivated as completely rainfed in dryland areas receiving rainfall 600–900 mm or as irrigated in rainfed areas rainfall more than 900–1500 mm. From the rainfall analysis it is concluded that the kharif crops are supposed to undergo the dry spell at least once in its total life cycle. Crop may suffer moisture stress at early season, mid-season or terminal stage due to dry spell. Critical period of soybean for water requirement is planting to emergence, reproductive growth stage, and pod filling stage. Supplemental irrigation should be given from the harvested and stored rainwater whenever there is a dry spell or moisture stress conditions occurs.

In rainfed areas, water stored from runoff during heavy rainfall is collected and stored in storage structures like farm ponds. This harvested and stored water can be utilized during drypell to save crop by giving protective irrigation. For protective irrigations in dryland areas the reuse of harvested rainwater can be efficiently used through micro irrigation system. Adaptation of the micro irrigation systems like sprinkler irrigation system will not only increase the efficiency of water application but will also increase the area under protective irrigation and can rescue more area from moisture stress. The depth of protective irrigation can be 1–2 cm for shallow rooted crops while it can be 3–5 cm for deep rooted crops (**Figure 1**).

Figure 1.
Sprinkler irrigation system for efficient reuse of harvested rain water.

3.13 Dry spell management

In Indian context dry spells occur in monsoon season invariably. The frequency of the dry spell is increasing in current scenario due to climate change. Moisture stress is a resultant of complex set of several climatic, edaphic, and agronomic factors. It can be characterized by major three varying parameters, i.e., occurrence timing, its intensity, and duration. Reduction of tissue water content, stomatal conductance, metabolic processes, and growth are subsequent effects of Dry spell induced a restricted water supply. Plants develop various adaptive mechanisms in response to moisture stress. These adaptations are drought tolerance and avoidance strategies to defy moisture stress. Under drought conditions, Plants by maintaining favorable water status may avoid moisture stress. Plant may avoid drought either by reducing water loss from leaves or by increasing the capacity of roots for water uptake.

Mitigation strategies of dry spell in soybean are:

i. Planting of more than three varieties as per maturity group.

ii. If the occurrence of drought is experienced more frequently, the short durational varieties should be selected for sowing.

iii. Organic mulching by applying crop residue at the rate 5 tons/ha after emergence

iv. Anti-transparent suitable for soybean crop are like $KNO3$ at the rate 1% or Glycerol at the rate 5% or $MgCO3$ at the rate 5%. These are recommended to spry under early period of dryspell conditions and when rainfall is likely to happen in upcoming few days. Avoid the conditions of crop failure and reduce the harmful effects on yield of soybean.

v. Mulching of removed weeds (*in-situ*) at 30 DAS other crops is also advised.

3.14 Soybean based cropping/intercropping systems

Areas where mean seasonal rainfall is 600–750 mm and the effective growing duration is 20–30 weeks, in that part intercropping is suggested, while the areas having mean annual rainfall more than 750 mm and duration of effective cropping season length is more than 30 weeks, double cropping of soybean is possible. In Madhya Pradesh and Maharashtra, particularly Marathwada region intercropping system of soybean + pigeon pea (4:2 row ratio) is very profitable and is recommended particularly for rainfed cultivation. Early maturing (short duration) pigeon pea varieties are more suitable for intercropping system. Soybean + sorghum or soybean + cotton intercropping is also recommended and practiced on large area in and around Madhya Pradesh (**Tables 2** and **3**).

The soybean-based cropping systems are not only productive but they have been profitable as good energy-efficient under various agro-climatic conditions. It is also advisable to farmers that continuous growing of soybean on the same piece of land should be avoided. Crop rotation tactics should be followed for sustainable soybean production (**Figure 2**) [6].

3.15 Conservation agriculture in soybean

Looking at the sustainability of the natural resources, CA is the promising system. CA is a broad term and it encompasses all conserving techniques that conserve resources anyway. It also involves the following technologies:

Zone	Cropping system	Intercropping system
Central (Madhya Pradesh, Bundelkhand region of U.P., Rajasthan, Gujarat, Northern and western parts of Maharashtra)	Soybean-wheat or chickpea soybean-wheat-corn fodder, soybean-potato, soybean-garlic/potato-wheat, soybean-rapeseed or mustard, soybean-pigeon pea or safflower or sorghum	Soybean + pigeon pea, soybean + corn, soybean + sorghum, soybean + sugarcane, soybean in mango/guava orchards, soybean in agro-forestry
Southern (Karnataka, Tamil Nadu, Andhra Pradesh, Kerala, Southern parts of Maharashtra)	Wheat-soybean-finger millet-peas, oat-cowpea-barley-soybean, soybean-finger millet-beans, soybean-wheat-groundnut	Soybean + pigeon pea, soybean + finger millet, soybean + sugarcane, soybean + sorghum, soybean + groundnut, soybean in coconut/mango/guava orchard and soybean in agro-forestry
Northern Plain (Punjab, Haryana, Delhi, North-Eastern plains of U.P., Western Bihar)	Soybean-wheat, soybean-potato, soybean-chickpea	Soybean + pigeon pea, soybean + corn, soybean + sorghum, soybean in mango/guava orchards, soybean in agro-forestry
Northern hill (Himachal Pradesh, North Hills of U.P.)	Soybean-wheat, soybean-pea, soybean-lentil, soybean-Toria	Soybean + corn, soybean + pigeon pea
Northeastern (Assam, Meghalaya, West Bengal, Bihar, Orissa)	Soybean-paddy, paddy-soybean	Soybean + finger millet, soybean + paddy, soybean + pigeon pea

Source: From [5, 6].

Table 2.
Soybean based remunerative cropping/intercropping system for different zones of India.

Treatments	Pooled mean				
	Soybean yield (kg/ha)	GMR (Rs./ha)	NMR (Rs./ha)	BC ratio	RWUE
Main plots: land configurations: (03)					
L_1: Flat bed	1148	34782	9263	1.36	2.77
L_2: BBF	1460	44191	18587	1.72	3.50
L_3: ridges and furrow	1321	39984	14232	1.54	3.16
SE ±	45.46	1145	341.6	—	—
CD at 5%	125.8	3168	976.4	—	—
Fertilizer cum stress management practices: (108)					
F_1: RDF (30:60:30 NPK kg/ha)	1140	34670	11457	1.48	2.81
F_2: RDF + KNO3 @ 1% and 2% (two sprays)	1421	42954	17545	1.68	3.35
F_3: RDF + (19:19:) @ 0.5%	1384	41884	16448	1.64	3.31
F_4: RDF + MoP @ 1% and 2%	1290	39099	14386	1.57	3.11
F_5: RDF + micronutrients mixture @ 0.5%	1440	43329	17624	1.68	3.34
F_6: RDF + straw mulch @ 3 tons/ha	1348	40738	13958	1.05	2.98
F_7: RDF + Anti-transparent Kaolin @ 7%	1224	37735	9105	1.31	2.98
F_8: RDF + water sprays	1211	36797	12570	—	—
SE +	37.75	1298	856.3	—	—
CD at 5%	104.4	3592	2401.6	—	—
Interaction SE+	65.39	2248	1031.5	—	—
CD at 5%	180.9	6223	1698.5	—	—

Table 3.
Soybean seed yield (kg/ha), gross monetary returns (GMR), net monitory returns (NMR), benefit cost (BC) ratio and rain water use efficiency (RWUE) as influenced by different treatments (2014–2016).

1. A minimum level of soil inversion (reduced or zero tillage)
2. Sensible and profitable crop rotation (inclusion of legume crops, cover crops)
3. Soil cover (through crop residues retention on the soil surface)

Resource conserving techniques confirm the optimal utilization of resources and improve input use efficiency. These practices include reduced tillage or zero tillage or no tillage, integrated crop management approaches, retaining crop residue and Use of beneficial microorganisms helpful in increasing both the biological fertility of soil and crop production. CA practices decrease resource degradation. Slow breakdown of surface residues which recovers soil organic matter status, soil microbial and biological activity, and diversity. These practices also contributes to the overall enhancement in soil quality. Sustainable improvements in efficient use of water and nutrients by improving nutrient balance and availability, infiltration, and retention by the soil, reducing water loss due to evaporation.

Figure 2.
Soybean + pigeon pea intercropping system in 4:2 row proportion is widely adopted system under rainfed conditions of Marathwada region.

3.16 Minimum/zero tillage

Minimum tillage is aimed at reducing tillage to the minimum necessary. It ensures a good seedbed, rapid seed germination, a satisfactory plant stand, and promising growing conditions. Minimum tillage will benefit in soil setting, improve soil infiltration, reduce soil resistance and soil compaction along with reduction in soil erosion. Although it has some adverse effects like lower seed germination, additional nitrogen requirement, hampers nodulation in some legumes, requires specially designed implements, development of herbicide residue and its pollution, and the perennial weeds become difficult to control.

Zero tillage is an ultimate form of CT in which mechanical soil manipulation is diminished to traffic and sowing only. In India at present area under zero tillage is 2.0 million ha and potential area is 2.0 million ha and 10 million ha. It is very supportive in the area of intensive cultivation where a turnaround period between two crops is very less and which can facilitate timely sowing. Zero tillage offers the benefit of reduced fuel consumption and labor expenses, reduction of total cost of cultivation, timely sowing is possible which offers yield advantage, reduced soil loss and erosion ultimately improvement in soil health.

3.17 Broad-bed and furrow system

BBF method comprised of the formation of broad-beds of 135 cm wide and 20 cm elevated beds and divided by 45 cm wide furrows, which are to a depth of 20 cm and graded across the contour to a 0.5% slope [7]. The main purpose of this system is to provide acceptable drainage during heavy rainfall events and draining excess rain water into grassy waterways or farm ponds. This stored rain water will be used to provide supplemental irrigation to Kharif and Rabi crops during dry spells. It will reduce runoff and soil loss, and for in-situ moisture conserving in the furrows. The movement of rain water in BBF system is in both the directions, i.e., vertically downward in furrows and laterally in micro and macro pores in the raised broad beds which enhances the efficient use of rain water.

National Innovations on Climate Resilient Agriculture Project (NICRA) adopted villages in Southern and Central parts of India, farmers who implemented broad bed furrow (BBF) sowing technique in soybean by BBF planter evaded injury to the crop owing to surplus rainfall during *Kharif* seasons of 2013, 2019, 2020, 2021 and gained about 40% yield benefit compared to flatbed sowing. BBF method for, soybean, cotton, pigeon pea and maize saved crop damage due to excess soil moisture by aiding quick drainage and avoiding water stagnation.

Figure 3.
Sowing of soybean with BBF planter.

BBF planter (**Figure 3**) facilitates the simultaneous preparation of broad beds and furrows (for soil and water conservation, in which furrows acting as safe drainage channels during heavy rainfall events (**Figure 4**) sowing of the seed with less seed

Figure 4.
Furrows acting as safe drainage channels during heavy rainfall events in soybean sown with BBF method.

Figure 5.
Soybean sown with BBF method.

requirement (kg/ha), deep placement of fertilizers below the seed enhancing the fertilizer use efficiency, which reduces the time, labor, energy and cost of cultivation (**Figure 5**).

3.18 Crop residue management

In India about 400 million tons of crop residues are produced annually. A large quantity of crop residues is left in the field when mechanical harvesting is practiced. This crop residue can be recycled which increases the organic carbon content in soil, as well as nutrient supply. About 25% of nitrogen (N) and phosphorus (P), 50% of sulfur (S), and 75% of potassium (K) uptake by cereal crops are retained in crop residues, making them valuable nutrient sources. Mulching with crop residues contributes to the conservation of soil and rainwater (**Figure 6**).

This technique reduces evaporative losses of soil moisture, nutrient loss along with runoff. Contrasting removal or burning of crop residue, incorporation of straw builds up soil from cropped fields. Crop residues modify soil biological activity resulting in improved soil fertility and better soil physical conditions.

Application or retention of crop residue at 2–5 tons/ha will conserve the soil moisture will also help in conservation of soil health.

Figure 6.
Crop residue management for conservation of soil and rain water in Soybean.

4. Conclusion

It may be concluded that with the use of appropriate improved climate resilient production technology, there is great scope to achieve the targeted average productivity. The adoption of the climate resilient practices like selecting suitable varieties, sowing in proper sowing window, BBF method of sowing, dry spell management practices, intercropping in soybean, resource conservation techniques can mitigate the effects of climate change and enhance the productivity of soybean in sustainable manner.

Author details

Bhagwan Asewar[1*], Megha Jagtap[1], Gopal Shinde[2], Shivaji Mehetre[3] and Madan Pendke[4]

1 Department of Agronomy, VNMKV, Parbhani, India

2 Farm Machinery and Power Engineering Department, VNMKV, Parbhani, India

3 Soybean Research Station, VNMKV, Parbhani, India

4 Soil and Water Conservation Engineering Department, All India Coordinating Project on Dryland Agriculture, VNMKV, Parbhani, India

*Address all correspondence to: asewarbv2007@gmail.com

References

[1] Krishnamurthy K, Shivashankar K. Soybean Production in Karnataka USA, Tech. Series No. 12. Bangalore: University of Agricultural Sciences; 1975

[2] Hymowitz T. Soybeans: The success story. In: Janick J, Simon J, editors. Advances in New Crops. Portland, Oregon: Timber Press; 1990. pp. 159-163

[3] Khoshoo TN. Census of India's biodiversity: Tasks ahead. Current Science. 1995;**69**:14-17

[4] Singh RJ, Hymowitz T. Soybean genetic resources and crop improvement. Genome. 1999;**42**(4). DOI: 10.1139/g99-039 [Reviewed abstract only]

[5] Available from: https://www.sopa.org

[6] Dupre BU, Kolhe S, Balasubramani N. Climate Smart Technologies & Practices for Increasing the Soybean Productivity [E-book]. ICAR-Indian Institute of Soybean Research & National Institute of Agricultural Extension Management (MANAGE): Hyderabad; 2021

[7] Asewar BV, Gore AK, Pendke MS, Gaikwad GK. Decadal Research Achievements in Dryland Farming for Marathwada Region, Technical Bulletin No: VNMKV/DOR/2/2018. Parbhani, Maharashtra, India: Vasantrao Naik Marathwada Krishi University; 2018

Section 4

Breeding and Biotechnology of Soybean

Chapter 8

Current Strategies and Future of Mutation Breeding in Soybean Improvement

Alp Ayan, Sinan Meriç, Tamer Gümüş and Çimen Atak

Abstract

Soybean, which has many foods, feed, and industrial raw material products, has relatively limited genetic diversity due to the domestication practices which mainly focused on higher yield for many centuries. Besides, cleistogamy in soybean plant reduces genetic variations even further. Improving genetic variation in soybean is crucial for breeding applications to improve traits such as higher yield, early maturity, herbicide, and pest resistance, lodging and shattering resistance, seed quality and composition, abiotic stress tolerance and more. In the 21st century, there are numerous alternatives from conventional breeding to biotechnological approaches. Among these, mutation breeding is still a major method to produce new alleles and desired traits within the crop genomes. Physical and chemical mutagen protocols are still improving and mutation breeding proves its value to be fast, flexible, and viable in crop sciences. In the verge of revolutionary genome editing era, induced mutagenesis passed important cross-roads successfully with the help of emerging supportive NGS based-methods and non-destructive screening approaches that reduce the time-consuming labor-intensive selection practices of mutation breeding. Induced mutagenesis will retain its place in crop science in the next decades, especially for plants such as soybean for which cross breeding is limited or not applicable.

Keywords: soybean, mutation breeding, mutagens, induced mutagenesis, next generation sequencing

1. Introduction

Soybean (*Glycine max* (L.) Merrill.) has a central position in agriculture along with barley, cassava, groundnut, maize, millet, potato, oil palm, rapeseed, rice, rye, sorghum, sugar beet, sugarcane, sunflower, and wheat which were considered as the most cultivated plants worldwide. Its central role is not only constituted due to the dense protein and high-quality oil contents but also industrial raw material supply. Tofu, soy milk, soy sauce, and miso are the main nutritious human soy products. Also, extracted soy oil, with over 75% oleic acid and under 10% polyunsaturated fatty acids, is one of the most preferred oils sold commercially in the United States today [1]. Long shelf-life required fry, spray, and ingredient oils should preferably contain higher oleic acid due to the better persistence to oxidation. Soy meal is also a major source of protein used in pig and poultry industries. The companion animal industry prefers soy meal as a protein source in animal diet, especially for dogs. High-quality amino acid composition and highly digestible protein content leads to the use of soy meal in aquaculture diets [2]. On the other hand, soy oil has various industrial uses as pharmaceuticals, plastics, papers, inks, paints, varnishes, and cosmetics.

In the verge of global warming effects, renewable energy sources as an alternative to fossil fuel are getting importance. Soybean is also an important biodiesel crop in many countries along with maize, especially in South America countries [3]. Besides the alternative bioenergy crop role, it has also environmental effects as being capable of utilizing atmospheric nitrogen through biological nitrogen fixation and is therefore less dependent on synthetic nitrogen fertilizers. While drought is one of the most plant growth and development limiting factors in present days, nitrogen deficiency is equally crippling for plants, as well, due to its structural, genetic, and metabolic functions in crop yield. Highly stable and non-reactive N_2 is the most abundant constituent of the Earth's atmosphere, still no eukaryotic organism can use it directly. Some members of *Leguminosae* (*Fabaceae*) family including soybean have adopted the ability to establish symbiotic interactions with diazotrophic bacteria known as rhizobia in evolutionary adaptations. By this means, a process called 'biological nitrogen fixation is a low-cost N source that sufficiently increases soybean yield with low environmental impact and avoids the use of synthetic N fertilizers [4].

Soybean (*G. max* (L.) Merr) as a member of the family *Fabaceae*/Leguminosae, subfamily *Papilionoideae,* and the tribe *Phaseoleae* contains two subgenera as *Glycine* which has 26 perennial species and *Soja* (Moench) F.J. Herm. having four annual species [5]. Domestication of cultivated soybean can be traced back to China in 5000 years ago, however, the geographical origin of *Glycine* genus can be traced back to putative ancestor (2n = 2x = 20) which was presumably migrated and formed unknown or extinct wild perennials (2n = 4x = 40) in China. Wild annuals (2n = 4x = 40; *Glycine soja*) and domesticated soybean (2n = 4x = 40; *G. max*) subsequently evolved [6]. The genetic diversity of *G. max* is assumed to regress due to man-made genetic bottlenecks through selection for high yielding lines in modern plant breeding applications. Indeed, yield is the backbone of the profitability and the feasibility. Varieties with other superior traits are not significant in industrial scale unless they have a high yield. As well as yield, maturity, herbicide, and pest resistance, lodging resistance, shattering resistance, seed quality and composition, abiotic stress tolerance are other breeding selection targets [7]. While the wild relative *G. soja* grows in various environmental conditions and have not been exposed to the selective bottlenecks, it retained significant genetic diversity over time.

On the other hand, soybean flowers represent cleistogamous characteristics. Cleistogamy, which is described as the production of both open (chasmogamous, CH) and closed (cleistogamous, CL) floral forms by one species, is very common among angiosperms. Soybean is pseudocleistogamous cleistogamy in which no morphological differences between CL and CH flowers occur other than a lack of expansion of petals and anthesis in CL flowers. It may also be induced by environmental stress factors, occasionally. Cleistogamy is observed both in cultivated soybean [*G. max* (L.) Merr.] and its wild relative [*G. soja* Sieb. & Zucc.]. Soybean usually produces both CH and CL flowers on the same plant. In these plants, fertilization occurs within closed petals of CL flowers [8–10]. The rates of natural cross-pollination have been observed between ranges of 0.03–1.14% in natural conditions for self-pollinating soybean plant [11]. Thus, cleistogamy may have influenced the genomic homogeneity and reduced genomic variation further in soybean along with domestication practices.

In this context, improving genomic variations is crucial in soybean breeding. This chapter will summarize present conventional and biotechnological methods in soybean breeding and emphasize on mutation breeding practices with the concluding discussion on future prospective.

2. Improving genomic variations

2.1 Conventional methods

In soybean breeding, oil and protein content, resistance to biotic and abiotic stresses have been the main breeding objectives in past decades. In conventional breeding practices variability of desired traits is based on the detection of novel genotypes which contains enhanced characteristic for the trait. Hybridization of these novel genotypes with the varieties which are already in use for commercial production is the base of the process. Subsequent, the selfing of progenies, which contain traits distributed according to basic genetical segregation rules, provide novel genotypes. Detection of the most favorable recombination in those progenies which is also referred as homozygosity by selection is based on numerous selection methods including pedigree selection, single-seed descent, bulk breeding, mass selection, selection among half-sib families, selection within half-sib families. However, the traditional pedigree method and the single-seed descent method (SSD) are the most successful and preferred in soybean breeding. The last step in the process is yield testing. Available genotypes and technical infrastructure (agricultural machines, greenhouses, and experienced stuff) as well as breeding objective are deciding factors in method selection. Breeding objectives generally depend on the local agroecological conditions, available acreage, production intensity, market demand, and economical value [11–13].

Pedigree selection is a highly labor-intensive method that depends on visual selection by the appearance in each generation. In this method, desirable genotypes are selected in each generation and the limited number of selected genotypes are advanced to the next generation by inbreeding/selfing. The labor intensity of the method is limiting for large scale breeding practices [14]. Single-seed descent (SSD) is the most preferred method with pedigree selection to increase homozygosity in soybean. Single pod descent (SPD) accelerates the SSD for harvesting process even further. This method is mostly preferred for high seed yield, oil content and quality, resistance to biotic and abiotic stresses and maturity duration breeding objectives [15].

2.2 Biotechnological approaches

Although, the improvement of plants by conventional breeding methods is one of the most preferred breeding strategies, the limited hybridization among species, transfer of undesirable genomic segments together with genes of interest (e.g., linkage drag) and the fact that diversity in species is based on spontaneous mutations with a very low frequency necessitated the development of new breeding strategies. Plant breeding has often benefited from new technologies to overcome such limitations. Molecular breeding as one of these strategies can be extensively defined as the utilization of genetic manipulation of DNA at the molecular level to improve of trait of interest in plants, including genetic engineering, molecular marker-assisted selection, marker-assisted backcrossing, marker-assisted recurrent selection, genome wide selection [16, 17]. Molecular breeding requires more complex equipment and molecular tools compared to conventional breeding approaches. The identification of functional genes and DNA markers associated with variation at the genomic level is an important part of molecular breeding. Marker-assisted breeding (MAB) which utilized marker-assisted selection involves the use of molecular markers in conjunction with linkage maps and genomics, and the improvement of crop plant traits based on genotypic analyses. Moreover, MAB requires minimum phenotypic information during the training phase. The convenience of use and analysis, low cost, a small amount of DNA requirement, co-dominance, reproducibility, high-rate polymorphism and genome-wide distribution are the most important factors for molecular tools used in marker-assisted breeding (MAB) in plants [18]. Along with the emergence of marker-assisted selection (MAS) after the mid-1980s, rapid improvement of plant yield and quality has been achieved thanks to the development of molecular maps by utilizing structural and functional genomics in plant breeding. MAS can be classified into five broad areas: marker-assisted evaluation of breeding material; marker-assisted backcrossing; marker-assisted pyramiding; early generation selection and combined MAS [19].

DNA markers have made significant contributions to increasing the efficiency of conventional and mutation breeding through marker-assisted selection and have been integrated into traditional schemes to develop novel varieties or used instead of traditional phenotypic selection. Many DNA marker techniques have been developed based on different polymorphism detection techniques or methods (such as nucleic acid hybridization, restriction enzyme digestion, PCR, DNA sequencing) such as RFLP, AFLP, RAPD, SSR, SNP. Advances in molecular marker techniques and the creation of large-scale marker datasets provide a reliable way to identify and trace the genetic basis of important agricultural traits. Molecular markers developed from functional genes have been used for the development of soybean varieties by improving important agricultural traits such as yield, disease resistance and abiotic stress tolerance [20]. Breeders can combine all the suitable alleles in a single variety to develop desired crops, thanks to molecular markers closely related to particular traits. However, although soybean yield remains the most important selection criterion for soybean breeders and the primary factor for profitability, it is very difficult to acquire complex traits such as yield, quality and abiotic stresses with marker-assisted selection. Genomic selection (GS) is a promising approach that leverages molecular genetic markers to design new breeding programs and develop new marker-based models for genetic evaluation. GS, which has high selection accuracy, reduced selection duration, greater gain per unit time, precise and accurate results provide breeders with opportunity faster development of improved crop varieties for complex traits. New marker technologies, such as NGS-based genotyping, have made the use of genomic selection as routine for crop improvement while increasing the efficiency of marker applications. The availability of genome-wide high-throughput, low-cost and flexible markers, usability for crop species with or without a reference genome sequence with a large population size are the most important factors for its successful and effective implementation in crop species [21].

Plant breeders have begun to take advantage of molecular breeding more through advances in the identification of QTLs/genes responsible for important agronomic traits. Numerous quantitative character loci (QTL) mapping studies performed for a variety of agricultural crops have resulted in the association of DNA markers and traits. The most notable high-throughput genotyping system is single-nucleotide polymorphisms (SNPs), which are heavily used in quantitative character locus (QTL) discovery. More than 10,000 QTLs using different marker systems have been reported in more than 120 studies involved 12 plant species aimed at improving quantitative properties with economic importance [22]. Linkage analysis for QTL mapping is frequently preferred in two-parent populations. Genotyping by next-generation sequencing become prominent as a promising technology and is also used for genome-wide association studies (GWAS) to identify useful genes to increase crop productivity. Soybean genome sequence information, as one of the most substantial resources, is the basis of genomic studies and has allowed the significant development of genomic applications for soybean breeding.

As in transgenesis, studies involving the transfer of a limited number of loci from one genetic background to another are also within the scope of molecular breeding. Especially in the last two decades, genetic engineering approaches that generate novel genetic variations in plant genome or enable the transfer of gene of interest for obtaining original traits to plants have been frequently preferred among the biotechnological approaches that have been successfully applied in plant breeding [23–25]. Along with recent developments in recombinant DNA technology, it has been paved the way for transferring the desired characteristics to plants within plant breeding in a short time. These genetic engineering and plant transformation approaches which make plant breeding faster, more predictable and improvable for a wide variety of species, include successful characterization, cloning, modification and transfer of DNA expressed the desired trait into cells. The gene pool utilized by plant breeders in conventional breeding since the mid-1990s has been considerably expanded by genetic transformation approaches and many different transgenic plants have been developed by transferring traits that are tough to transfer [26–28]. Genetically Modified Organisms (GMOs), whose agricultural traits have been improved through inter-species gene transfer by utilizing genetic engineering techniques, have been increasingly planted, globally. The total cultivation areas of approved GM plants have increased approximately 113 times, from 1.7 million hectares in 1996 to 191.7 million hectares in 2018. This increase reveals that transgenic technology is the fastest adopted technology in recent years. A total of 2.5 billion hectares or 6.3 billion acres GM crops have been planted in the first 23 years (1996–2018) of commercialization of transgenic plants [29]. Especially soybean (95.9 million hectares) which comprises 50% of the global area of GM crops, corn (58.9 million hectares), cotton (24.9 million hectares) and canola (10.1 million hectares) are the four main transgenic crops cultivated. Transgenic crops, which were initially developed for only producers/farmers on the purpose of agriculture such as insect resistance and herbicide tolerance, afterwards were developed for other traits such as disease resistance, abiotic stress tolerance, modified product quality for both the producers/farmers and consumers. Especially cultivation of stacked events which are GM crops with more than one genetic modification, gather momentum.

During the 23-year period from 1996 to 2018, herbicide tolerance has accounted for the majority of transgenic crops area planted. Only herbicide tolerance cultivation areas of transgenic crops have been gradually decreasing over the years with the increasing importance of stacked cultivars with multiple traits (e.g., both insect resistance and herbicide tolerance; IR/HT). In 2018, stacked (IR/HT) traits used in soybean, maize and cotton have accounted for 42% of the total transgenic acreage, up 4% annually. Traits such as herbicide tolerance, insect resistance, disease resistance, pollination control, modified crop quality, anti-allergy, delayed fruit softening, delayed ripening, enhancement of vitamin A content, modified alpha-amylase, modified amino acid, modified oil/fatty acid, modified starch/carbohydrate, nicotine reduction, non-browning phenotype, phytase production, reduced acrylamide potential, reduced black spot bruising have been transferred to plants and many of these have been combined in various combinations [29]. Thanks to these features brought to agricultural plants, the product yield obtained from the cultivation areas increases significantly. Along with the acceleration of the transfer of the appropriate gene combinations to plants with high added value, products that can provide significant gains in the agricultural economy have been developed. In this process, about 30 different types of transgenic plants such as particularly G. max (soybean), Zea mays (corn), Gossypium hirsutum (cotton), Bassica napus (canola) and including fruits and vegetables such as Phaseolus vulgaris (bean), Prunus domestica (plum), Beta vulgaris (sugar beet), Solanum melongen (eggplant), Cucumis melo (melon), Carica papaya (papaya) have been approved [30]. Stacked traits such as Intacta™Roundup Ready™ 2 Pro, Enlist E3™ and Vistive Gold™ soybeans are favored by farmers for their cost-saving technologies. In 2018, the planting of crops with novel stacked traits in various combinations, including herbicide-tolerant and high-oleic acid soybean, herbicide-tolerant and salt-tolerant soybean varieties were approved. The global acreage of soybeans in 2018 was 123.5 million hectares, of which 78% (95.9 million hectares) were GM soybeans. GM soybeans have been planted on 95.9 million hectares, 50% of the global cultivated area for GM crops; USA (34.1 million hectares), Brazil (34.9 million hectares), Argentina (18.0 million hectares), Paraguay (3.35 million hectares), Canada (2.42 million hectares), Uruguay (1.26 million hectares), Bolivia (1.26 million hectares) and Southern Africa (694,000 hectares). In the USA, soybean is the second most important crop with a total cultivated area of 36.26 million hectares in 2018, with 94% GM. These GM soybeans contain herbicide-tolerant traits that control a variety of weed species depending on the genes deployed. Other features incorporated into HT soybeans include consumer properties such as high monounsaturated oleic acid and enriched omega-fatty acid. In Brazil which has the second-largest GM crop cultivation area with 51.3 million hectares in 2018, GM soybean was planted in an area of 34.86 million hectares. As for Argentina which was the third country to plant the most GM crops in 2018, 18 million hectares of soybeans were planted [29, 30].

2.3 Mutation breeding

Term of mutation was first introduced by de Vries as the sudden and unexpected emergence of hereditary alterations in defining traits apart from recombination in Mutation Theory Vol. I [31]. In 1920s, following Stadler's experiments on genetic effects of X-rays on maize, plant breeders started to use physical and chemical mutagens to induce heritable mutations in plants [32]. As a term, mutation breeding is introduced to the scientific world by Freisleben and Lein defined as the deliberate exposure of biological materials to mutagens for induction of mutation frequency exceeding the natural mutation frequency to develop new varieties [33].

Mutations that cause genetic variation among living organisms can be categorized under spontaneous and induced mutation terms. Spontaneous mutations, which occur in low frequency and accumulate for a long time, allow plants to adapt very distinct environments apart from their original habitat [34–36].

The spontaneous mutation may occur due to the exposure to physical (cosmic radiation, natural background radiation of earth), chemical (alkylating agents, base analogs, antibiotics) mutagens and biological factors (transposon activation) during the reproductive stage. Spontaneous mutation frequency is calculated as

10^{-6} in plants during DNA replication, repair, or genomic element activities [37]. In vitro and in vivo propagation processes may also trigger gene methylation and cause epigenetic alterations while transposon mobility may trigger somoclonal variation and increase spontaneous mutations. Loss or activation of gene through transposable elements (TEs) regulate many biological processes. There are various studies on somoclonal variation-based trait improvement in plants. However, low mutation frequency is a real draw back for considering this method as common breeding alternative [35, 37]. Mutations can also be induced through physical and chemical mutagens. The use of mutagens may induce 103-fold more mutants comparing to the spontaneous mutations. Ossowski et al. [38] calculated spontaneous mutation frequency as $7x10^{-9}$ substitutions per site per generation for *Arabidospsis* plant in 30 generations. This frequency was increased by ethyl methanesulfonate (EMS) treatment to $3x10^{-5}$ substitutions per site per generation. EMS is a mutagenic, teratogenic, and carcinogenic organic compound with formula $C_3H_8SO_3$ which produces random mutations, mostly G:C to A:T transitions induced by guanine alkylation, in genetic material by nucleotide substitution. EMS typically produces only point mutations. Genetic alterations due to physical and chemical mutagens can be classified as genome, chromosome, and gene mutations [31, 35, 39–42].

Genome mutations not only affect genome size (ploidy) but also genome rearrangement in plants. Many plant species as bread wheat: 6X; durum wheat: 4X, cotton: 4X, potato: 4X have polyploidy in nature. Polyploidy leads various advantages as enhanced nucleus size, enlargement on cell and organism basis, yield, increase in gene variations. Polyploidy can be induced as genome duplication (autoploidy) and increase in genome size (alloploidy) through use of mutagens [34, 43].

Chromosome mutations occur during meiotic cell division in very low frequencies. In euploidy state of plants, one set of chromosomes are present, while radiation exposure may result whole or partial chromosome deletions, insertions or translocations and cause aneuploidy. Besides, chromosome inversions, which are characterized as a chromosome rearrangement in which a segment of a chromosome is reversed 180 degrees end-to-end, cause very high gene recombination. In chromosome translocations, break off chromosome parts may attach to the same chromosome (intra-chromosomal) or different chromosome (inter-chromosomal). Both, inter- and intra-chromosomal translocations lead to devastating effects on gene expression.

Gene mutations can be either as gene copy number alterations or as point-mutations, insertions, deletions on nucleotides of gene sequence. Plants may increase gene copy numbers to enhance protein expression during metabolic functions. Mutagens can affect gene expression profiles through either by increasing or decreasing gene copy numbers. Point-mutations occur particularly in chemical mutagen applications. Single or set of nucleoid alterations cause silent mutations if they do not occur in genic regions. Alternatively, they can also cause nonfunctional gene products or nonsense mutations. Nucleotide insertions or deletions can alter codon structure and cause shift in open reading frames. These alterations can also occur on promotor regions, coding sequences or intron regions of genes, therefore, significantly effect protein expression [34, 35].

Single nucleotide changes as deletions generally cause functional gene mutations by the leading formation of novel alleles. Hence, they are particularly important for plant breeding studies for inducing genetic variations. There are numerous examples of plant height, abiotic stress tolerance, pesticide and herbicide resistance improvement cases in rice, wheat, barley, soybean plants, and more [44–47]. Nucleotide mutations can also occur in non-genic regions and cause silent mutations which have no apparent effect on gene expression. Silent mutations generally occur following the alkylating chemical applications and do not affect translation [37]. Deletions among intergenic regions remain silent as long as they do not affect regulating sequences. Still, the possibility of open reading frame shift is present and may lead to nonfunctioning peptide formation [35].

In mutation breeding studies, whole plant, meristem tips, pollens, in vitro explants, embryos, microspores, callus cultures can be selected as initial materials. However, seeds are mostly preferred materials by plant breeders due to the advantages as metabolic inactivity, easy transport, ease of application, low space requirement, ease of storage comparing to others.

2.3.1 Mutagens

Choice of appropriate mutagen is one of the deciding factors on succession of the mutation breeding program. Physical, chemical, or biological agents are viable alternatives. Among physical mutagens ionizing radiation sources, particle (electrons, protons, neutrons, alpha and beta particles) or electromagnetic (X-rays, gamma rays), are widely used. Ionizing radiation interacts with genetic material and cause mutations on DNA sequences. Magnitude of mutagenic effect is proportional to the radiation dose. It is crucial to determine and optimize the effective radiation dose based on experimental plant variety, plant part, and radiation source. 80% of mutation breeding studies prefer physical mutagens and of 60% of this use gamma radiation [35].

Chemical mutagens offer much larger alternative choices. However, the most widespread use of chemical mutagens is among alkylating agents. Ethyl methane sulphonate (EMS), diethyl sulphate (DES), ethylene imine (EI), N-ethyl-N-nitroso urea (ENU), ethyl nitrite urethane (ENU), N-methyl-N-nitrosourea (MNU) are the most generally preferred chemicals. O^6-alkylguanine, N^3-alkyladenine, N^3-alkylcitosine leads to alternative allele formation. Besides methylating agents, nitric acid, nucleic acid analogs, some antibiotics (streptozotocin, mitomycin C, azaserine) are other important chemical mutagens. 60% of registered chemically induced mutant plants are developed by use of EMS, MNU and EMU. One-third of these mutants are obtained by EMU which has ease of supply among others.

Among physical mutagens, gamma radiation has the most frequent use. In nature, there are various gamma-emitting isotopes such as potassium-40 (^{40}K), however, in plant breeding applications cobalt-60 (^{60}Co) and cesium-137 (^{137}Cs) are the common choices.

In the last 20 years, there are 599 different developed mutant plants belonging to 78 different plant species registered to International Atomic Energy Agency (IAEA) Mutant Variety Database [48]. Soybean is in the third place among these plants with 46 registered mutants (8%) after 247 rice (42%) and 55 wheat (9%) mutants. In the category referred as others, chickpea, carnation, tomato, mung bean, Hibiscus, rapeseed, sesame, orchid, pepper, cowpea, glory bush and sunflower have the most mutants (**Figure 1**).

Among the soybean mutants, there are 15 different improved traits. They can be listed as; high yield, high protein content, resistance to soybean mosaic virus (SMV), early maturity, resistance to leaf rust, resistance to purple seed stain, resistance to cyst nematode (SCN), resistance to lodging, drought tolerance, super nodulation, absence of lipoxygenase, temperature tolerance, low allergenicity and higher nitrogen fixation (**Figure 2**). Thirty-six of these traits were improved by the use of gamma rays

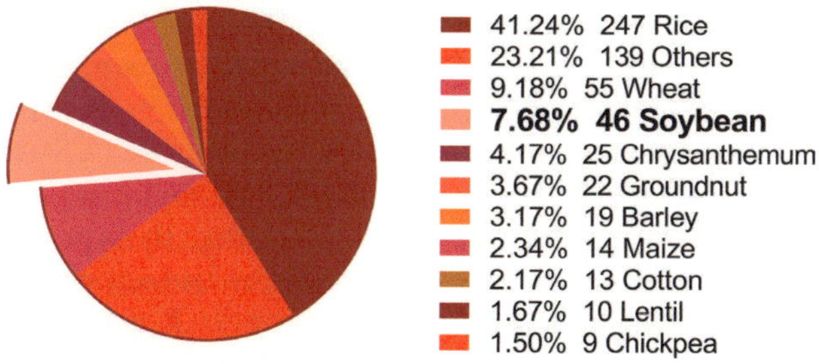

Figure 1.
Mutation variety database of IAEA registered mutant plants in last 20 years [48].

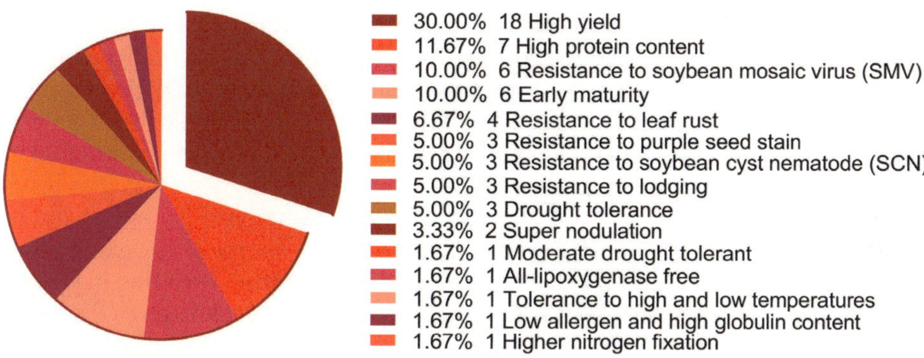

Figure 2.
Radiation-induced trait improvements achieved and registered to MVD in last 20 years.

as physical mutagens, while 7 of them were developed by chemical mutagens. In this period, China is the leading country with 9 registered soybean mutants while Japan (9), Viet Nam (5), Bulgaria (3), India (3), Indonesia (3), Republic of Moldova (3) Republic of Korea (1) and Thailand (1) are the followers.

2.3.2 Present applications of mutation breeding in soybean

In the last decades of mutation breeding, radiosensitivity of different plant species and tissues were investigated and dose limits were determined for various plants. In present days, molecular marker-based techniques were widely applied to estimate genetic diversity and population structure. Among these techniques restriction fragment length polymorphisms (RFLPs), random amplified polymorphic DNAs (RAPDs), amplified fragment length polymorphisms (AFLPs), and inter-simple sequence repeats (ISSRs) are viable options depending on the advantages and limitations of each technique. SNPs, which are spread across in both non-coding and coding regions of the genome, are also preferred in many mutation studies [49]. Present applications of the marker-based techniques include even transposable elements (TEs). The target region amplification polymorphism (TRAP) is a novel, polymerase chain reaction (PCR)-based marker system which exploits the available EST database sequence data to generate polymorphic markers targeting candidate genes. This method utilizes an 18-mer primer derived from the EST sequence and pairs it with an arbitrary primer that targets the intron and/or exon region. TRAP method is useful for germplasm genotyping and producing markers associated with desirable agronomic traits in mutation breeding. Hung et al. [50] employed this simple rapid method by using the consensus terminal inverted repeat sequences of PONG, miniature inverted-repeat transposable element (MITE)-Tourist (M-t) and MITE-Stowaway (M-s) as target region amplification polymorphism (TE-TRAP) markers to investigate the mobility of TEs in a gamma-irradiated soybean mutant pool. They concluded that MITEs were significant enough to confirm their practical utility as molecular markers for investigating mutant populations which were induced by random variations caused through physical mutagenesis (X-ray or gamma-ray). Also, the TE-TRAP marker system was suggested as it provides a simple, rapid, and cost-effective alternative for investigating genetic diversity and identifying mutant lines in irradiated soybean mutant breeding. Kim et al. [51] conducted a genetic diversity and association analysis of soybean mutants to assess elite mutant lines which were induced by 250 Gy of gamma rays using a ^{60}Co gamma-irradiator. They have chosen 208 soybean mutants by phenotypic traits to mutant diversity pool (MDP) and investigated the genetic diversity and inter-relationships of these MDP lines using TRAP markers. MDP has been suggested to have great potential for soybean genetic resources. TRAP markers were found useful for the selection of soybean mutants in mutation breeding applications [51].

Besides the genetic diversity and population structure analysis, genetic characterization of improved mutants and the determination of the source of the gained trait in sequence basis studies have taken over the course of mutation breeding in present days. Before the genomic era which was ignited through the breakthrough discovery of DNA sequencing by Sanger et al. [52], the heteroduplex mismatch cleavage assay which is based on mismatch-specific endonuclease Cel I, was the standard method to detect point mutations. As a simple, rapid, and cheaper mutant discovery method, high resolution melting (HRM) analysis was applied to many agronomical crops. Following the Sanger sequencing, the final step of mutation screening was changed to Sanger to evaluate the changes in the genome and effects of mutation on amino acid substitutions. Today, next generation sequencing (NGS) technologies are the gold standard in the mutation detection field with various options as Roche 454 pyrosequencing, sequencing-by-synthesis, SOLiD sequencing and the HiSeq 2000, which is the gold standard of high-throughput sequencing. Tsuda et al. [53] reported the construction of a high-density mutant library in soybean and the development of a mutant retrieval method referred as amplicon sequencing which is an alternative, cheaper method for sequencing the PCR amplicons in targeted regions. The library of DNA and seeds of EMS-induced plants revealed large morphological and physiological variations. They retrieved the mutants through HRM and indexed amplicon sequencing analysis and confirmed by Sanger sequencing in the final step. They concluded that indexed amplicon sequencing allows researchers to scan a longer sequence range and skip screening steps and also, to know the sequence information of mutation due to the utilization of systematic DNA pooling and the index of NGS reads, which simplifies the discovery of mutants with amino acid substitutions comparing to the HRM screening [53].

MutMap method which utilizes the sequencing technique for mapping the mutated genes responsible for the desired trait was introduced for mutation breeding studies. The first application of the method has been developed by Abe et al. [54] to identify the mutated gene responsible for the change in leaf color from dark green to light green in rice [54]. Thereafter, it has been commonly used for mapping the monogenic recessive genes. In this method, a cultivar with a known reference sequence can be mutagenized by either chemical or physical mutagens. After the selfing and homozygosity experiment for the desired trait between M_3 to M_6 generations, mutants are crossed with their parental or wild type varieties. F_2 population is obtained by selfing of F_1. If the desired trait is inherited through a single recessive gene, the segregation ratio should be of 3:1 in wild and mutant phenotype in F_2 population. In MutMap method, DNA of homozygous mutant plants are extracted and subjected to whole genome sequencing. The mutant genomes are compared to the publicly accessible reference sequences to determine single nucleotide polymorphic (SNPs) variations. The linkage between mutants and wild type plants can be evaluated according to SNP ratios in which the ratio infer that the SNP variation is not linked to the mutation if ranged between 0.1 and 0.5, while it can be linked to the mutation when ranges are between 0.51 to 1 [54, 55]. Kato et al. [56] introduced Lumi-Map, which is a high-throughput platform for identifying causative SNPs for studying pathogen-associated molecular patterns (PAMP) triggered immunity (PTI) signaling components, in combination with MutMap. In Lumi-Map method, they generated nine transgenic *Arabidopsis* reporter lines expressing the LUC gene fused to multiple promoter sequences of defense-related genes, that generates luminescence upon activation of FLAGELLIN-SENSING 2 (FLS2) by flg22, a PAMP derived from bacterial flagellin treatment. Mutagenesis of the line as achieved through EMS treatment and the mutants with altered luminescence patterns were screened by a high-throughput real-time bioluminescence monitoring system. They subjected MutMap method on selected mutants to identify the causative SNP responsible for the luminescence pattern alterations. WRKY29-promoter reporter line was selected to identify mutants in the signaling pathway downstream of FLS2. Twenty-two mutants with altered WRKY29 expression upon flg22 treatment among 24,000 EMS-induced mutants of the reporter line were isolated. In this mutagenesis study, Lumi-Map method combined with MutMap revealed three genes not previously associated with PTI and suggested as a potential alternative to identify novel PAMPs and their receptors as well as signaling components downstream of the receptors [56]. Takagi et al. [57] exploited the rapid and versatile properties of MutMap for more than 20,000 ha of rice paddy field which was inundated with seawater, resulting in salt contamination of the land in Japan following the 2011 earthquake and tsunami that affected Japan. They needed an improved rice variety at short notice as local rice landraces were not tolerant of high salt concentrations caused by seawater. They obtained 6000 EMS-induced mutant lines of a local elite cultivar, 'Hitomebore'. MutMap method was used to rapidly identify a loss-of-function mutation responsible for the salt tolerance of hst1 rice. The detected salt-tolerant hst1 mutant was used to breed a salt-tolerant Kaijin variety which differs from Hitomebore by only 201 SNPs. Conducted field trials presented that improved variety had the equal growth and yield performance as the parental line under normal growth conditions. The whole process was completed only in 2 years which proves the efficiency of MutMap in mutation breeding studies [57]. Fekih et al. [58] improved the method even further and introduced the MutMap+ which is a modified version of MutMap developed for the cases in which obtaining F_2 mapping population is impossible due to the lethal mutations or sterility. MutMap+ has advantages over MutMap as it is less complex, time-consuming, and costly especially in large mapping population. Also, hybridization step of MutMap can be relatively compelling especially in small flower plant species and in crops that are recalcitrant to artificial crosses, therefore, MutMap+, which notably does not necessitate artificial crossing between mutants and the wild-type parental line, is advantageous. In MutMap+ method, again, a cultivar with known reference sequence can be mutagenized by either chemical or physical mutagens. M_1 plants are selfed to develop

M_2 generation. However, in MutMap+ mutants are not crossed with their parental or wild type varieties. The heterozygous M_2 mutant plants are selfed to develop M_3 generation in which the segregation ratio of 3:1 for wild and mutant phenotype is expected. DNAs of tagged mutants and parental varieties are extracted, and pooled. Following the whole genome sequencing, data is compared to the reference genome and SNP profiles are determined. They identified causal nucleotide changes of rice mutants of NAP6 gene that is responsible for change in leaf color and consequent lethality after germination. This versatile extension of MutMap method, also allow determination of recessive lethal alleles [58].

In soybean, Liu et al. [59] investigated two types of resistant sources which are widely used against soybean cyst nematode (SCN, *Heterodera glycines* Ichinohe). Peking-type soybean requires both rhg1-a and Rhg4 alleles, while PI 88788-type soybean requires only the rhg1-b allele for resistance. Instead of MutMap, they preferred the region-specific extraction sequencing (RSE-Seq) method which is developed to enrich a targeted chromosomal segment for genome sequencing to identify SCN resistance genes within the identified 300 kb chromosomal segment carrying the rhg1 locus, due to the requirement of MutMap to an additional procedure of backcross of phenotypic mutants with the wild-type. They suggested GmSNAP18 gene as a candidate for the resistance of two various resistant types of soybeans for SCN [59]. RSE is a cost-effective, long-range DNA target capture methodology that relies on the specific hybridization of short (20–25 base) oligonucleotide primers to selected sequence motifs within the DNA target region This target enrichment method can produce sequencing templates more than 20 kbp in length. These capture primers are then enzymatically extended on the 3′-end, incorporating biotinylated nucleotides into DNA. Streptavidin-coated beads are subsequently used to pull down the original, long DNA template molecules through synthesized, biotinylated DNA that is bound to them [60]. QTL-seq is another method adapted from MutMap to identify quantitative trait loci. In presence of pooled two segregating progeny populations with opposite traits as resistant and susceptible and single whole-genome resequencing of either of the parental cultivars, it utilizes pooled sequences. Also, modified QTL-seq using high-resolution mapping has been developed to cover the weakness of original QTL-seq which do not assume a highly heterozygous genome [61]. Direct whole genome re-sequencing (WGRS) is also utilized effectively to identify candidate genes involved in resistance to SCN in soybean due to the requirement of time-consuming backcrosses in MutMap and QTL-seq methods. Two EMS-induced soybean mutants and six relevant whole genomes were re-sequenced to determine genomic variants as SNPs and InDels. Comparison by this method eliminated many genomic variants from the mutant lines that overlapped non-phenotypic but mutant progeny plants. Therefore, the method was suggested as simple but effective to the identify other trait genes in soybean, even in other organisms [62]. Likewise, comparative genomic analyses of two segregating soybean mutants which were selected among 500 EMS-induced candidates revealed seven genes potentially involved in resistance to *Fusarium equiseti* through WGRS. These genes were suggested to facilitate the breeding of resistant germplasm resources and the identification of resistance to *Fusarium spp*. in soybean [63].

3. Future Prospect and conclusion

Soybean genetic variation improvement is important for the development of superior cultivars. One of the greatest challenges in mutation breeding is random

İntended Development	Target Gene/Site/ Protein	Soybean Line	Transformation Method	Other Model Organizm	gRNAs	Referance
Soybean cyst nematode (SCN) (*Heterodera glycines*) resistance	Glyma.15G191200/γ-SNAP Protein	LD10–30110 Resistance / LD10–30080 and LD10–30092 Susceptible	*Agrobacterium rhizogenes* strain ARqua 1	*N. benthamiana*	TCCGCTGCTGCTCTCGCAA and GTATTCTGTTGCAGCTAAT	[64]
Soybean cytoplasmic malesterile (CMS)	The Aborted Microspores (AMS) Gene	Williams 82	*Agrobacterium tumefaciens* strain EHA105	—	NA	[65]
Heat Stress Tolerance	Glyma.16G178800.1/ Hsp90A2	TianLong No. 1	*A. tumefaciens* strain EHA101	Yeast strain NMY51	NA	[66]
Salt Stress Tolerance	Glyma.06 g21020.1/ NAC06	Williams 82	*A. rhizogenes* strain K599	Yeast strain AH109	NA	[67]
Fatty Acid Content	Glyma.10G278000/ FAD2–1A and Glyma.20G111000/ FAD2–1B	Williams 82 and Maverick	*A. rhizogenesis* strain K599	—	CCAAACACAAAGCCACCATTCAC and GATGAAGGAACATCCGAGAA	[68]
Flowering time and plant height	Glyma.16G091300/ APETALA1 (AP1)	Williams 82 and HX3	*A. tumefaciens* strain EHA101	*Nicotiana benthamiana*	NA	[69]
Flowering Time	Glyma.16 g26660/FT2a	Jack	A. tumefaciens strain EHA105	—	GTAGGGATCCTCTCGTTGTTGGG	[70]
Lipoxygenase-Free	Glyma.13 g347600, Glyma.13 g347500, and Glyma.15 g026300/ LOX1, LOX2, and LOX3	Huachun 6/ Lipoxygenase Free Cultivar Wuxing 4	*A. tumefaciens* strains GV3101	—	GGAAAGGATACGTTCTTG GAAGG(sgRNA(GmLox1/2)/ CCTTTCCTTATCCTCGTAGGGGG(sgRNA-GmLox3)	[71]

İntended Development	Target Gene/Site/ Protein	Soybean Line	Transformation Method	Other Model Organizm	gRNAs	Referance
Decreased Allergenic Genes	Glyma.U020300.1/ Bd 28 K and Glyma.08G116300.1/ Bd 30 K	Enrei and Kariyutaka	*A. tumefaciens* EHA105	—	CCACTCAGCGAACCGGATATTGG and ACCCAAGTAAAGTACCAAGGGGG	[72]
Increases İsoflavone Content	Glyma.11G253000/ Phytoene Desaturase (PDS) and Glyma.10G278000/FAD2	Jack	*A. rhizogenes* strain K599	—	GAAGCAAGAGACGTTCTAGGTGG and AGTTGGCCAACAGTGAATGGTGG	[73]
Soybean Mosaic Virus (SMV)	Glyma.04G196100/ Asetolaktat sentaz (ALS)	Williams82	*A. tumefaciens* strain EHA105	—	CGTCGGCGAGGCCGCTCACGAGG	[73]
Reduced Saturated Fatty Acids	Glyma.05G012300/ FATB1a and Glyma.17G012400/ FATB1b	Williams 82	*Agrobacterium strain tumefaciens* LBA4404	*Arabidopsis thaliana*	GTTAAAAGTGCTGGGCTTCTTGG and GTTAAAAGTGCTGGGCTTCT	[74]
Early Flowering	Glyma.06G207800/E1 Protein	Jack	*A. tumefaciens* strain EHA105	—	CCCTTCAGATGAAAGGGAGCAGT and CCACCATATGCGAAGCCTCTAAC	[75]
Seed Storage	Glyma.20 g148400/ Conglycinins (7S) and Glyma.03 g163500/ Glycinins (11S)	Harosoy 63	*A. rhizogenes* strain K599	—	CCTTCTGAT GAGGTG GGC GT and GATAAC CGTATAGAGTCAGA	[76]
Architecture in Soybean	Glyma.02G177500/ Squamosa Promoter Binding-like Protein 9 (Spl9a)	Williams 82	*A. tumefaciens* strain EHA105	—	TCCCTTGATGGCTTGAAGTTTGG	[76]
Drought Stress	Glyma.16G151500/NAC8	Tianlong No.1	*A. tumefaciens* strain EHA101	*Nicotiana benthamiana*	CCATCTTATCTGAGAACCACTCC	[77]
Herbicide Resistant	DD20 and DD43	93B86	Particle Bombardment	—	GGAACTGACACACGACATGATGG	[78]

Table 1.
Targeted mutagenesis application examples for soybean.

(uncontrolled) nature of induced mutagenesis. Large population requirement for desired mutant selection brings intensive labor. The emergence of clustered regularly interspaced short palindromic repeats/CRISPR associated protein 9 (CRISPR/Cas9) technology has brought wider insight to the field through allowing targeted mutagenesis. It has been widely used in numerous plants as rice, wheat, maize, oilseed rape, barley, cotton, tomato and soybean as well. However, utilization of CRISPR/Cas9 system in soybean is still limited due to the transformation challenges in soybean. As summarized in **Table 1**, most of the targets which were successfully applied to soybean were single gene edits. Paleopolyploid genome of soybean in which approximately 75% of the genes have multiple copies, requires multiple genes or paralogous genes to regulate many important traits. Therefore, these traits may only be targeted by editing which requires the engineering of homologous sequences using more than one sgRNA for recognition. Introducing multiple constructs simultaneously to soybean is relatively limiting in terms of genome editing associated soybean breeding approaches. Recently, Zhang et al. [73] successfully optimized one sgRNA CRISPR/Cas9 system in soybean for the target-specific mutations at multiple loci of GmFAD2 and GmALS. They evaluated the efficiency, type, specificity, and patterns of multiple targeted mutations by selecting three different genes with known functions in soybean and suggested that CRISPR/Cas9 could specifically and efficiently induce targeted mutations at one locus or multiple loci in the T_0 generation. Moreover, they demonstrated the necessity of simultaneous modification of different homoeologous gene copies in polyploid soybean for successful CRISPR-Cas9-mediated breeding [73]. Therefore, induced mutagenesis is still a major method to produce new alleles and new desired traits within the crop genomes. Physical and chemical mutagen protocols are still improving and mutation breeding proves its value to be fast, flexible, and viable in crop sciences.

The second most limiting prospect of induced mutagenesis was the requirement of at least three generation before any stable selection of desired traits in mutants which leads to 7–9 years of average mutation breeding study, previously. However, as described in previous sections NGS based approaches as MutMap accelerated the selection periods significantly. Novel non-destructive measurement methods allow automated imaging and optical measurements of the same plants for desired periods. These approaches provide high measurement densities and fill the gap between genotype and phenotype in mutation breeding studies which is still another limitation in this field. Repeated imaging of particular genotypes under different environmental conditions leads to the generation of development models for biologically relevant parameters. In the present omics era, future procedures may shorten the selection procedures even further [79].

In conclusion, mutation breeding passed important cross-roads successfully during recent advances in plant biotechnology, transformation and targeted mutagenesis by its particular great advantages. Mutagenesis will retain its place in crop science in next decades especially for the plants as soybean for which cross breeding is limited or not applicable.

Author details

Alp Ayan*, Sinan Meriç, Tamer Gümüş and Çimen Atak
Faculty of Science and Letters, Department of Molecular Biology and Genetic, Istanbul Kultur University, Istanbul, Turkey

*Address all correspondence to: a.ayan@iku.edu.tr

References

[1] Knowlton S. High-oleic soybean oil. In: Flider FJ, editor. High Oleic Oils. IL, USA: AOCS Press; 2022. pp. 53-87. DOI: 10.1016/B978-0-12-822912-5.00007-1

[2] Willis S. The use of soybean meal and full fat soybean meal by the animal feed industry. In: 12th Australian Soybean Conference. Bundaberg: Soy Australia; 2003

[3] Costantini M, Bacenetti J. Soybean and maize cultivation in South America: Environmental comparison of different cropping systems. Cleaner Environmental Systems. 2021;**2**:100017. DOI: 10.1016/j.cesys.2021.100017

[4] de Freitas VF, Cerezini P, Hungria M, Nogueira MA. Strategies to deal with drought-stress in biological nitrogen fixation in soybean. Applied Soil Ecology. 2022;**172**:104352. DOI: 10.1016/j.apsoil.2021.104352

[5] Schoch CL, Ciufo S, Domrachev M, Hotton CL, Kannan S, Khovanskaya R, et al. NCBI taxonomy: A comprehensive update on curation, resources and tools. Database. 2020;**2020**:1-21

[6] Hymowitz T. The history of the soybean. In: Soybeans. IL, USA: AOCS Press; 2008. pp. 1-31. DOI: 10.1016/B978-1-893997-64-6.50004-4

[7] Bilyeu K, Ratnaparkhe MB, Kole C. Genetics, Genomics, and Breeding of Soybean. Genet Genomics, Breed Soybean. London, UK: CRC Press; 2016. pp. 1-362

[8] Valliyodan B, Qiu D, Patil G, Zeng P, Huang J, Dai L, et al. Landscape of genomic diversity and trait discovery in soybean. Scientific Reports. 2016;**6**(1):1. DOI: 10.1038/srep23598

[9] Valliyodan B, Brown AV, Wang J, Patil G, Liu Y, Otyama PI, et al. Genetic variation among 481 diverse soybean accessions, inferred from genomic re-sequencing. Scientific Data. 2021;**8**(1):1-9. DOI: 10.1038/s41597-021-00834-w

[10] Takahashi R, Kurosaki H, Yumoto S, Han OK, Abe J. Genetic and linkage analysis of cleistogamy in soybean. Journal of Heredity. 2001;**92**(1):89-92. DOI: 10.1093/jhered/92.1.89

[11] Pratap A, Gupta SK, Kumar J, Solanki RK. Soybean. In: Gupta S, editor. Technological Innovations in Major World Oil Crops. Vol. 1. New York, NY: Springer; 2012. pp. 293-321. DOI: 10.1007/978-1-4614-0356-2_12

[12] Miladinovic J, Burton JW, Tubic SB, Miladinovic D, Djordjevic V, Djukic V. Soybean breeding: Comparison of the efficiency of different selection methods. Turkish Journal of Agriculture and Forestry. 2011;**35**(5):469-480. DOI: 10.3906/tar-1011-1474

[13] Yuhong G, Rasheed A, Zhao ZH, Gardiner JJ, Ilyas M, Akram M, et al. Role of conventional and molecular techniques in soybean yield and quality improvement: A critical review. Notulae Botanicae Horti Agrobotanici Cluj-Napoca. 2021;**49**(4):12555. DOI: 10.15835/nbha49412555

[14] Beaver JS, Osorno JM. Achievements and limitations of contemporary common bean breeding using conventional and molecular approaches. Euphytica. 2009;**168**(2):145-175. DOI: 10.1007/s10681-009-9911-x

[15] Lavanya C, Ushakiran B, Sarada C, Manjunatha T, Senthilvel S, Ramya KT,

et al. Use of single seed descent versus pedigree selection for development of elite parental lines in castor (*Ricinus communis* L.). Genetic Resources and Crop Evolution. 2021;**68**(1):295-305. DOI: 10.1007/s10722-020-00985-6

[16] Jiang GL. Molecular markers and marker-assisted breeding in plants. Plant Breeding from Laboratories to Fields. 2013;**3**:45-83. DOI: 10.5772/52583

[17] Ribaut JM, De Vicente MC, Delannay X. Molecular breeding in developing countries: Challenges and perspectives. Current Opinion in Plant Biology. 2010;**13**(2):213-218. DOI: 10.1016/j.pbi.2009.12.011

[18] Jiang GL. Molecular markers and marker-assisted breeding in plants. Plant Breeding from Laboratories to Fields. 2013;**3**:45-83. DOI: 10.5772/52583

[19] Collard BC, Mackill DJ. Marker-assisted selection: An approach for precision plant breeding in the twenty-first century. Philosophical Transactions of the Royal Society, B: Biological Sciences. 2008;**363**(1491):557-572. DOI: 10.1098/rstb.2007.2170

[20] Anderson EJ, Ali ML, Beavis WD, Chen P, Clemente TE, Diers BW, et al. Soybean [*Glycine max* (L.) Merr.] breeding: History, improvement, production and future opportunities. In: Advances in Plant Breeding Strategies: Legumes. Cham: Springer; 2019. pp. 431-516. DOI: 10.1007/978-3-030-23400-3_12

[21] Bhat JA, Ali S, Salgotra RK, Mir ZA, Dutta S, Jadon V, et al. Genomic selection in the *era* of next generation sequencing for complex traits in plant breeding. Frontiers in Genetics. 2016;**7**:221. DOI: 10.3389/fgene.2016.00221

[22] Crossa J, Pérez-Rodríguez P, Cuevas J, Montesinos-López O, Jarquín D, De Los CG, et al. Genomic selection in plant breeding: Methods, models, and perspectives. Trends in Plant Science. 2017;**22**(11):961-975. DOI: 10.1016/j.tplants.2017.08.011

[23] Alok A, Sharma S, Kumar J, Verma S, Sood H. Engineering in plant genome using *Agrobacterium*: Progress and future. In: Metabolic Engineering for Bioactive Compounds. Singapore: Springer; 2017. pp. 91-111. DOI: 10.1007/978-981-10-5511-9_5

[24] Belhaj K, Chaparro-Garcia A, Kamoun S, Nekrasov V. Plant genome editing made easy: Targeted mutagenesis in model and crop plants using the CRISPR/Cas system. Plant Methods. 2013;**9**(1):1. DOI: 10.1186/1746-4811-9-39

[25] Benning C, Sweetlove L. Synthetic biology for basic and applied plant research. The Plant Journal. 2016;**87**(1):3-4. DOI: 10.1111/tpj.13245

[26] Çelik Ö, Meriç S, Ayan A, Atak Ç. Biotic stress-tolerant plants through small RNA technology. In: Plant Small RNA. Massachusetts, USA: Academic Press; 2020. pp. 435-468. DOI: 10.1016/B978-0-12-817112-7.00020-1

[27] Meriç S, Ayan A, Atak Ç. Molecular abiotic stress Tolerans strategies: From genetic engineering to genome editing era. In: Abiotic Stress in Plants. London, UK: IntechOpen; 2020. p. 91. DOI: 10.5772/intechopen.94505

[28] Ayan A, Meriç S, Gümüş T, Atak Ç. Next generation of transgenic plants: From farming to pharming. In: Genetically Modified Plants and beyond. London, UK: IntechOpen; 2022. DOI: 10.5772/intechopen.102004

[29] James C. Brief 54: Global Status of Commercialized Biotech/GM Crops in 2018: Biotech Crops Continue to

Help Meet the Challenges of Increased Population and Climate Change. Ithaca, NY: ISAAA; 2018

[30] ISAAA. GM Approval Database. GMO Database. Available on: https://www.isaaa.org/gmapprovaldatabase/

[31] Oladosu Y, Rafii MY, Abdullah N, Hussin G, Ramli A, Rahim HA, et al. Principle and application of plant mutagenesis in crop improvement: A review. Biotechnology and Biotechnological Equipment. 2016;**30**(1):1-6. DOI: 10.1080/13102818.2015.1087333

[32] Stadler LJ. Genetic effects of X-rays in maize. Proceedings of the National Academy of Sciences of the United States of America. 1928;**14**(1):69. DOI: 10.1073/pnas.14.1.69

[33] Freisleben R, Lein A. Über die Auffindung einer mehltauresistenten Mutante nach Röntgenbestrahlung einer anfälligen reinen Linie von Sommergerste. Die Naturwissenschaften. 1942;**30**(40):608. DOI: 10.1007/BF01488231

[34] Shu QY, Forster BP, Nakagawa H, Nakagawa H. Plant Mutation Breeding and Biotechnology. Wallingford, UK: Cabi; 2012

[35] Bado S, Forster BP, Nielen S, Ali AM, Lagoda PJ, Till BJ, et al. Plant mutation breeding: Current progress and future assessment. Plant Breeding Reviews. 2015;**39**:23-88. DOI: 10.1002/9781119107743.ch02

[36] Jankowicz-Cieslak J, Tai TH, Kumlehn J, Till BJ. Biotechnologies for Plant Mutation Breeding: Protocols. Berlin, Germany: Springer Nature; 2017

[37] Lehnert S. Biomolecular Action of Ionizing Radiation. London, UK: CRC Press; 2007

[38] Ossowski S, Schneeberger K, Lucas-Lledó JI, Warthmann N, Clark RM, Shaw RG, et al. The rate and molecular spectrum of spontaneous mutations in *Arabidopsis thaliana*. Science. 2010;**327**(5961):92-94. DOI: 10.1126/science.1180677

[39] Lagoda PJ. Effects of mutagenic agents on the DNA sequence in plants. Plant Breeding and Genetics Newsletter. 2007;**19**:13-14

[40] Jain SM, Brar DS, Ahloowalia BS. Somaclonal Variation and Induced Mutations in Crop Improvement. Berlin, Germany: Springer Science & Business Media; 2013. DOI: 10.1007/978-94-015-9125-6_11

[41] Azman AS, Mhiri C, Tam S. Transposable elements and the detection of somaclonal variation in plant tissue culture: A review. Malaysian Applied Biology. 2014;**43**(1):1-2

[42] Raina A, Laskar RA, Khursheed S, Amin R, Tantray YR, Parveen K, et al. Role of mutation breeding in crop improvement-past, present and future. Asian Research Journal of Agriculture. 2016;**2**(2):1-3. DOI: 10.9734/ARJA/2016/29334

[43] Dhawan OP, Lavania UC. Enhancing the productivity of secondary metabolites via induced polyploidy: A review. Euphytica. 1996;**87**(2):81-89. DOI: 10.1007/BF00021879

[44] Rostoks N, Mudie S, Cardle L, Russell J, Ramsay L, Booth A, et al. Genome-wide SNP discovery and linkage analysis in barley based on genes responsive to abiotic stress. Molecular Genetics and Genomics. 2005;**274**(5):515-527. DOI: 10.1007/s00438-005-0046-z

[45] Jehan T, Lakhanpaul S. Single nucleotide polymorphism (SNP)–methods

and applications in plant genetics: A review. Indian Journal of Biotechnology. 2006;**5**:435-459

[46] Tondelli A, Francia E, Barabaschi D, Aprile A, Skinner JS, Stockinger EJ, et al. Mapping regulatory genes as candidates for cold and drought stress tolerance in barley. Theoretical and Applied Genetics. 2006;**112**(3):445-454. DOI: 10.1007/s00122-005-0144-7

[47] Tuberosa R, Salvi S. Genomics-based approaches to improve drought tolerance of crops. Trends in Plant Science. 2006;**11**(8):405-412. DOI: 10.1016/j.tplants.2006.06.003

[48] IAEA. Mutanat Variety Database (MVD) [Internet]. 2022. Available from: https://mvd.iaea.org/#!Home [Accessed: 07-02-2022]

[49] Atak Ç, Alikamanoğlu S, Açık L, Canbolat Y. Induced of plastid mutations in soybean plant (*Glycine max* L. Merrill) with gamma radiation and determination with RAPD. Mutation Research: Fundamental and Molecular Mechanisms of Mutagenesis. 2004;**556**(1-2):35-44. DOI: 10.1016/j.mrfmmm.2004.06.037

[50] Hung NN, Kim DG, Lyu JI, Park KC, Kim JM, Kim JB, et al. Detecting genetic mobility using a transposon-based marker system in gamma-ray irradiated soybean mutants. Plants. 2021;**10**(2):373. DOI: 10.3390/plants10020373

[51] Kim DG, Lyu JI, Lee MK, Kim JM, Hung NN, Hong MJ, et al. Construction of soybean mutant diversity pool (MDP) lines and an analysis of their genetic relationships and associations using TRAP markers. Agronomy. 2020;**10**(2):253. DOI: 10.3390/agronomy10020253

[52] Sanger F, Nicklen S, Coulson AR. DNA sequencing with chain-terminating inhibitors. Proceedings of the National Academy of Sciences. 1977;**74**(12):5463-5467. DOI: 10.1073/pnas.74.12.5463

[53] Tsuda M, Kaga A, Anai T, Shimizu T, Sayama T, Takagi K, et al. Construction of a high-density mutant library in soybean and development of a mutant retrieval method using amplicon sequencing. BMC Genomics. 2015;**16**(1):1-8. DOI: 10.1186/s12864-015-2079-y

[54] Abe A, Kosugi S, Yoshida K, Natsume S, Takagi H, Kanzaki H, et al. Genome sequencing reveals agronomically important loci in rice using MutMap. Nature Biotechnology. 2012;**30**(2):174-178. DOI: 10.1038/nbt.2095

[55] Tribhuvan KU, Kumar K, Sevanthi AM, Gaikwad K. MutMap: A versatile tool for identification of mutant loci and mapping of genes. Indian Journal of Plant Physiology. 2018;**23**(4):612-621. DOI: 10.1007/s40502-018-0417-1

[56] Kato H, Onai K, Abe A, Shimizu M, Takagi H, Tateda C, et al. Lumi-map, a real-time luciferase bioluminescence screen of mutants combined with MutMap, reveals *Arabidopsis* genes involved in PAMP-triggered immunity. Molecular Plant-Microbe Interactions. 2020;**33**(12):1366-1380. DOI: 10.1094/MPMI-05-20-0118-TA

[57] Takagi H, Tamiru M, Abe A, Yoshida K, Uemura A, Yaegashi H, et al. MutMap accelerates breeding of a salt-tolerant rice cultivar. Nature Biotechnology. 2015;**33**(5):445-449. DOI: 10.1038/nbt.3188

[58] Fekih R, Takagi H, Tamiru M, Abe A, Natsume S, Yaegashi H, et al. MutMap+: Genetic mapping and mutant identification without crossing in rice.

PLoS One. 2013;**8**(7):e68529. DOI: 10.1371/journal.pone.0068529

[59] Liu S, Kandoth PK, Lakhssassi N, Kang J, Colantonio V, Heinz R, et al. The soybean *GmSNAP18* gene underlies two types of resistance to soybean cyst nematode. Nature Communications. 2017;**8**(1):1. DOI: 10.1038/ncomms14822

[60] Dapprich J, Ferriola D, Mackiewicz K, Clark PM, Rappaport E, D'Arcy M, et al. The next generation of target capture technologies-large DNA fragment enrichment and sequencing determines regional genomic variation of high complexity. BMC Genomics. 2016;**17**(1):1-4. DOI: 10.1186/s12864-016-2836-6

[61] Sugihara Y, Young L, Yaegashi H, Natsume S, Shea DJ, Takagi H, et al. High-performance pipeline for MutMap and QTL-seq. PeerJ. 2022;**10**:e13170. DOI: 10.7717/peerj.13170

[62] Liu S, Ge F, Huang W, Lightfoot DA, Peng D. Effective identification of soybean candidate genes involved in resistance to soybean cyst nematode via direct whole genome re-sequencing of two segregating mutants. Theoretical and Applied Genetics. 2019;**132**(9):2677-2687. DOI: 10.1007/s00122-019-03381-6

[63] Zhang L, Huang W, Peng D, Liu S. Comparative genomic analyses of two segregating mutants reveal seven genes likely involved in resistance to *Fusarium equiseti* in soybean via whole genome re-sequencing. Theoretical and Applied Genetics. 2019;**132**(11):2997-3008. DOI: 10.1007/s00122-019-03401-5

[64] Butler KJ, Fliege C, Zapotocny R, Diers B, Hudson M, Bent AF. Soybean cyst nematode resistance quantitative trait locus *cqSCN-006* alters the expression of a γ-SNAP protein. Molecular Plant-Microbe Interactions. 2021;**34**(12):1433-1445. DOI: 10.1094/MPMI-07-21-0163-R

[65] Chen X, Yang S, Zhang Y, Zhu X, Yang X, Zhang C, et al. Generation of male-sterile soybean lines with the CRISPR/Cas9 system. The Crop Journal. 2021;**9**(6):1270-1277. DOI: 10.1016/j.cj.2021.05.003

[66] Huang Y, Xuan H, Yang C, Guo N, Wang H, Zhao J, et al. GmHsp90A2 is involved in soybean heat stress as a positive regulator. Plant Science. 2019;**285**:26-33. DOI: 10.1016/j.plantsci.2019.04.016

[67] Li M, Chen R, Jiang Q, Sun X, Zhang H, Hu Z. *GmNAC06*, a NAC domain transcription factor enhances salt stress tolerance in soybean. Plant Molecular Biology. 2021;**105**(3):333-345. DOI: 10.1007/s11103-020-01091-y

[68] Do PT, Nguyen CX, Bui HT, Tran LT, Stacey G, Gillman JD, et al. Demonstration of highly efficient dual gRNA CRISPR/Cas9 editing of the homeologous *GmFAD2-1A* and *GmFAD2-1B* genes to yield a high oleic, low linoleic and α-linolenic acid phenotype in soybean. BMC Plant Biology. 2019;**19**(1):1-4. DOI: 10.1186/s12870-019-1906-8

[69] Chen L, Nan H, Kong L, Yue L, Yang H, Zhao Q, et al. Soybean *AP1* homologs control flowering time and plant height. Journal of Integrative Plant Biology. 2020;**62**(12):1868-1879. DOI: 10.1111/jipb.12988

[70] Cai Y, Chen L, Liu X, Guo C, Sun S, Wu C, et al. CRISPR/Cas9-mediated targeted mutagenesis of *GmFT2a* delays flowering time in soya bean. Plant Biotechnology Journal. 2018;**16**(1):176-185. DOI: 10.1111/pbi.12758

[71] Wang J, Kuang H, Zhang Z, Yang Y, Yan L, Zhang M, et al. Generation of seed lipoxygenase-free soybean using CRISPR-Cas9. The Crop Journal.

2020;**8**(3):432-439. DOI: 10.1016/j.cj.2019.08.008

[72] Sugano S, Hirose A, Kanazashi Y, Adachi K, Hibara M, Itoh T, et al. Simultaneous induction of mutant alleles of two allergenic genes in soybean by using site-directed mutagenesis. BMC Plant Biology. 2020;**20**(1):1-5. DOI: 10.1186/s12870-020-02708-6

[73] Zhang L, Wang Y, Li T, Qiu H, Xia Z, Dong Y. Target-specific mutations efficiency at multiple loci of CRISPR/Cas9 system using one sgRNA in soybean. Transgenic Research. 2021;**30**(1):51-62. DOI: 10.1007/s11248-020-00228-5

[74] Ma J, Sun S, Whelan J, Shou H. CRISPR/Cas9-mediated knockout of *GmFATB1* significantly reduced the amount of saturated fatty acids in soybean seeds. International Journal of Molecular Sciences. 2021;**22**(8):3877. DOI: 10.3390/ijms22083877

[75] Han J, Guo B, Guo Y, Zhang B, Wang X, Qiu LJ. Creation of early flowering germplasm of soybean by CRISPR/Cas9 technology. Frontiers in Plant Science. 2019;**1446**. DOI: 10.3389/fpls.2019.01446

[76] Bao A, Chen H, Chen L, Chen S, Hao Q, Guo W, et al. CRISPR/Cas9-mediated targeted mutagenesis of *GmSPL9* genes alters plant architecture in soybean. BMC Plant Biology. 2019;**19**(1):1-2. DOI: 10.1186/s12870-019-1746-6

[77] Yang C, Huang Y, Lv W, Zhang Y, Bhat JA, Kong J, et al. GmNAC8 acts as a positive regulator in soybean drought stress. Plant Science. 2020;**293**:110442. DOI: 10.1016/j.plantsci.2020.110442

[78] Li Z, Liu ZB, Xing A, Moon BP, Koellhoffer JP, Huang L, et al. Cas9-guide RNA directed genome editing in soybean. Plant Physiology. 2015;**169**(2):960-970. DOI: 10.1104/pp.15.00783

[79] Schunk CR, Eberius M. Phenomics in plant biological research and mutation breeding. Plant Mutation Breeding and Biotechnology. 2012:535-560. DOI: 10.1079/9781780640853.0535

Chapter 9

Breeding of *Rj* Gene-Accumulated Soybean Genotypes and Their Availability for Improving Soybean Productivity

Sokichi Shiro and Yuichi Saeki

Abstract

Some soybean varieties harbor the *Rj* genes, which regulate nodulation by preventing infection and nodulation by specific rhizobial strains. Soybean genotypes carrying several *Rj* genes may enhance the occupancy of useful bradyrhizobia, which exhibit potent nitrogen fixation ability and exhibit nodulation compatible with the *Rj* genotype of soybean. Therefore, we bred soybean lines presenting the $Rj_2Rj_3Rj_4$ genotype by crossing the Japanese cultivars "Bonminori" (Rj_2Rj_3) and "Fukuyutaka" (Rj_4) and studied the effects of *Rj* gene accumulation on productivity. To investigate yield components, three *Rj* gene-accumulated soybean lines (B × F – E, B × F – M, and B × F – L) and three soybean cultivars ("Enrei," "Sachiyutaka," and "Fukuyutaka") were cultivated in 2016 and 2017. Pod and seed number and yield were the highest in B × F – M. The occupancy rate of isolates in cluster of *Bradyrhizobium diazoefficiens* USDA 110^T carrying the hydrogen uptake genes tended to be lower in the $Rj_2Rj_3Rj_4$ soybean lines than in "Sachiyutaka" and "Fukuyutaka." Additionally, the occupancy rate of this cluster was positively correlated with yield. Therefore, promoting infection by bradyrhizobial strains carrying the hydrogen uptake genes may improve soybean productivity. Moreover, the $Rj_2Rj_3Rj_4$ genotype of soybean may be inoculated with *B. diazoefficiens* USDA 110^T, which is not restricted by the Rj_2 gene, to further enhance soybean productivity.

Keywords: soybean, *Rj* gene, breeding, yield components, infection tendency

1. Introduction

Soybean (*Glycine max* (L.) Merr.) is one of the most important legume crops in the world, including Japan. According to the information on soybean production and consumption published by the Japanese Ministry of Agriculture, Forestry and Fisheries (MAFF), soybean yield in the country is 166 kg 10 a^{-1}, which is lower than those in major producing countries, including the United States (358 kg 10 a^{-1}), Brazil (342 kg 10 a^{-1}), Argentina (309 kg 10 a^{-1}), and China (188 kg 10 a^{-1}) [1]. To improve this lower productivity, breeding of high-yielding soybean and improvement of cultivation techniques, such as pest control, field management, and plantation methods, have been extensively studied. One of the cultivation techniques is the inoculation of rhizobia, which exhibit potent nitrogen-fixing capacity, during soybean plantation.

As a leguminous plant, soybean roots bear nodules formed as a result of infection by nodulating rhizobia, which perform symbiotic nitrogen fixation, and the plant acquires atmospheric nitrogen in the form of ammonia through these root nodules. Major soybean-nodulating rhizobia include *Bradyrhizobium japonicum*, *Bradyrhizobium diazoefficiens*, *Bradyrhizobium elkanii*, and *Sinorhizobium* (=*Ensifer*) *fredii* [2–7]. In addition to these, *Bradyrhizobium yuanmingense*, *Bradyrhizobium liaoningense, Sinorhizobium xinjiangense*, and *Mesorhizobium tianshanense* have been reported as soybean-nodulating rhizobial species [8–14]. *B. diazoefficiens* USDA 110^T is a symbiont possessing a hydrogen uptake (Hup) system that recycles H_2 produced as a by-product of nitrogenase activity, thereby increasing nitrogen fixation efficiency [15–17]. The inoculation of bradyrhizobia possessing this system, such as *B. diazoefficiens* Hup^+ strains, enhances the productivity of legume crops [18]. However, the efficiency of inoculated rhizobia with high nitrogen fixation ability remains poor in the field, because they cannot compete with indigenous soybean-nodulating rhizobia in the soil. To solve this problem, the ecology of indigenous soybean-nodulating rhizobia in terms of genetic diversity and compatibility with the host soybean must be elucidated.

Rj or *rj* are the well-known host genes that regulate soybean nodulation, and non-*Rj*, rj_1, Rj_2, Rj_3, Rj_4, and *Rfg1* genotypes of soybean have been confirmed to exist naturally [19–24]. In addition to these, *Rj* genotypes, including rj_5, rj_6, and rj_7, have been developed through experimental mutagenesis [25–31]. The Rj_2, Rj_3, Rj_4, and *Rfg1* genotypes are known to restrict nodulation by specific strains of *Bradyrhizobium* or *Sinorhizobium* species. Meanwhile, the Rj_2, Rj_3, Rj_4, and *Rfg1* genotypes restrict nodulation by *B. diazoefficiens* USDA 122, *B. elkanii* USDA 33, *B. elkanii* USDA 61, and *Sinorhizobium fredii* USDA 257. Furthermore, *B. japonicum* Is–1 and Is–34 exhibit incompatibility with the Rj_2 and Rj_4 genotypes, respectively [32]. The rj_1, rj_5, and rj_6 genotypes restrict nodulation by all soybean-nodulating rhizobial strains. The rj_7 genotype developed through ethyl methane sulfonate (EMS)-induced mutagenesis is a "hypernodulation" genotype, which can form abundant nodules [33]. The Rj_2/*Rfg1* gene encodes a member of the Toll-interleukin receptor–nucleotide-binding site–leucine-rich repeat (TIR–NBS–LRR) class of plant resistance (R) proteins, which confer resistance against microbial pathogens through an effector-triggered immune (ETI) response [34]. Furthermore, the amino acid determinant of the Rj_2 genotype in cultivated and wild soybeans has been reported [35]. The Rj_4 gene encodes a thaumatin-like protein (TLP), classified as pathogenesis-related protein 5 (PR5). PR proteins are induced by pathogen attack and involved in host resistance [36, 37]. In addition, the type III secretion system (T3SS) structural gene in *B. elkanii* USDA 61 and *B. japonicum* Is–34 is involved in the restriction of nodulation in the Rj_4 genotype of soybean [38, 39].

The compatibility and preference for nodulation by bradyrhizobial strains of soybean cultivars and varieties exhibiting the *Rj* genotype have been investigated [32], and the $Rj_2Rj_3Rj_4$ genotype lines, in which the *Rj* genes are accumulated, have been bred by crossing the soybean cultivars "IAC-2" (Rj_2Rj_3) and "Hill" (Rj_4) [40]. The $Rj_2Rj_3Rj_4$ genotype is superior to the non-*Rj*, Rj_2Rj_3, and Rj_4 genotypes in terms of the efficiency of nodulation by inocula with potent nitrogen fixation ability [41]. In addition, the community structure of indigenous soybean-nodulating bradyrhizobia was significantly different across five *Rj* genotypes (non-*Rj*, Rj_2Rj_3, Rj_3, Rj_4, and $Rj_2Rj_3Rj_4$) [42]. Furthermore, the Rj_2Rj_3 and $Rj_2Rj_3Rj_4$ genotypes presented a higher occupancy of the indigenous soybean-nodulating bradyrhizobial cluster of *B. diazoefficiens* USDA 110^T than the non-*Rj*, Rj_3, and Rj_4 genotypes, regardless of the cultivation temperature [43]. Thus, the availability of the $Rj_2Rj_3Rj_4$ genotype of soybean has been reported. Since the $Rj_2Rj_3Rj_4$ genotype of soybean has been produced by crossing foreign cultivars, *Rj* gene-accumulated soybean cultivars that match the needs of Japanese consumers and producers must be developed. Therefore, we bred the $Rj_2Rj_3Rj_4$ genotype of soybean by crossing the Japanese soybean cultivars "Bonminori" (Rj_2Rj_3) and "Fukuyutaka" (Rj_4). According to the information on the development and diffusion of new soybean cultivars published by the Japanese MAFF, "Fukuyutaka" is the most cultivated soybean cultivar in the country [44], and this cultivar was registered in 1980 [45].

In this chapter, we describe breeding and selection processes, shoot growth, yield components, and infection tendency of useful bradyrhizobia of *Rj* gene-accumulated soybean genotypes produced by crossing Japanese cultivars.

2. Breeding and selection of *Rj* gene-accumulated soybean

2.1 Phenotypic analysis of "Bonminori" × "Fukuyutaka" F3 seeds

To select *Rj* gene-accumulated soybean lines with the $Rj_2Rj_3Rj_4$ genotype, 157 F3 seeds from the experimental field of the Faculty of Agriculture, University of Miyazaki (31°49′41″N, 131°24′45″E), were subjected to the inoculation test. *B. japonicum* Is–1 and *B. japonicum* Is–34, which exhibit incompatibility with the Rj_2 and Rj_4 genotypes of soybean, respectively, were used as the inoculum strains. The strains were cultured in 1 mL of HEPES-MES (HM) medium [46] supplemented with 0.1% L-arabinose [47] for 3–5 days in the dark at 28°C. The bacterial cultures were then diluted with sterile distilled water to 10^6 cells mL^{-1} and mixed with the respective strain. Culture pots (1 L) were filled with vermiculite containing aqueous N-free nutrient solution (40% v/v) [48] and autoclaved at 121°C for 20 min. F3 seeds were sterilized using 70% ethanol for 30 s and diluted sodium hypochlorite solution (0.25% available chlorine) for 3 min and then washed with sterile distilled water. Five seeds of each line were sown per pot, inoculated with 1 mL of diluted bacterial culture per seed, and cultivated for 3–4 weeks in a growth chamber (day, 28°C for 16 h; night, 25°C for 8 h), with weekly supply of sterile distilled water. After cultivation for 3–4 weeks, the presence or absence of nodulation on soybean roots was observed.

2.2 Selection of non-nodulation phenotype using the inoculation test

Theoretically, all F1 lines obtained by crossing "Bonminori" and "Fukuyutaka" should present the $Rj_2Rj_2Rj_4Rj_4$ genotype, and according to Mendel's law, the probability of F2 lines exhibiting this genotype is 1/16. In the F3 lines, individuals with the $Rj_2Rj_2Rj_4Rj_4$ genotype can be obtained by self-fertilization of F2 individuals with the $Rj_2rj_2Rj_4Rj_4$, $Rj_2Rj_2Rj_4rj_4$, and $Rj_2rj_2Rj_4rj_4$ genotypes. To select individuals with the $Rj_2Rj_2Rj_4Rj_4$ genotype, we subjected 153 lines of F3 seeds, excluding four lines of seeds that had decayed during storage in the refrigerator, to inoculation test. The appearance of soybean roots during the inoculation test is shown in **Figure 1**. Dominant homozygous plants, such as those with the $Rj_2Rj_2Rj_4Rj_4$ genotype, did not form root nodules (**Figure 1a**). We proceeded with the screening based on this phenotype and selected eight lines from the 153 lines (**Table 1**). These eight lines were grown for

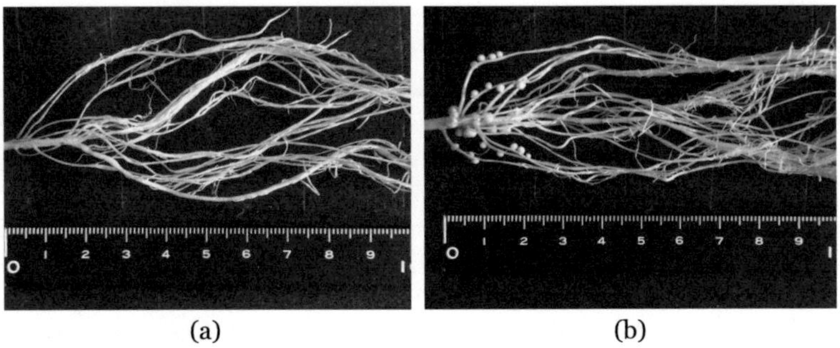

Figure 1.
Appearance of nodulation following the inoculation of Bradyrhizobium japonicum Is-1 and Is-34. Soybean roots with (a) and without (b) nodulation restriction.

Phenotypes of F3 seeds	Number of soybean lines
non-nodulation	8
non-nodulation or nodulation	63
nodulation	82
not tested	4
Total	157

Among the five seeds sown for selection, lines that did not form root nodule on all plants were classified as "non-nodulation," lines that formed root nodule more than one of the five plants were classified as "non-nodulation or nodulation," and lines that formed root nodule all plants classified as "nodulation."

Table 1.
Nodulation phenotypes of F3 seeds.

several years. Three plants differing in terms of the flowering and ripening periods by approximately 1 week each were detected and selected for further cultivation.

3. Evaluation of shoot growth and yield components of *Rj* gene-accumulated soybean

3.1 Experimental site, design, and tested soybean variety

A 2-year field trial was conducted in 2016 and 2017 in the experimental field of the Agricultural Science Section, Education and Research Center for Biological Resources, Faculty of Life and Environmental Science, Shimane University, Japan (35°30′55″N, 133°06′36″E). The experimental sites were located at 35°30′60″N, 133°06′35″E in 2016 and 35°31′02″N, 133°06′40″E in 2017. Both experimental fields had gray lowland soil (paddy conversion fields). Soil pH (H_2O) and electrical conductivity (mS cm^{-1}) were respectively 6.42 and 0.10 in 2016 and respectively 6.72 and 0.06 in 2017. Before sowing, nitrogen, potassium, and phosphorus were applied at doses of 40, 100, and 100 kg ha^{-1}, respectively. To correct soil pH, magnesium lime was applied at the dose of 1000 kg ha^{-1}. The experiment followed the split-plot design with three replicates. Three soybean cultivars, namely "Enrei," "Sachiyutaka," and "Fukuyutaka," as well as F10 or F11 plants of three *Rj* gene-accumulated soybean lines with different flowering and ripening periods, namely B × F – E, B × F – M, and B × F – L, were tested. "Enrei" and "Sachiyutaka" were registered in 1971 and 2001, respectively [45]. "Enrei" presents the *rj*$_4$ genotype [37]. "Sachiyutaka" may present the *Rj*$_4$ genotype, because it is bred through backcrossing "Enrei" with F2 plants from a cross "Enrei" and "Fukuyutaka" [45]. All soybean seeds were sown at a depth of 3–4 cm on June 20, 2016, and June 21, 2017, respectively, and the planting density was 11.1 and 10.1 plants m^{-2} in 2016 and 2017, respectively.

3.2 Data collection and analysis

Soybean growth was evaluated during the flowering and harvest periods. Samples were collected by from 10 consecutive plots per replicate. During the flowering period, plant height, node number, branch number, stem and leaf dry weight, and main culm dry weight (2017 only) were measured. During the harvest period, plant height, node number, shoot dry weight, pod number, seed number, 100-seed weight, and yield were measured. Plant dry weight was measured after drying at 70°C for over 72 h in a drying apparatus. All statistical analyses were performed using R version 4.0.3 [49]. Soybean growth parameters during the flowering period were analyzed using Tukey's honestly significant difference (HSD) test for multiple comparisons using the R package "multcomp." Soybean yield components were subjected to two-way analysis of variance using anovakun version 4.8.5 [50]. Meteorological data during soybean cultivation were collected from past information provided by the Japan Weather Association (**Table 2**).

3.3 Growth and yield of *Rj* gene-accumulated soybean

The results of soybean growth during the flowering period in 2017 are presented in **Table 3**. The measurements during the flowering period were conducted on August 10, 2017, for B × F – E and "Enrei"; August 17, 2017, for B × F – M and "Sachiyutaka"; and August 26, 2017, for B × F – L and "Fukuyutaka." The plant height of *Rj* gene-accumulated soybean lines tended to increase in the order B × F – E < B × F – M < B × F – L, indicating dependence on the lateness of the flowering period. Similarly, the plant height of other soybean cultivars tended to increase in the order of "Enrei" < "Sachiyutaka" < "Fukuyutaka," indicating dependence on the lateness of the flowering period. Branch number was the highest in "Fukuyutaka" and the lowest in "Enrei." In addition, among the *Rj* gene-accumulated soybean lines, plant height tended to be higher in B × F – L than in the other lines, although the difference was not significant. There were no significant differences in shoot dry weight among the cultivars, although "Fukuyutaka," B × F – M, and "Enrei" showed higher values in that order.

The results of yield components of soybean cultivars during the harvest period in 2016 and 2017 are presented in **Table 4**. In ANOVA, all yield components, except 100-seed weight, significantly differed between years and among cultivars. Specifically, pod and seed number and yield were significantly higher in 2016 than in 2017. Conversely, plant height, node number, and shoot dry weight were significantly higher in 2017 than in 2016. Based on the average values of the 2 years, pod and seed number in B × F – M was significantly higher than that in the other cultivars. Moreover, the yield of B × F – M and "Sachiyutaka" was significantly higher than that of B × F – E, B × F – L, and "Enrei." Furthermore, 100-seed weight of "Sachiyutaka" was significantly higher than that of the other cultivars, except "Fukuyutaka." Plant height and shoot dry weight of B × F – L tended to be higher

Year	month	Temperature (°C)			Precipitation	Sunshine
		minimum	maximum	mean	(mm)	duration (h)
2016	Jun.	19.0	26.8	22.3	166.0	4.9
	Jul.	23.5	30.8	26.6	77.0	5.7
	Aug.	23.6	32.2	27.2	140.5	7.7
	Sep.	20.7	26.7	23.3	293.0	2.8
	Oct.	15.3	22.3	18.5	103.5	3.5
	Nov.	8.8	16.2	12.2	120.0	3.2
2017	Jun.	16.5	25.9	20.9	86.5	7.3
	Jul.	24.8	31.5	27.6	168.5	5.6
	Aug.	24.4	31.5	27.4	141.5	6.8
	Sep.	18.5	26.3	22.1	214.5	4.9
	Oct.	14.3	21.0	17.5	358.0	3.6
	Nov.	6.7	15.7	11.1	93.0	4.6

Values indicate monthly averages, and each value was calculated based on meteorological data provided by the Japan Weather Association.

Table 2.
Meteorological data during soybean cultivation in 2016 and 2017.

Cultivar	Plant height (cm plant^{-1})	Node number (No. plant^{-1})	Branching number (No. plant^{-1})	Main culm dry weight (g m^{-2})	Stem and leaf dry weight (g m^{-2})	Shoot dry weight (g m^{-2})
BxF–E	66.2 ab	13.5	6.1 ab	13.0	23.1	36.1
BxF–M	67.9 ab	13.1	6.5 ab	15.3	25.1	40.4
BxF–L	74.9 b	13.6	7.2 bc	16.7	22.4	39.0
Enrei	53.7 a	11.5	4.5 a	12.2	27.0	39.2
Sachiyutaka	57.3 a	12.3	6.4 ab	12.7	22.6	35.4
Fukuyutaka	64.5 ab	12.2	9.2 c	17.4	23.8	41.2
ANOVA	*	ns	***	ns	ns	ns

Values are presented as the means of three replicates. $^{*}p < 0.05$, $^{***}p < 0.001$, *and ns = not significant. Different letters indicate significant differences (Tukey's HSD test) at* $p < 0.05$ *for different soybean cultivars.*

Table 3.
Growth of soybean cultivars during the flowering period in 2017.

than those of the other cultivars. The interaction between year and cultivar was detected for all test parameters, except seed number and yield. Therefore, multiple comparison analysis was performed among 12 cohorts for each test item, and the results are shown in **Figure 2**. Briefly, pod and seed number and yield were lower in all soybean cultivars in 2017 than in 2016. Furthermore, pod and seed number of B × F – E, B × F – M, "Enrei," and "Sachiyutaka" decreased significantly. While the yield of "Sachiyutaka" decreased significantly, that of B × F – E, B × F – M, B × F – L, "Enrei," and "Fukuyutaka" tended to decrease, albeit without significant differences.

Year	Cultivar	Plant height (cm plant^{-1})	Node number (No. plant^{-1})	Shoot dry weight (g m^{-2})	Pod number (No. m^{-2})	seed number (No. m^{-2})	100-seed weight (g)	Yield (g m^{-2})
2016	BxF–E	50.3	11.1	251.6	1426.7	937.0	23.9	214.4
	BxF–M	51.1	11.3	298.5	1927.0	2276.8	22.9	514.5
	BxF–L	71.1	12.1	336.6	1036.0	1372.9	23.5	325.8
	Enrei	34.3	9.5	215.7	542.6	421.4	22.6	96.1
	Sachiyutaka	46.1	12.1	300.3	1531.1	1742.9	33.8	584.6
	Fukuyutaka	56.3	12.1	267.1	1057.3	1377.3	29.8	407.4
2017	BxF–E	67.4	13.4	251.8	318.8	218.9	24.6	53.8
	BxF–M	69.4	13.8	432.8	1180.7	1477.4	23.6	348.2
	BxF–L	71.0	13.3	424.1	784.0	1037.7	20.7	214.6
	Enrei	56.2	12.8	340.0	101.0	56.2	29.7	16.7
	Sachiyutaka	55.2	12.1	344.2	704.1	974.4	29.3	285.8
	Fukuyutaka	70.5	16.1	467.9	934.1	1245.6	25.0	311.7
2016		51.5	11.4	278.3	1253.4	1354.7	26.1	357.1
2017		64.9	13.6	376.8	670.4	835.0	25.5	205.1
	BxF–E	58.8 bc	12.2 a	251.7 a	872.7 b	578.0 a	24.2 a	134.1 a
	BxF–M	60.3 bc	12.5 ab	365.7 b	1553.8 c	1877.1 c	23.2 a	431.4 c
	BxF–L	71.0 d	12.7 ab	380.4 b	910.0 b	1205.3 b	22.1 a	270.2 b
	Enrei	45.2 a	11.1 a	277.8 a	321.8 a	238.8 a	26.1 a	56.4 a
	Sachiyutaka	50.6 ab	12.1 a	322.3 ab	1117.6 b	1358.6 b	31.6 b	435.2 c
	Fukuyutaka	63.4 cd	14.1 b	367.5 b	995.7 b	1311.5 b	27.4 ab	359.6 bc
ANOVA	Year (Y)	***	***	***	***	***	ns	***
	Cultivar (C)	***	***	***	***	***	***	***
	Y x C	*	**	**	**	ns	**	ns

Values are presented as the means of three replicates. $^{*}p < 0.05$, $^{**}p < 0.01$, $^{***}p < 0.001$, *and ns = not significant. Different letters indicate significant differences (Tukey's HSD test) at* $p < 0.05$ *for different soybean cultivars.*

Table 4.
Yield components of soybean cultivars in 2016 and 2017.

"Enrei," "Sachiyutaka," and "Fukuyutaka" are soybean cultivars that are suitable or possible to cultivate in the Chugoku region of Japan, including Shimane prefecture, where the cultivation test was conducted in the present study [51, 52]. Additionally, pod and seed numbers are the most important soybean yield components, which are primarily determined during the period from before and after flowering to pod set, including the beginning of the seed filling period [53]. However, increasing temperature during the growing season can negatively affect soybean leaf photosynthesis, growth, flowering, pod and seed number, and yield [54, 55]. In the present study, the monthly mean maximum temperature in August during the flowering period of soybean was 32.2°C in 2016 and 31.5°C in 2017 (**Table 2**). Specifically, in early August of 2017, when B × F – E and "Enrei" were flowering, the temperature remained above

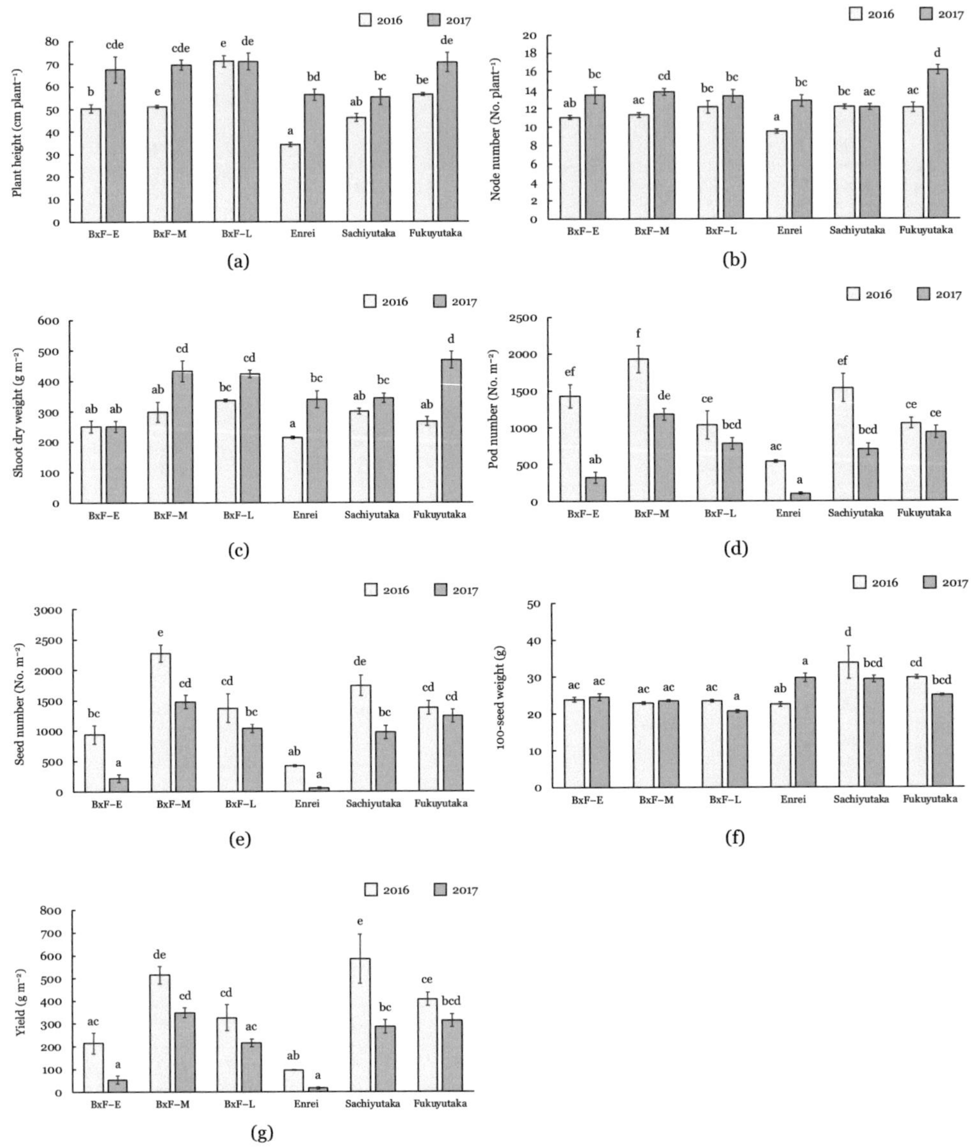

Figure 2.
Yield components for each soybean cultivar in 2016 and 2017 growing seasons. Values are presented as the mean ± SE of three replicates. Different letters indicate significant differences (Tukey's HSD test) at $p < 0.05$. (a) Plant height. (b) Node number. (c) Shoot dry weight. (d) Pod number. (e) Seed number. (f) 100-seed weight. (g) Yield.

35°C for 3 consecutive days. Furthermore, in late August of 2017, the temperature remained above 32°C for 4 consecutive days. Additionally, in October 2017, nearly 3.5 times the amount of precipitation in 2016 was recorded (**Table 2**). Soybean pod and seed number and yield in 2017 were significantly lower than the values in 2016 due to the effects of these meteorological factors (**Table 4**). Moreover, the 100-seed weight of B × F – M was lower than that of "Sachiyutaka" and "Fukuyutaka" (**Table 4**). Therefore, backcrossing with these cultivars is expected to produce soybean cultivars with larger seeds and higher yield.

4. Occupancy of indigenous soybean-nodulating bradyrhizobia

4.1 Sampling of soybean root nodules and isolation of nodulating bradyrhizobia

The nodules used in the present experiment were collected from soybean roots at the flowering stage in 2017, as described in Section 3.2. The nodules were surface-sterilized with 70% ethanol for 3 min and diluted sodium hypochlorite solution (0.25% available chlorine) for 30 min, followed by washing with sterile distilled water. After washing, 24 nodules were randomly collected and transferred to 1.5 mL microcentrifuge tubes. Each nodule was homogenized in sterile distilled water and streaked onto a yeast extract–mannitol agar (YMA) plate [56]; to isolate a single colony per nodule, the plates were incubated for 5–7 days in the dark at 28°C. A total of 144 isolates were obtained from six soybean plants and used for PCR-restriction fragment length polymorphism (RFLP) analysis of the 16S–23S rRNA gene internal transcribed spacer (ITS) region.

4.2 PCR-RFLP analysis of the 16S: 23S rRNA gene ITS region

For DNA extraction, each isolate was cultured in 1.5 mL HM medium supplemented with 0.1% L-arabinose for 5–7 days in the dark at 28°C. Total DNA for use as the PCR template was extracted from the isolates in BL extraction buffer, as described previously [42] based on the method reported by Hiraishi *et al.* [57]. As reference strains, *B. japonicum* USDA 6^T; *B. diazoefficiens* USDA 110^T; and *B. elkanii* USDA 46, 76^T, and 94 were used [16]. Total DNA of the reference strains for use as the PCR templates was also extracted using the same method [42, 57].

The 16S–23S rRNA gene ITS region was PCR-amplified using *TaKaRa Ex Taq*® Hot Start Version (TaKaRa Bio, Shiga, Japan) and the ITS primer set (BraITS-F: 5′-GACTGGGGTGAAGTCGTAAC-3′ and BraITS-R1: 5′-ACGTCCTTCATCGCCTC-3′) [58]. The PCR cycle comprised a pre-run at 94°C for 5 min, denaturation at 94°C for 30 s, annealing at 55°C for 30 s, and extension at 72°C for 1 min. This temperature control sequence was repeated for 30 cycles, followed by a final run at 72°C for 10 min. RFLP analysis of the 16S–23S rRNA gene ITS region was performed using the restriction enzyme *Msp*I (TaKaRa Bio, Shiga, Japan). The PCR product was digested with the restriction enzyme at 37°C for 16 h, and the restricted fragments were separated using 3% agarose gel electrophoresis.

4.3 Occupancy of infected indigenous bradyrhizobia carrying *hupSL* genes

Eighty-seven indigenous soybean-nodulating bradyrhizobial isolates belonging to the cluster *B. diazoefficiens* USDA 110^T were investigated for the presence of *hupSL* genes. PCR templates of the test isolates, obtained as described in Section 4.2, were used. The PCR amplification for *hupSL* was performed using KAPA *Taq*® Extra Hot Start ReadyMix with dye (Kapa Biosystems, USA) and the *hupSL* primer set (hupS-F261: 5′-TCGAACAGGCGTTGTAAGTG-3′, hupS-R830: 5′-TCGACTACGACGACACCATC-3′, hupL-F962:5′-TCGGGCAGATAGACCATTTC-3′ and hupL-R1632: 5′-GGGATCGAAGTGATCCTGAA-3′). The PCR cycle comprised a pre-run at 95°C for 3 min, denaturation at 95°C for 15 s, annealing at 55°C for 15 s, and extension at 72°C for 1 min. This temperature control sequence was repeated for 30 cycles, followed by a final run at 72°C for 1 min. The PCR products were electrophoresed on a 2% agarose gel to confirm amplification.

4.4 Occupancy of soybean-nodulating bradyrhizobia carrying *hupSL* genes

The occupancy rate of indigenous bradyrhizobia infecting each soybean cultivar is presented in **Table 5**. Based on the fragment patterns obtained from PCR–RFLP analysis, the indigenous isolates with the similar patterns as the five reference strains, namely *B. japonicum* USDA 6^T; *B. diazoefficiens* USDA 110^T; and *B. elkanii* USDA 46, 76^T, and 94 were defined as Bj6, Bd110, Be46, Be76, and Be94, respectively. Since cluster Bd110 includes isolates carrying the *hupSL* genes, such as *B. diazoefficiens* USDA 110^T and 122 [16], the occupancy rate of these isolates was also determined. PCR analysis targeting the *hupSL* genes revealed amplicons exhibiting zero to two bands. The isolates exhibiting two amplification products corresponding to *hupS* and *hupL* were defined as $hupS^+L^+$, those exhibiting a single amplification product corresponding to *hupL* were defined as $hupS^-L^+$, and those exhibiting no amplification products were defined as $hupS^-L^-$. In B × F – E, Bd110 isolates exhibiting $hupS^-L^+$ were the most dominant (62.5%), followed by Bd110 isolates exhibiting $hupS^+L^+$ (12.5%) and Bj6 isolates (12.5%). In B × F – M, Bj6 isolates were the most dominant (70.8%), followed by Bd110 isolates exhibiting $hupS^-L^+$ (16.7%). In B × F – L, Bd110 isolates exhibiting $hupS^-L^+$ (45.8%) were the most dominant, followed by Bd110 isolates exhibiting $hupS^+L^+$ (25.0%). In "Enrei," Bj6 isolates were the most dominant (83.4%), followed by Bd110 isolates exhibiting $hupS^-L^+$ (8.3%) or $hupS^+L^+$ (8.3%). In "Sachiyutaka," Bd110 isolates exhibiting $hupS^+L^+$ (39.1%) were the most dominant, followed by Bd110 isolates exhibiting $hupS^-L^+$ (34.8%). In "Fukuyutaka," Bd110 isolates exhibiting $hupS^+L^+$ (37.5%) were the most dominant, followed by Bj6 (25.0%) isolates. "Sachiyutaka" and "Fukuyutaka" with the Rj_4 genotype tended to present a higher occupancy rate of Bd110 isolates exhibiting $hupS^+L^+$. In contrast, soybean lines with the $Rj_2Rj_3Rj_4$ genotype tended to present a lower occupancy rate of Bd110 isolates exhibiting $hupS^+L^+$. These results may be explained by the effect of the presence of Rj_2. Rj_2 restricts *B. diazoefficiens* USDA 122 [32]. Indigenous bradyrhizobial isolates, such as *B. diazoefficiens* USDA 110 ^{T}, which are not restricted by the Rj_2 gene, can infect *Rj* gene-accumulated soybean lines. To solve this problem, the occupancy rate of inocula carrying the *hupSL* genes may be improved by inoculating *Rj* gene-accumulated soybean with *B. diazoefficiens* USDA 110^T during cultivation. To test this hypothesis, we are currently investigating the effect of *B. diazoefficiens* USDA

Cultivar	Bj6	Bd110			Be46	Be76	Be94
		$hupS^-L^-$	$hupS^-L^+$	$hupS^+L^+$			
BxF–E	12.5	8.3	62.5	12.5	0.0	0.0	4.2
BxF–M	70.8	0.0	16.7	12.5	0.0	0.0	0.0
BxF–L	0.0	8.3	45.8	25.0	4.2	16.7	0.0
Enrei	83.4	0.0	8.3	8.3	0.0	0.0	0.0
Sachiyutaka	13.0	8.7	34.8	39.1	0.0	4.4	0.0
Fukuyutaka	25.0	20.8	16.7	37.5	0.0	0.0	0.0

Bj6, Bd110, Be46, Be76, and Be94 showed RFLP patterns similar to those of B. japonicum USDA 6^T, B. diazoefficiens USDA 110^T, B. elkanii USDA 46, B. elkanii USDA 76^T, and B. elkanii USDA 94, respectively. $hupS^-L^-$, $hupS^-L^+$, and $hupS^+L^+$ indicate isolates carrying or not carrying the hupS and/or hupL genes.

Table 5.
Occupancy rate (%) of indigenous soybean-nodulating bradyrhizobia in each soybean cultivar in 2017.

110^{T} inoculation on the growth and yield of various soybean genotypes, including *Rj* gene-accumulated ones.

4.5 Correlation between the occupancy rate of indigenous bradyrhizobial strains and yield components of soybean

Correlation analysis was used to evaluate the association between the occupancy rate of indigenous bradyrhizobial strains and yield components. Correlation coefficients were computed based on data obtained from the measurement of yield components and occupancy rate of indigenous soybean-nodulating bradyrhizobia. The R package "psych" was used to compute and plot the correlations. Additionally, the significance of the correlations was tested using the "cor.test" function in R.

The results of correlation analysis between the occupancy rate of indigenous bradyrhizobial strains and yield components of soybean are presented in **Figure 3**. The correlation coefficients of the occupancy rate of Bj6 isolates with plant height, node number, shoot dry weight, pod number, seed number, 100-seed weight, and yield were – 0.30, −0.04, 0.09, −0.13, −0.15, 0.41, and – 0.16, respectively. The correlation

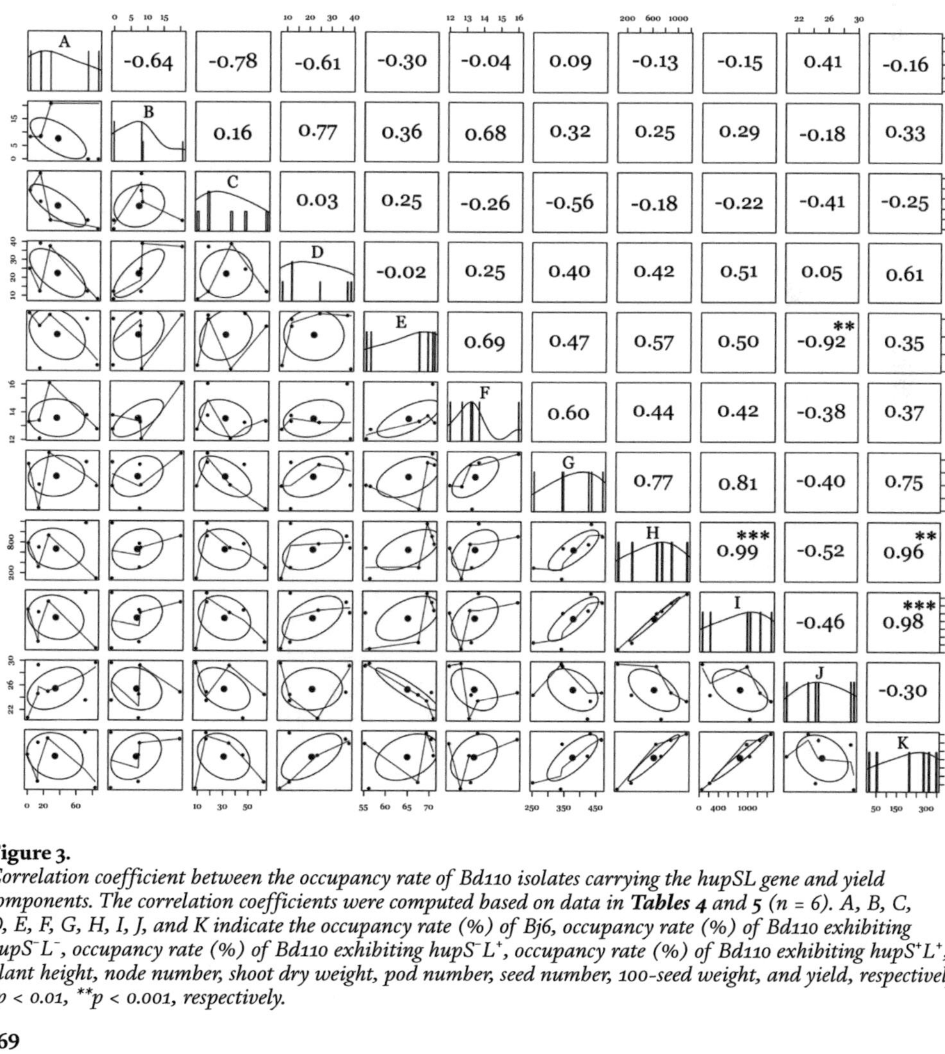

Figure 3.
Correlation coefficient between the occupancy rate of Bd110 isolates carrying the hupSL gene and yield components. The correlation coefficients were computed based on data in ***Tables 4*** *and* ***5*** *(n = 6). A, B, C, D, E, F, G, H, I, J, and K indicate the occupancy rate (%) of Bj6, occupancy rate (%) of Bd110 exhibiting hupS*$^{-}$*L*$^{-}$*, occupancy rate (%) of Bd110 exhibiting hupS*$^{-}$*L*$^{+}$*, occupancy rate (%) of Bd110 exhibiting hupS*$^{+}$*L*$^{+}$*, plant height, node number, shoot dry weight, pod number, seed number, 100-seed weight, and yield, respectively.* $^{*}p < 0.01$*,* $^{**}p < 0.001$*, respectively.*

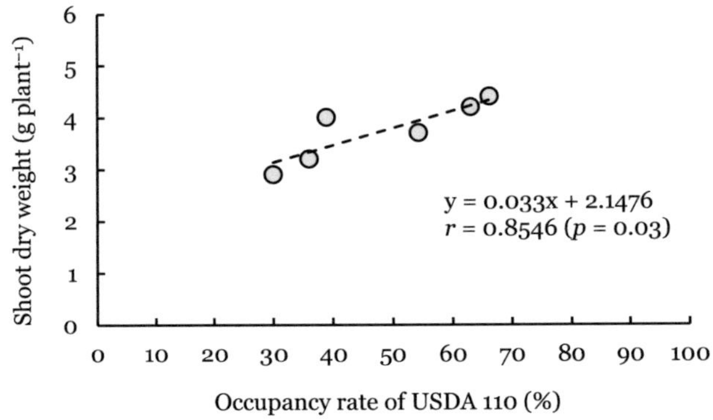

Figure 4.
Correlation between the occupancy rate of Bradyrhizobium diazoefficiens USDA 110^T and shoot dry weight of soybean. The values were obtained based on the growth investigation of soybean inoculated with B. diazoefficiens USDA 110^T, B. japonicum USDA 6^T, B. japonicum USDA 123, and Bradyrhizobium elkanii USDA 31 at the same bacterial density (10^6 cells mL^{-1}) and then cultivated for 5 weeks. Values are presented as the mean of three replicates.

coefficients of the occupancy rate of Bd110 isolates exhibiting *hupS⁻L⁻* with plant height, node number, shoot dry weight, pod number, seed number, 100-seed weight, and yield were 0.36, 0.68, 0.32, 0.25, 0.29, −0.18, and 0.33, respectively. The correlation coefficients of the occupancy rate of Bd110 isolates exhibiting $hupS^-L^+$ with plant height, node number, shoot dry weight, pod number, seed number, 100-seed weight, and yield were 0.25, −0.26, −0.56, −0.18, −0.22, −0.41, and − 0.25, respectively. The correlation coefficients of the occupancy rate of Bd110 isolates exhibiting $hupS^+L^+$ with plant height, node number, shoot dry weight, pod number, seed number, 100-seed weight, and yield were – 0.02, 0.25, 0.40, 0.42, 0.51, 0.05, and 0.61, respectively. Among correlations between the occupancy rate of each indigenous bradyrhizobial strain and soybean yield, Bd110 isolates exhibiting $hupS^+L^+$ presented a strong positive correlation (r = 0.61), albeit without significant differences. Additionally, in another experiment, which revealed different results from the present findings, showed that the occupancy rate of *B. diazoefficiens* USDA 110^T was correlated with the shoot dry weight of soybean (**Figure 4**). This result was obtained in the correlation analysis between the growth and the occupancy rate of *B. diazoefficiens* USDA 110^T, *B. japonicum* USDA 6^T and 123, and *B. elkanii* USDA 31 in soybean inoculated at the same bacterial density (10^6 cells mL^{-1}) and cultivated for 5 weeks. Although this experiment was set in a greenhouse using cultivation pots, a significant positive correlation between the occupancy rate of USDA 110 and shoot dry weight of soybean plants was noted (r = 0.86, p = 0.03). A positive correlation between soybean growth and yield has been reported in previous study [59–62]. In the present study, yield was positively correlated with plant height (r = 0.35) and shoot dry weight (r = 0.75), albeit without significant differences (**Figure 3**). Therefore, enhancing the infection rate of bradyrhizobial strains, such as *B. diazoefficiens* USDA 110^T carrying the *hupSL* genes, may promote the growth of soybean and consequently increase yield.

5. Conclusion

In the present chapter, we described the breeding and selection processes, shoot growth, yield components, and infection tendency of useful bradyrhizobial strains carrying the *hupSL* genes of *Rj* gene-accumulated soybean lines exhibiting the

Rj$_2$*Rj*$_3$*Rj*$_4$ genotype, obtained by crossing the Japanese soybean cultivars "Bonminori" (*Rj*$_2$*Rj*$_3$) and "Fukuyutaka" (*Rj*$_4$).

First, we selected eight lines exhibiting the characteristics of the *Rj*$_2$*Rj*$_2$*Rj*$_4$*Rj*$_4$ genotype from 153 lines of F3 seeds following inoculation with *B. japonicum* Is–1 and *B. japonicum* Is–34 (**Table 1**). These eight lines were grown for several years, and three groups (B × F – E, B × F – M, and B × F – L) differing in terms of the flowering and ripening periods by approximately a week each were detected, which were cultivated further. Second, to investigate yield components, three *Rj* gene-accumulated soybean lines (B × F – E, B × F – M, and B × F – L) of F10 or F11 plant and three soybean cultivars ("Enrei," "Sachiyutaka," and "Fukuyutaka") were cultivated in the 2016 and 2017 growing seasons. The yield of B × F – M was equivalent to that of "Sachiyutaka," and this genotype likely possesses a greater yield potential than the parent soybean cultivar "Fukuyutaka," among the *Rj* gene-accumulated soybean lines. However, the 100-seed weight of B × F – M was lower than that of "Sachiyutaka" and "Fukuyutaka." Therefore, backcrossing with these cultivars is expected to produce soybean cultivars with larger seeds and higher yield potential. Third, to assess the occupancy rate of infected indigenous soybean-nodulating bradyrhizobia carrying the *hupSL* genes, we collected nodules from soybean roots and performed PCR-RFLP analysis of the 16S–23S rRNA gene ITS region. Furthermore, 87 indigenous soybean-nodulating bradyrhizobial strains belonging to the *B. diazoefficiens* USDA 110^T cluster were investigated for the presence of the *hupSL* genes using PCR. The occupancy rate of Bd110 isolates carrying the *hupSL* genes tended to be lower in the *Rj* gene-accumulated soybean lines than in "Sachiyutaka" and "Fukuyutaka." In addition, among the *Rj* gene-accumulated soybean lines, B × F – L presented the highest occupancy rate of Bd110 isolates carrying the *hupSL* genes. Based on these results, during the cultivation of *Rj* gene-accumulated soybean, the occupancy rate of inocula carrying the *hupSL* genes can be improved by inoculating *B. diazoefficiens* USDA 110^T, which is not restricted by the *Rj*$_2$ gene.

Finally, to evaluate the association between the occupancy rate of indigenous bradyrhizobial strains and yield components of soybean, correlation analysis was performed. Correlation coefficients of the occupancy rate of Bd110 isolates exhibiting *hupS*$^+$*L*$^+$ with plant height, node number, shoot dry weight, pod number, seed number, 100-seed weight, and yield were – 0.02, 0.25, 0.40, 0.42, 0.51, 0.05, and 0.61, respectively, and the occupancy rate of Bd110 isolates exhibiting *hupS*$^+$*L*$^+$ was strongly and positively correlated with yield components ($r = 0.61$), albeit without significant differences. Furthermore, soybean yield was positively correlated with plant height ($r = 0.35$) and shoot dry weight ($r = 0.75$), albeit without significant differences. Therefore, enhancing the infection rate of bradyrhizobial strains, such as *B. diazoefficiens* USDA 110^T carrying the *hupSL* genes, may promote the growth of soybean and consequently increase its yield. In the future, we intend to conduct further inoculation tests with useful strains, such as *B. diazoefficiens* USDA 110^T carrying the *hupSL* genes, to evaluate in greater detail the availability of *Rj* gene-accumulated soybean lines with the *Rj*$_2$*Rj*$_3$*Rj*$_4$ genotype for improving productivity.

Acknowledgements

The authors thank the members of the laboratories of Shimane University and University of Miyazaki involved in the present study. Additionally, the authors thank the technical staff of the Agricultural Science Section, Education and Research Center for Biological Resources, Faculty of Life and Environmental Science, Shimane

University for their support in managing soybean cultivation. The authors also thank the Faculty of Life and Environmental Sciences at Shimane University for financial support to publish this chapter.

Conflict of interest

The authors declare no conflict of interest.

Author details

Sokichi Shiro[1*] and Yuichi Saeki[2]

1 Institute of Agricultural and Life Sciences, Academic assembly, Shimane University, Shimane, Japan

2 Faculty of Agriculture, Department of Biochemistry and Applied Biosciences, University of Miyazaki, Miyazaki, Japan

*Address all correspondence to: skc-shiro@life.shimane-u.ac.jp

References

[1] Ministry of Agriculture, Forestry, and Fisheries of Japan (MAFF). The circumstances on soybean (in Japanese) [Internet]. 2021. Available from: https://www.maff.go.jp/j/seisan/ryutu/daizu/attach/pdf/index-144.pdf [Accessed: November 29, 2021]

[2] Jordan DC. Transfer of *Rhizobium japonicum* Buchanan 1980 to *Bradyrhizobium* gen. Nov., a genus of slow-growing, root nodule bacteria from leguminous plants. International Journal of Systematic Microbiology. 1982;**32**:136-139. DOI: 10.1099/00207713-32-1-136

[3] Scholla MH, Elkan GH. *Rhizobium fredii* sp. nov. a fast-growing species that effectively nodulates soybeans. International Journal of Systematic Bacteriology. 1984;**34**:484-486. DOI: 10.1099/0207713-34-4-484

[4] Chen WX, Yan GH, Li JL. Numerical taxonomic study of fast-growing soybean rhizobia and a proposal that *Rhizobium fredii* be assigned to *Sinorhizobium* gen. nov. International Journal of Systematic Microbiology. 1988;**38**:392-397. DOI: 10.1099/00207713-38-4-392

[5] Kuykendall LD, Saxena B, Cevine TE, Udell SE. Genetic diversity in *Bradyrhizobium* Jordan 1982 and a proposal for *Bradyrhizobium elkanii* sp. nov.. Canadian journal. Microbiology. 1992;**38**:501-505. DOI: 10.1139/m92-082

[6] Young JM. The genus name *Ensifer* Casida 1982 takes priority over *Sinorhizobium* Chen *et al*. 1988, and *Sinorhizobium morelense* Wang *et al*. 2002 is a later synonym of *Ensifer adhaerens* Casida 1982. Is the combination '*Sinorhizobium adhaerens*' (Casida 1982) Willems *et al*. 2003 legitimate? Request for an opinion. International Journal of Systematic and Evolutionary Microbiology. 2003;**53**:2107-2110. DOI: 10.1099/ijs.0.02665-0

[7] Delamuta JRM, Riberio RA, Ormeño-Orrillo E, Melo IS, Maritínez-Romero E, Hungria M. Polyphasic evidence supporting the reclassification of *Bradyrhizobium japonicum* group Ia strains as *Bradyrhizobium diazoefficiens* sp. nov. International Journal of Systematic and Evolutionary Microbiology. 2013;**63**:3342-3351. DOI: 10.1099/ijs.0.049130-0

[8] Chen W, Wang E, Wang S, Li Y, Chen X, Li Y. Characteristics of *Rhizobium tianshanense* sp. nov., a moderately and slowly growing root nodule bacterium isolated from an arid saline environment in Xinjiang, People's republic of China. International Journal of Systematic Bacteriology. 1995;**45**:153-159. DOI: 10.1099/00207713-45-1-153

[9] Xu LM, Ge C, Cui Z, Li J, Fan H. *Bradyhizobium liaoningense* sp. nov., isolated from the root nodules of soybeans. International Journal of Systematic Bacteriology. 1995;**45**:706-711. DOI: 10.1099/00207713-45-4-706

[10] Tan ZY, Xu XD, Wang ET, Gao JL, Martinez-Romero E, Chen WX. Phylogenetic and genetic relationships of *Mesorhizobium tianshanense* and related rhizobia. International Journal of Systematic Bacteriology. 1997;**47**:874-879. DOI: 10.1099/00207713-47-3-874

[11] Peng GX, Tan ZY, Wang ET, Reinhold-Hurek B, Chen WF, Chen WX. Identification of isolates from soybean nodules in Xinjiang region as *Sinorhizobium xinjiangense* and genetic differentiation of *S. xinjiangense* from *Sinorhizobium fredii*. International

Journal of Systematic and Evolutionary Microbiology. 2002;**52**:457-462. DOI: 10.1099/00207713-52-2-457

[12] Yao ZY, Kan FL, Wang ET, Wei GH, Chen WX. Characterization of rhizobia that nodulate legume species of the genus *Lespedeza* and description of *Bradyrhizobium yuanmingense* sp. nov. International Journal of Systematic and Evolutionary Microbiology. 2002;**52**:2219-2230. DOI: 10.1099/ijs.0.01408-0

[13] Vinuesa P, Rojas-Jiménez K, Contreras-Moreira B, Mahna SK, Prasad BN, Moe H, et al. Multilocus sequence analysis for assessment of the biogeography and evolutionary genetics of four *Bradyrhizobium* species that nodulate soybeans on the Asiatic continent. Applied and Environmental Microbiology. 2008;**74**:6987-6996. DOI: 10.1128/AEM.00875-08

[14] Adhikari D, Kaneto M, Itoh K, Suyama K, Pokharel BB, Gaihre YK. Genetic diversity of soybean-nodulating rhizobia in Nepal in relation to climate and soil properties. Plant and Soil. 2012;**357**:131-145. DOI: 10.1007/s11104-012-1134-6

[15] Evans HJ, Harker AR, Papen H, Russell SA, Hanus FJ, Zuber M. Physiology, biochemistry, and genetics of the uptake hydrogenase in rhizobia. Annual Review of Microbiology. 1987;**41**:335-361. DOI: 10.1146/annurev.mi.41.100187.002003

[16] Saeki Y, Aimi N, Hashimoto M, Tsukamoto S, Kaneko A, Yoshida N, et al. Grouping of *Bradyrhizobium* USDA strains by sequence analysis of 16S rDNA and 16S-23S rDNA internal transcribed spacer region. Soil Science and Plant Nutrition. 2004;**50**:517-525. DOI: 10.1080/00380768.2004.10408508

[17] Masuda S, Saito M, Sugawara C, Itakura M, Eda S, Minamisawa K. Identification of the hydrogen uptake gene cluster for chemolithoautotrophic growth and symbiosis hydrogen uptake in *Bradyrhizobium diazoefficiens*. Mirobes and Environments. 2016;**36**:76-78. DOI: 10.1264/jsme2.ME15182

[18] Drevon J, Kalia VC, Heckmann M, Salsac I. Influence of the *Bradyrhizobium* hydrogenase on the growth of *Glycine* and *Vigna* species. Applied and Environmental Microbiology. 1987;**53**:610-612. DOI: 10.1128/aem.53.3.610-612.1987

[19] Williams LF, Lynch DL. Inheritance of a non-nodulating character in the soybean. Agronomy Journal. 1954;**46**:28-29. DOI: 10.2134/agronj1954.00021962004600010008x

[20] Caldwell BE. Inheritance of a strain-specific ineffective nodulation in soybean. Crop Science. 1966;**6**:427-428. DOI: 10.2135/cropsci1966.0011183X000600050010x

[21] Vest G. *Rj*$_3$ - A gene conditioning ineffective nodulation in soybean. Crop Science. 1970;**10**:34-35. DOI: 10.2135/cropsci1970.0011183X001000010013x

[22] Vest G, Caldwell BE. *Rj*$_4$ - A gene conditioning ineffective nodulation in soybean. Crop Science. 1972;**12**:692-693. DOI: 10.2135/CROPSCI1970.0011183X001000010013X

[23] Weiser GV, Skipper HD, Wollum AG. Exclusion of inefficient *Bradyrhizobium japonicum* serogroups by soybean genotypes. Plant and Soil. 1990;**121**:99-105. DOI: 10.1007/BF00013102

[24] Trese AT. A single dominant gene in McCall soybean prevents effective nodulation with *Rhizobium fredii* USDA257. Euphytica. 1995;**81**:279-282. DOI: 10.1007/BF00025618

[25] Madsen EB, Madsen LH, Radutoiu S, Olbryt M, Rakwalska M, Szczyglowski K,

et al. A receptor kinase gene of the LysM type is involved in legume perception of rhizobial signals. Nature. 2003;**425**:637-640. DOI: 10.1038/nature02045

[26] Arrighi JF, Barre A, Amor BB, Bersoult A, Soriano LC, Mirabella R, et al. The *Medicago truncatula* lysine motif-receptor-like kinase gene family includes *NFP* and new nodule-expressed genes. Plant Physiology. 2006;**142**:265-279. DOI: 10.1104/pp.106.084657

[27] Indrasumunar A, Gresshoff PM. Duplicated nod-factor receptor 5 (NFR5) genes are mutated in soybean (*Glycine max* L. Merr.). Plant Signaling and Behavior. 2010;**5**:535-536. DOI: 10.4161/psb.11028

[28] Nishimura R, Hayashi M, Wu GJ, Kouchi H, Imaizumi-Anraku H, Murakami Y, et al. HAR1 mediates systemic regulation of symbiotic organ development. Nature. 2002;**420**:426-429. DOI: 10.1038/nature01231

[29] Krusell L, Madsen LH, Sato S, Aubert G, Genua A, Szczyglowski K, et al. Shoot control of root development and nodulation is mediated by a receptor kinase. Nature. 2002;**420**:422-426. DOI: 10.1038/nature01207

[30] Searle IR, Men AE, Laniya TS, Buzas DM, Iturbe-Ormaetxe I, Carroll BJ, et al. Long-distance signaling in nodulation directed by a CLAVATA1-like receptor kinase. Science. 2003;**299**:109-112. DOI: 10.1126/science.1077937

[31] Schnabel E, Journet EP, de Carvalho-Niebel F, Duc G, Frugoli J. The *Medicago truncatula SUNN* gene encodes a *CLV1*-like leucine-rich repeat receptor kinase that regulates nodule number and root length. Plant Molecular Biology. 2005;**58**:809-822. DOI: 10.1007/s11103-005-8102-y

[32] Ishizuka J, Suemasu Y, Mizogami K. Preference of *Rj*-soybean cultivars for *Bradyrhizobium japonicum* for nodulation. Soil Science and Plant Nutrition. 1991;**37**:15-21. DOI: 10.1080/00380768.1991.10415005

[33] Hayashi M, Saeki Y, Haga M, Harada K, Kouchi H, Umehara Y. *Rj* (*rj*) genes involved in nitrogen-fixing root nodule formation in soybean. Breeding Science. 2012;**61**:544-553. DOI: 10.1270/jsbbs.61.544

[34] Yang S, Tang F, Gao M, Krishnan HB, Zhu H. R gene-controlled host specificity in the legume-rhizobia symbiosis. Proceeding of the National Academy of Sciences of the United States of America. 2010;**107**:18735-18740. DOI: 10.1073/pnas.1011957107

[35] Sugawara M, Umehara Y, Kaga A, Hayashi M, Ishimoto M, Sato S, et al. Symbiotic incompatibility between soybean and *Bradyrhizobium* arises from one amino acid determinant in soybean Rj2 protein. PLoS One. 2019;**14**:e0222469. DOI: 10.1371/journal.pone.0222469

[36] van Loon LC, Rep M, Pieterse CMJ. Significance of inducible defense-related proteins in infected plants. Annual Review of Phytopathology. 2006;**44**:135-162. DOI: 10.1146/annurev.phyto.44.070505.143425

[37] Hayashi M, Shiro S, Kanamori H, Mori-Hosokawa S, Sasaki-Yamagata H, Sayama T, et al. A thaumatin-like protein, Rj4, controls nodule symbiotic specificity in soybean. Plant and Cell Physiology. 2004;**55**:1679-1689. DOI: 10.1093/pcp/pcu099

[38] Okazaki S, Zehner S, Hempel J, Lang K, Göttfert M. Genetic organization and functional analysis of the type III secretion system of *Bradyrhizobium elkanii*. FEMS Microbiology Letters. 2009;**295**:88-95. DOI: 10.1111/j.1574-6968.2009.01593.x

[39] Tsurumaru H, Hashimoto S, Okizaki K, Kanesaki Y, Yoshikawa H, Yamakawa T. A putative type III secretion system effector encoded by the *MA20_12780* gene in *Bradyrhizobium japonicum* Is–34 causes incompatibility with *Rj*4 genotype soybeans. Applied and Environmental Microbiology. 2015;**81**:5812-5819. DOI: 10.1128/AEM.00823-15

[40] Ishizuka J, Kim SD, Hussain AKMA, Yamakawa T. Soybean preference for *Bradyrhizobium japonicum* for modulation. Isolation of Rj_2Rj_4-lines from the cross of soybean cvs. IAC-2 (Rj_2) and Hill (Rj_4). Soil Science and Plant Nutrition. 1993;**39**:79-86. DOI: 10.1080/00380768.1993.10416977

[41] Yamakawa T, Eriguchi M, Hussain AKMA, Ishizuka J. Soybean preference for *Bradyrhizobium japonicum* for nodulation. Nodulation by $Rj_2Rj_3Rj_4$-genotypes isolated from the progenies of cross of soybean cvs. IAC-2 (Rj_2Rj_3) and Hill (Rj_4). Soil Science and Plant Nutrition. 1999;**45**:461-469. DOI: 10.1080/00380768.1999.10409360

[42] Minami M, Yamakawa T, Yamamoto A, Akao S, Saeki Y. Estimation of nodulation tendency among *Rj*-genotype soybeans using the bradyrhizobial community isolated from an andosol. Soil Science and Plant Nutrition. 2009;**55**:65-72. DOI: 10.1111/j.1747-0765.2008.00333.x

[43] Shiro S, Yamamoto A, Umehara Y, Hayashi M, Yoshida N, Nishiwaki A, et al. Effect of *Rj* genotype and cultivation temperature on the community structure of soybean-nodulating bradyrhizobia. Applied and Environmental Microbiology. 2012;**78**:1243-1250. DOI: 10.1128/AEM.06239-11

[44] Ministry of Agriculture, Forestry, and Fisheries of Japan (MAFF). The development and diffusion of new soybean cultivars (in Japanese) [Internet]. 2015. Available from: https://www.maff.go.jp/j/seisan/ryutu/daizu/d_ziten/pdf/fukyu.pdf [Accessed: December 1, 2021]

[45] Togashi A, Oikawa S. Leaf productivity and persistence have been improved during soybean (*Glycine max*) domestication and evolution. Journal of Plant Research. 2021;**134**:223-233. DOI: 10.1007/s10265-021-01263-x

[46] Cole MA, Elkan GH. Transmissible resistance to penicillin G, neomycin and chloramphenicol in *Rhizobium japonicum*. Antimicrobial Agents and Chemotherapy. 1973;**4**:248-253. DOI: 10.1128/aac.4.3.248

[47] Sameshima R, Iwasa T, Sadowsky J, Hamada T, Kasai H, Shutsrirung A, et al. Phylogeny and distribution of extra-slow-growing *Bradyrhizobium japonicum* harboring high copy numbers of RSα, RSβ and IS*1631*. FEMS Microbiology Ecology. 2003;**44**:191-202. DOI: 10.1016/S0168-6496(03)00009-6

[48] Saeki S, Akagi I, Takaki H, Nagatomo Y. Diversity of indigenous *Bradyrhizobium* strains isolated from three different *Rj*-soybean cultivars in terms of randomly amplified polymorphic DNA and intrinsic antibiotic resistance. Soil Science and Plant Nutrition. 2000;**46**:917-926. DOI: 10.1080/00380768.2000.10409157

[49] R Core Team R. Available from. In: A Language and Environment for Statistical Computing. Vienna, Austria: R Foundation for Statistical Computing; 2020 https://www.R-project.org/ [Accessed: December 23, 2021]

[50] Iseki R. The page of Iseki Ryuta. ANOVAKUN/Release Notes (in Japanese) [Internet]. 2020. Available from: http://

riseki.php.xdomain.jp/index.php?ANOVA%E5%90%9B%2F%E3%83%AA%E3%83%AA%E3%83%BC%E3%82%B9%E3%83%8E%E3%83%BC%E3%83%88 [Accessed: December 23, 2021]

[51] Ministry of Agriculture, Forestry, and Fisheries of Japan (MAFF). The characteristics of Japanese soybean cultivars (in Japanese) [Internet]. 2019. Available from: https://www.maff.go.jp/j/seisan/ryutu/daizu/d_ziten/attach/pdf/index-21.pdf [Accessed: January 24, 2022]

[52] Konno T, Takahashi T, Nakano K, Nira R, Ohashi Y, Kudo T, et al. Survey of wet and dry stress in soybean in Japanese farmers' fields using dry and wet indices based on FAO56 model (in Japanese). Japanese Journal of Crop Science. 2020;**89**:337-345. DOI: 10.1626/jcs.89.337

[53] Kantolic A, Slafer GA. Reproductive development and yield components in indeterminate soybean as affected by post-flowering photoperiod. Field Crops Research. 2005;**93**:212-222. DOI: 10.1016/j.fcr.2004.10.001

[54] Kumagai E, Sameshima R. Genotypic differences in soybean yield responses to increasing temperature in a cool climate are related to maturity group. Agricultural and Forest Meteorology. 2014;**198-199**:265-272. DOI: 10.1016/j.agrformet.2014.08.016

[55] Shiro S, Mazuka S, Kadowaki M, Saeki Y. The effect of cultivation temperature on the soybean (*Glycine max*) growth and community structure of bradyrhizobia (in Japanese). Japanese Journal of Soil Science and Plant Nutrition. 2021;**92**:255-262. DOI: 10.20710/dojo.92.3_255

[56] Vincent JM. A Manual for the Practical Study of the Root-Nodule Bacteria. International Biological Program Handbook No. 15. Oxford: Blackwell Scientific Publishers; 1970. p. 164. DOI: 10.1002/jobm.19720120524

[57] Hiraishi A, Kamagata Y, Nakamura K. Polymerase chain reaction amplification and restriction fragment length polymorphism analysis of 16S rRNA genes from methanogens. Journal of Fermentation and Bioengineering. 1995;**79**:523-529. DOI: 10.1016/0922-338X(95)94742-A

[58] Saeki Y, Aimi N, Tsukamoto S, Yamakawa T, Nagatomo Y, Akao S. Diversity and geographical distribution of indigenous soybean-nodulating bradyrhizobia in Japan. Soil Science and Plant Nutrition. 2006;52:418-426. DOI: 10.1111/j.1747-0765.2006.00050.x

[59] Shirakawa T, Ueno N, Shimada S, Horie T. Correlation between yielding ability and dry matter productivity during initial seed filling stage in various soybean genotypes. Plant Production Science. 2004;**7**:138-142. DOI: 10.1626/pps.7.138

[60] Kakiuchi J, Kobata T. The relationship between dry matter increase of seed and shoot during the seed-filling period in three kinds of soybeans with different growth habits subjected to shading and thinning. Plant Production Science. 2006;**9**:20-26. DOI: 10.1626/pps.9.20

[61] Kasu-Bandi BT, Kidinda LK, Kasendue GN, Longanza LB, Emery KL, Lubobo AK. Correlations between growth and yield parameters of soybean (*Glycine max* (L.) Merr.) under the influence of *Bradyrhizobium japonicum* in Kipushi (the Democratic Republic of Congo). American Journal of Agricultural and Biological Sciences. 2019;**14**:86-94. DOI: 10.3844/ajabssp.2019.86.94

[62] Borowska M, Prusinski J. Effect of soybean cultivars sowing dates on seed yield and its correlation with yield parameters. Plant, Soil and Environment. 2021;**67**:360-366. DOI: 10.17221/73/2021-PSE

Chapter 10

Role of CRISPR/Cas9 in Soybean (*Glycine max* L.) Quality Improvement

Summra Siddique

Abstract

Biotechnology has made significant advances in recent years, and the area of genetic engineering is progressing day by day, generating several advantages. Through the new ability to precisely change and modify the genomes of living organisms, genome editing technology has transformed genetic and biological research. Genome editing technology first appeared in the 1990s, and different approaches for targeted gene editing have subsequently been created. The fields of functional genomics and crop improvement have been transformed by advances in genome editing tools. CRISPR/Cas9 (clustered regularly interspaced short palindromic repeat)-Cas9 is a versatile genetic engineering tool based on the complementarity of the guide RNA (gRNA) to a specific sequence and the endonuclease activity of the Cas9 endonuclease. This RNA-guided genome editing tool has produced variations in plant biology fields. CRISPR technology is continually improving, allowing for more genetic manipulations such as creating knockouts, precise changes, and targeted gene activation and repression. Soybean is a leguminous crop, high in protein and oil contents that are used for poultry and livestock feed industry. In this chapter, we focus on the recent advances in CRISPR/Cas9-based gene editing technology and discuss the challenges and opportunities to harnessing this innovative technology for targeted improvement of traits in soybean and other crops.

Keywords: clustered regularly interspaced short palindromic repeats, genome editing (GE), guide RNA (gRNA), nonhomologous end joining (NHEJ), homology-directed repair (HDR), Cas9

1. Introduction

Nowadays, almost one billion people suffer from malnourishment due to increasing population, and our agricultural system is degrading by the loss of biodiversity and climate change [1]. To overcome the malnourishment, there is a need to improve the crop plants. To achieve this goal, conventional breeding approach is labor-intensive, and it takes several years to form the commercial varieties. Genome editing tools are advanced biotechnological techniques to modify an organism's genome efficiently and precisely. Although recently developed genome editing technologies, such as zinc finger nucleases (ZFN) and transcription activator-like effector nucleases (TALENs), have many advantages but also has some drawbacks too. CRISPR (clustered regularly interspaced short palindromic repeats)/Cas9 technology has site-specific genome editing with precision, efficiency, and ease of use.

The type II CRISPR/Cas system is a prokaryotic adaptive immune response system that guides the Cas9 nuclease to induce site-specific DNA cleavage using noncoding RNAs as a template. The CRISPR/Cas type II system is a flexible genome editing tool for crop improvement [2]. It is a simple, effective, and cost-effective approach that can target several genes. Many plants have advantage from the CRISPR/Cas9 system, including rice, maize, wheat, soybean, sorghum, and barley [3].

The CRISPR/Cas9 system has been utilized for genome editing in all mammalian cells, which may be used to make gene knockouts (through insertion/deletion). A single-guided RNA (sgRNA) is used to guide the Cas9 nuclease to a specific genomic region in order to disrupt genes (**Figure 2**). Double-strand breaks caused by Cas9 are repaired by the NHEJ DNA repair mechanism. Because the repair is prone to errors, insertions and deletions (INDELs) might occur, causing gene function to be disrupted. Cellular DNA repair processes, either the nonhomologous end joining DNA repair pathway (NHEJ) or the homology-directed repair (HDR) pathway, fix the DNA damage or DNA repair pathway (i.e., HDR).

Mechanism of CRISPR/Cas9-mediated gene disruption is as follows: (1) A single-guide RNA (sgRNA) binds to a recombinant form of Cas9 protein with DNA endonuclease activity, consisting of a crRNA sequence specific to the DNA target and a tracrRNA sequence that interacts with the Cas9 protein. (2) The resultant complex will cleave double-stranded DNA that is particular to the target. (3) Then, cleavage efficiency of sgRNAs will be tested.

Crop development techniques should enable to increase production, biotic and abiotic stress resistance, as well as quality and nutritional value. Over several decades, innovative agricultural technology has considerably enhanced crop productivity. Consumers are more concerned about crop quality since it provides many nutrients such as proteins, fiber, vitamins, minerals, and bioactive substances, all of which are directly linked to human health [4]. In addition, scientists and breeders have switched their focus from increasing production to enhancing quality. Traditional crossing breeding, chemical and radiation-mediated mutation breeding, molecular marker-assisted breeding (MAB), and genetic engineering breeding have all proven successful in improving various crop qualities [5–8]. Traditional mutagenesis-based breeding techniques are time-consuming and labor-intensive, especially for polyploid crop production [9]. Recently, crop breeding has advantage from genome editing (GE) technology, which alters plant genomes in a precise and predictable manner [10].

Genome editing can produce predictable and inheritable mutations in specified regions of the genome, with minimal off-target effects and no external gene sequence integration. Deletions, insertions, and single-nucleotide substitutions (SNPs) are all examples of GE-mediated alterations. There are four SDN (site-directed nuclease) families in a nucleotide excision process, i.e., homing endonucleases or mega-nucleases (HEs) [11], zinc-finger nucleases (ZFNs) [12], transcription activator-like effector nucleases (TALENs) [13], and CRISPR-associated protein (Cas) [14]. The majority of SDNs can precisely target double-stranded template DNA and produce a double-strand break (DSB). The DSBs are naturally repaired by a plant's endogenous repair system, which uses one of two major DNA damage repair mechanisms: nonhomologous end joining (NHEJ) or homologous-directed recombination (HDR). A FokI cleavage domain and a particular DNA-binding domain from TALE proteins make up TALENs. TALENs technology has a greater target binding specificity and decreased off-target effects when compared with ZFNs [15]. In rice [16], wheat [17], maize [16], and tomato [18], it was widely used as a gene-editing method. However, ZFN and TALENs have long construction procedure, which has limited their use in plants on a wide scale. CRISPR (clustered regularly interspaced short palindromic repeat) was first discovered in *Escherichia coli* in 1987 and described as an immunological response to viral and plasmid DNA invasion [19]. CRISPR/Cas systems have become the most popular GE technology in recent years. Because the specificity of editing is governed by nucleotide complementarity of the guided RNA to a specific sequence without protein engineering, the CRISPR/Cas systems are more efficient for genome editing than other SDNs [20].

Soybean is a leguminous crop, has a great economic value, and is high in protein and oil. With the growing demand for soybeans around the world, it is more important to understand gene function and speed up functional gene research and breeding to increase yield and improve quality. Traditional soybean breeding procedures are insufficient to meet the growing demand for soybean products and the problems posed by the agricultural environment. As a result, it is critical to implement quick, precise, and effective breeding procedures in order to develop improved varieties, particularly with improved yield, quality, and stress tolerance or resistance [21, 22]. Genome editing technology is a highly desired technology given the advantages listed above, and it is also an excellent tool for improving soybean genetics. The number of crops engineered by genome editing has increased day by day. Crop quality is one of the most important objectives among the different target traits for crop improvement. Here is a brief description of different quality traits improvement through CRISPR/Cas9-mediated tool.

2. CRISPR/Cas9 gene-editing system in plants

CRISPR/Cas systems are split into two classes and five kinds based on the Cas protein classification. The *Streptococcus pyogenes* type II CRISPR/SpCas9 system has

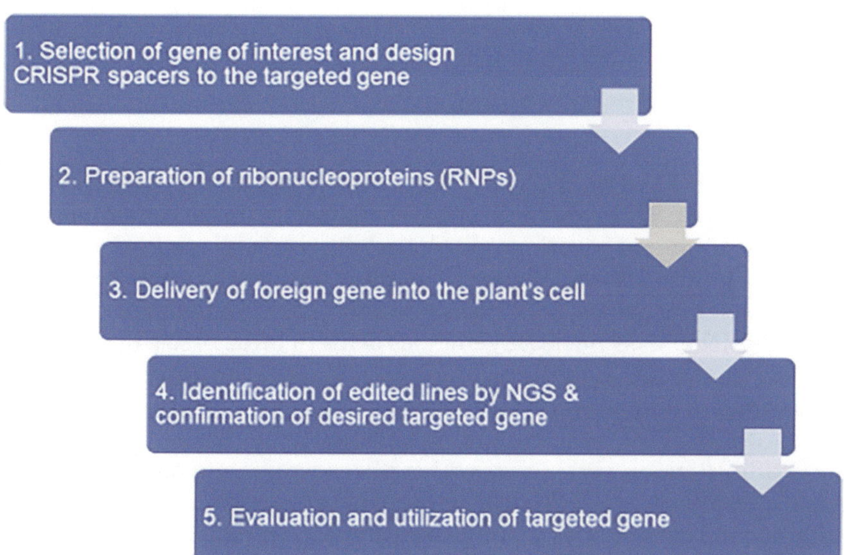

Figure 1.
The workflow of CRISPR/Cas9-based gene editing in plants.

Application	Crop	Gene target
Physical appearance and quality	Rice	GS3, Gn1a, GW2, GW5, TGW6, GL2/OsGRF4, OsGRF3, GS9, GW5, OsGS3, OsGW2, and OsGn1a, ANT1, SlMYB12, SlMYB12, Psy1, CrtR-b2
	Tomato	OVATE, Fas, Fw2.2, fas, lc, ENO, CLV3,
	Wheat	TaGW7, TaGW2
	Maize	Psy1
Texture, palatability, quality	Tomato	ALC, PL, PG2a, TBG4
	Banana	MaACO1, OsGBSSI, OsGBSSI
	Rice	OsAAP6, OsAAP10, OsBADH2, SH2, GBSS
	Maize	Wx1
Nutritional quality	Rice	OsGAD3, OsNramp5, OsFAD2-1, OsPLD*a*1
	Tomato	SlGAD2, SlGAD3, slyPDS, BnFAD2
	Wheat	*a*-gliadin genes

Table 1.
List of research on crop quality improvement by using CRISPR/Cas gene-editing technology.

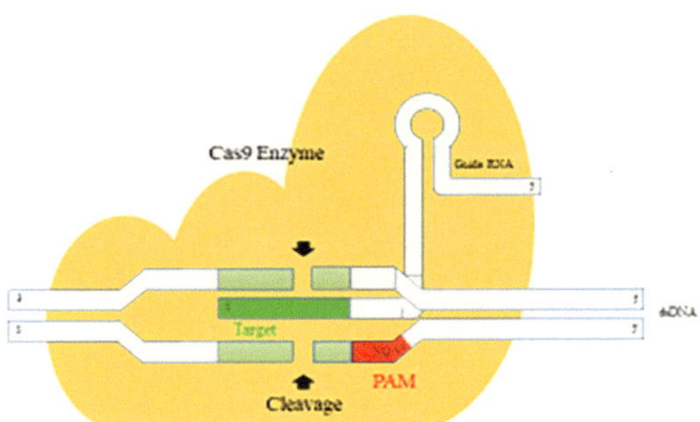

Figure 2.
Mechanism of CRISPR/Cas9 system.

been adapted and evolved as a powerful GE tool for several purposes [2]. The Cas9 protein and the guide RNA (gRNA or sgRNA) are the two main components. CRISPR RNA (crRNA) and trans-activating crRNA are both made up of gRNA (tracrRNA). A 20-nt fragment (also known as spacer, complementary to a certain region of target genes) is followed by a protospacer adjacent motif (PAM) in the target genes of interest in the gRNA. Cas9 nuclease generates DSBs to three base pairs upstream of the PAM motif under the supervision of gRNA [23]. Gene deletion or loss of protein function is common outcome of NHEJ cleavage repair [24]. **Figure 1** indicates the general steps of CRISPR/Cas9 gene editing technique.

Using CRISPR/Cas9, several scientists have been working to improve crop yield, quality, and stress resistance. CRISPR/Cas9-mediated genome editing has been reported in 41 food crop species, 15 industrial crops, 6 oil crops, 8 ornamental crops, 1 fiber crop, and 1 feed crop (**Table 1**) [25]. Mechanism of CRISPR/Cas9 system breakage is presenting in **Figure 2**.

3. CRISPR/Cas9-mediated molecular breeding enhances the crop quality

Crop quality has been an important factor in determining market value of crop. External and internal traits, in general, determine crop quality. Physical and visual aspects such as size, color, texture, and aroma are examples of external quality attributes. Internal quality factors, on the other hand, consist of nutrients (such as protein, carbohydrate, and fats etc.) as well as bioactive chemicals (carotenoids, lycopene, −aminobutyric acid, flavonoid, etc.).

3.1 Improvement in the physical appearance of crop

3.1.1 Modification of shape and size

CRISPR/Cas9 technology was utilized to alter the shape and size of the crops to the consumer's demand. Several genes/quantitative trait loci (QTLs) have been proposed to be responsible for crop appearance quality. Rice and tomato supplied the most information about fruit shape and fruit size. The first QTL found to increase grain length, GS3 (GRAIN SIZE 3), has been successfully knocked out in five japonica rice varieties. The grain length of T1 lines has been increased in all genetic backgrounds compared with wild type [26, 27]. Grain shape affects grain weight (GW) as well as quality [28]. For example, disruption of many grain weight negative regulators, GW2, GW5, and GW6, enhanced the rice grain weight. The disruption of TaGW7 in wheat using CRISPR/Cas9 technique helps in increasing grain width and weight [29]. Researchers can manipulate tomato fruit form and size in horticulture species by changing the expression of OVATE, CLV [30], fas and lc [31], and ENO [32]. OVATE and SUN, for example, are involved in asymmetric and symmetric fruit elongation [33, 34], whereas SlWUS and SlCLV3 are genes that modulate tomato locule quantity. CLV3 mutations that cause gain-of-function and WUS mutations that cause partial loss of function are referred to as the fas and lc locus, respectively. Both mutations increase the size of the fruit [35–37]. This was further validated by disrupting the CLV-WUS cis-regulatory genes [31].

3.1.2 Color modification

Plant pigments made up of carotenoids, anthocyanins, and polyphenols determine plant color. The color of the fruit, leaves, and flower buds, especially in plant edible parts, affects the consumer's taste. Europeans and Americans like red tomatoes, whereas Asian customers prefer pink tomatoes [38]. It was reported that the absence of flavonoid pigments in the peel caused the pink phenotype. CRISPR/Cas9 can thus be used to manipulate the color of fruits by interrupting genes involved in the pigment production pathway. MYB12 regulates the accumulation of flavonoids and controls the pink skin phenotype as a flavonoid biosynthesis pathway transcription factor. SlMYB12 has been successfully knocked out, resulting in pink-fruited tomatoes [39]. In addition, by targeting PSY1 and ANT1, researchers were able to develop yellow and purple tomatoes. The PSY1 gene controls the early stages of carotenogenesis by encoding phytoene synthase. PSY1 mutations lowered total lycopene levels, resulting in yellow flesh tomato fruit [40, 41], whereas ANT1-modified tomatoes increased anthocyanin accumulation, resulting in purple plant tissue [42]. The anthocyanin biosynthetic structural genes are predominantly controlled by R2R3-MYB, bHLH, and WD-repeat proteins in all crop species investigated. Yellow roots occurred from CRISPR/Cas9 knockout of DcMYB7, an R2R3-MYB, in the solid purple carrot [43]. Flower color influences the market value of ornamental crops, and plant breeders are continually looking for new colors. Several groundbreaking flower color alteration research studies have previously been completed. Flavanone 3-hydroxylase (F3H), a major enzyme involved in flavonoid production, is required for anthocyanin accumulation. Disrupting F3H with CRISPR/Cas9 has resulted in pale blue flower torenia variations and pale purplish-pink flowered petunia varieties [44, 45].

3.2 Improving crop texture

3.2.1 Prolonging shelf life

Fruit texture is another important factor in commercial crop production. CRISPR/Cas9 technology has a pivotal role for extending the shelf life of tomatoes and bananas. Several naturally occurring mutant genes, such as Nr, alc, rin, nor, and Cnr [46], have the potential to extend shelf life. These modifications, however, are accompanied by a lack of color, an unpleasant flavor, and a low nutritional value [47]. According to one study, alc mutation not only extended the shelf life of fruits but also preserved their color and smell [48]. To induce tomato ALC gene mutations, HDR-mediated gene replacement was used, and the intended alc homozygous mutants in T1 generation displayed good storage performance [49]. Cell wall degrading enzymes can alter the texture of fruits [50]. During fruit softening, the pectate lyase enzyme can degrade the cell wall [51]. Pectate lyase enzyme interferes with RNA for prolonging life in tomato to exhibit a good fruit phenotype [52]. Similarly, SlPL gene knockout mutations based on CRISPR/Cas9 showed a harder phenotype and longer shelf life without compromising organoleptic or nutritional quality [53, 54]. Downregulating endogenous ethylene production, in addition to silencing genes involved in cell wall disintegration, can be another effective technique for delaying the fruit softening process [55]. Ethylene is the most important element affecting banana post-harvest preservation and shelf life. MA-ACO1 is involved in the ethylene production process and has an impact on the after-ripening process [55]. After ethephon treatment, the after-ripening process in MA-ACO1-mutant lines was slowed by around 2 days. More intriguingly, the amount of vitamin C and sugar in the fruit was increased without any negative effects on the fruit's quality [56].

3.3 Improving palatability

3.3.1 Improving eating and cooking quality

Consumer acceptance and market value are both determined by the eating and cooking quality (ECQ). Amylose production requires the Waxy (Wx) gene, which codes for granule-bound starch synthase I (GBSSI). After cooking, rice varieties with a somewhat low amylose concentration (7–10%) have a soft and sticky texture, making them more popular among Asian buyers. Several genetic improvement studies have effectively used the CRISPR/Cas9 system to alter the Wx gene in japonica background rice accessions, resulting in accessions with grain amylose content of 5–12% without sacrificing other desirable features [57, 58]. A number of rice mutants with fine-tuned amylose contents have been created by precise alteration of particular bases of Wx genes to fulfill the different demands on ECQ [59]. Meanwhile, by disrupting the Wx gene with CRISPR/Cas9 [60], waxy maize mutants have been generated in 12 elite inbred lines. Furthermore, rice with low palatability has a high grain protein content (GPC), which is inversely connected to ECQ. As a result, many elite rice cultivars with good ECQ have a low GPC content (often less than 7%) [61]. The GPC-related QTL qPC1 was discovered in rice for the first time. In rice, a positive regulator of GPC was found in an amino acid transporter (OsAAP6) found in qPC1 loci [62]. Rice cultivars with high ECQ can be quickly reduced in GPC and improved in ECQ using targeted mutagenesis of OsAAP6 and OsAAP10 [63].

3.3.2 Improving flavor

Next to ECQ, aroma is a preferred quality feature. Rice-eating communities in Asia and Europe both prefer rice cultivars with aroma [64]. According to research, most fragrant rice cultivars are particularly high in the 2-acetyl-1-pyrroline (2AP) chemical [65], which is also found in fresh bread and popcorn and gives food products a popcorn or cracker-like aroma [66]. BADH2 (encoding a betaine aldehyde dehydrogenase) has been linked to fragrance generation in genetic research [67, 68].

Functional BADH2 was found to engage in the conversion of -aminobutyraldehyde (GABald) to GABA, whereas nonfunctional BADH2 mutants convert GABald to 2AP [69]. As a result, RNAi technology has been employed to disrupt OsBADH2 and boost 2AP production [70]. In 2015, the first fragrant rice was generated by employing TALENs to target the OsBADH2 gene [71]. Researchers have recently made a breakthrough in the creation of novel OsBADH2 alleles using CRISPR/Cas9, successfully converting an unscented rice variety, ASD16, into a unique aromatic rice [72].

3.4 Biofortification of nutrient elements

Consumer preferences are changing toward healthier, more nutritionally rich foods. As a result, researchers have been struggling to develop new goods to meet the needs of this expanding industry. Many nutrients found in fruits and vegetables have anti-inflammatory, anticancer, and antioxidant properties. Biofortification of several nutrients, such as carotene, γ-aminobutyric acid (GABA), iron, and zinc, has been implemented in many crops through breeding programs. Through gene editing for biofortification, it has been attempted to satisfy the “hidden hunger” with high-quality nutrients.

3.4.1 Increasing carotenoid content

Carotenoids have been linked to antioxidant mechanisms and the prevention of eye diseases. Humans, on the other hand, are unable to produce carotenoids and must obtain them from their food. Lycopene and phytoene also aid in the prevention of cancer and cardiovascular disease. Previously, researchers used traditional genetic engineering to add CrtI and PSY genes into rice, as well as manufacture β-carotene. Many anti-GMO researchers believe that golden rice may not offer enough β carotene to eliminate vitamin A deficiency; and allergies and antibiotic resistance are potential dangers of planting and eating golden rice. GMO crops may also have an adverse effect on the ecosystem and biodiversity [73]. Carotenoid biofortification in rice, tomato, and banana has been achieved using CRISPR/Cas9 genome editing. Due to the lack of external gene integration in host genomes, those created by this technique have a good chance of avoiding GM regulation. Carotenoid biofortification was generally accomplished using one of two methods. First, carbon input into the carotenoid biosynthesis pathway is imposed by overexpression of phytoene synthase genes using CRISPR/Cas9-mediated knock-in. A carotenogenesis cassette including CrtI and PSY genes was successfully integrated into the target site in rice, yielding marker-free gene-edited mutants with 7.9 g/g β carotene in dry weight [74]. Another technique is to prevent their precursors from being converted or to silence associated genes, such as LCYe, BCH, ZEP, and CCD4. The loss of the LCYe gene, for example, resulted in a golden fruit banana mutant with a sixfold increase in β-carotene content [75]. Similarly, a fivefold increase in lycopene content was achieved by disrupting five carotenoid metabolic-related genes (SGR1, LCYe, BLC, LCY-B1, and LCY-B2), resulting in a lycopene-enriched tomato [76].

3.4.2 Increasing γ-aminobutyric acid content

GABA is a nonprotein amino acid inhibitory neurotransmitter that regulates blood pressure and acts as an antianxiety agent [77]. As a result, the food sector has turned its attention to generating new GABA-rich foods. Glutamate decarboxylase (GAD) is a crucial enzyme that catalyzes glutamate decarboxylation to GABA. GAD has a C-terminal autoinhibitory region that inhibits GAD action. The C-terminal has been fully removed using CRISPR/Cas9 in order to boost GABA content. GABA accumulation increased sevenfold in mutant tomatoes [78]. Furthermore, researchers generated GABA-rich rice by truncating the C-terminal of the OsGAD3 gene using the CRISPR/Cas9 system, resulting in a sevenfold increase in GABA content [79]. GABA-rich vegetables clearly have a positive impact on human health. However, pursuing a high GABA content without regard for the phenotype of the fruit could result in not only a reduction in glutamate but also a faulty phenotype [80]. Li et al. [81] employed a multiplex CRISPR/Cas9 approach to remove SlGABA-Ts and SlSSADH, resulting in a 20-fold increase in GABA levels but significant reductions in tomato fruit size and yield [82].

3.4.3 Biofortification of micronutrients

Micronutrient deficiencies, such as selenium, zinc, iron, and iodine, affect around two billion people worldwide. For those who have an imbalanced diet, biofortification of crop plants with micronutrients would be a long-term solution. Knocking down vacuolar iron transporter (VIT) genes, such as OsVIT2, to increase Fe content in rice grain is a potential use of the CRISPR/Cas9 technology. In a recent study, OsVIT2 mutation resulted in increased Fe distribution to the grains' embryo and endosperm, as well as higher Fe content in the polished grain without affecting yield [83]. Furthermore, the rice gain-of-function arsenite tolerant 1 (astol1) mutant enhanced the grain content of selenium (Se), an essential mineral with antioxidant properties for humans. Gene editing can also help in the formation of micronutrient-rich rice and wheat grains by regulating the expression of genes involved in ion homeostasis [84].

3.4.4 Improving fatty acid composition

Olive oil is high in monounsaturated fatty acids (MUFA), such as oleic acid (18:1). Diets high in oleic acid have been shown to improve cardiovascular health. Saturated fatty acids and trans-fatty acids, on the other hand, are frequently referred to as "unhealthy" fats and have been associated to cardiovascular disease [85, 86]. Soybean oil, the most extensively produced and used edible oil, contains just 20% oleic acid, compared to 65–85% in olive oil [87]. For controlling the fatty acid composition in soybean, several fatty acid desaturase genes, such as FAD2 and FAD3, were targeted and altered. By altering two homologous genes of GmFAD2, researchers have already boosted oleic acid levels from 20% to 80% in 2019, while lowering linoleic acid levels from 50% to 4.7% [88]. Similar breeding tactics have been used in rapeseed and camelina, with increases in oleic acid content of 7% and 34%, respectively [89, 90]. The first gene-edited high oleic soybean line, with 80% oleic acid and up to 20% less saturated fatty acid, was recently marketed for sale in the US market [91].

3.4.5 Eliminating antinutrients

Phytic acid, gluten protein, and cadmium (Cd) are only a few of the chemicals that have a negative impact on crop nutritional quality. Genome editing can also be used to reduce the amount of unwanted compounds in the body. Due to a lack of comparable degrading enzymes, humans are unable to metabolize phytic acid. Because phytic acid can interact with minerals and proteins to form complexes, absorption of minerals and protein is inhibited when large amounts of it are consumed by people [92]. CRISPR/Cas9 was used to knock out an ITPK gene encoding an enzyme that catalyzes the penultimate phase of phytate production in rapeseeds in order to lower phytic acid concentration [93]. The ITPK mutants had a 35% decrease in phytic acid without affecting plant performance [94]. Furthermore, gluten proteins in wheat can cause celiac disease in people who are gluten-intolerant [95]. Due to the more than 100 loci in the wheat genome that code for gluten protein, traditional breeding strategies are unlikely to reduce gluten concentration. Low-gluten, transgene-free wheat lines have been generated using CRISPR/Cas9 to target a conserved area of the α-gliadin genes [96]. Furthermore, the CRISPR/Cas9 technology has aided in the development of heavy metal pollution-resistant rice varieties. Cd is a human carcinogen, and long-term use of Cd-contaminated rice can result in chronic diseases such renal failure and cancer [97]. As a result, scientists face a problem in developing low-heavy-metal rice in Cd-contaminated areas [98]. Researchers created new Indica rice lines with reduced Cd accumulation in grain by altering OsNramp5, which mediates Cd root uptake. Furthermore, when cultivated in high Cd circumstances, osnramp5 mutants' agronomic characteristics and grain yield were unaffected [99].

4. Challenges and future perspectives

Gene editing in crops is now progressing at a considerably faster rate than in other disciplines. The CRISPR/Cas9 technology has successfully transformed and improved several quality-related features in various crops, as shown in **Table 1**. Although several gene-edited crops, such as the TALEN-fad2 soybean, TALEN-ppo potato, and CRISPR-wx1 maize, have been commercialized, we are still at the beginning of the gene-editing revolution.

To begin with, gene-edited crop rules and regulations different countries have distinct regulatory frameworks. Most countries' regulatory frameworks for genetically modified organisms (GMOs) govern the development and commercialization of novel gene-edited crops. The United States and some South American countries, such as Argentina, Brazil, Chile, and Colombia, have used product-based regulations that exempt gene-edited products from GMO supervision if the final products do not contain exogenous DNA [100, 101], whereas the European Union (EU) and New Zealand have strict process-based regulations for genome-edited crops, resulting in costly and time-consuming GM safety tests. China also has a process-based GMO regulation framework, as any gene-edited crops are scrutinized closely and no gene-edited crop has yet to be sold. The benefits of genome editing have been negated as a result of such rigorous restriction. As a result, establishing a worldwide unified and specialized regulatory system for genome-edited crops is vital. Thirteen WTO countries recently released a declaration supporting the use of gene editing in agricultural innovation, marking the first step toward building a global regulatory framework [102].

Furthermore, the delivery of CRISPR/Cas9 would be the most difficult challenge in using plant gene-editing technology. The recipient genotype has a big impact on biolistic bombardment and Agrobacterium-mediated transformation efficiency, especially in monocots. Some elite rice cultivars, for example, are notoriously difficult to change due to a lack of culture and regeneration-friendly traits [103]. Furthermore, T-DNA incorporation is unavoidable, and following plant regeneration methods are frequently technical and time-consuming. As a result, creating delivery systems that do not require tissue culture and can be used to a variety of plant species is important. Exogenous genes were delivered into pollen grains of many model crops using "pollen magnetofection," a unique approach that uses magnetic NPs as DNA transporters. About 1% of transgenic plants were produced after pollination with magnetofected pollen [104]. Some scientists, however, questioned pollen magnetofection's reproducibility [105]. It will be a shortcut to establish heritable gene alteration in transgenic seeds without tissue culturing if CRISPR/Cas9 can be delivered to reproductive cells and stably expressed via the pollen magnetofection approach [106]. Furthermore, because nano delivery technologies are nonintegrating and nonpathogenic, nanomaterial-mediated gene-edited crops may be exempt from GMO [107].

The specificity of plant CRISPR/Cas9 systems for targeted gene editing is another issue. According to several research studies, CRISPR/Cas systems have a high potential for off-target activity, and sgRNA/Cas9 complexes could create mismatched DNA sequences in mammals [108, 109]. Despite this, whole-genome sequencing demonstrated that the incidence of off-target mutations caused by CRISPR/Cas9 in plants is extremely low [110]. Off-targeting can be a problem in gene functional investigations since it might influence the phenotype of interest and lead to incorrect results interpretation [111]. Off-target effects can, however, be avoided when utilizing CRISPR technologies in crop breeding. Beneficial off-target mutations can be retained in progeny because off-target mutations with detrimental phenotype consequences are rejected throughout the breeding process [112]. As a result, in the breeding of gene-edited crops, screening advantageous mutations is more crucial than discovering off-target variants. To reduce off-targeting, several solutions have been proposed [113]. First, by developing highly precise sgRNAs with the fewest projected off-targets, the majority of off-targeting can be avoided. Second, high-fidelity Cas9 enzymes such as eSpCas9 [114] and SpCas-HF [115] can improve the specificity of CRISPR systems. Finally, the ribonucleoprotein (RNP) delivery approach can be employed to shorten the time that genomic DNA is exposed to CRISPR reagents, reducing off-targeting rates [116].

5. General procedure of genome editing in soybean and factors for success

The first successful genome editing in soybean was done in hairy roots, where ZFNs were used to target the GmDcl4a and GmDcl4b genes. ZFNs were also used to develop the first viable GE soybean plants with a GmDcl4 gene mutation (either GmDcl4a or GmDcl4b). The first TALENs-mediated GE events with two target sites were reported by Haun et al. [117]. Jacobs et al. [118] reported the first successful CRISPR GE in soybean. The majority of CRISPR/Cas9 research first concentrated on developing a GE system and analyzing its targeting effectiveness in hairy roots, and the multiplex property of CRISPR to target pairs of genes at the same time was also confirmed. Meanwhile, target gene knockout [119] and homology-directed recombination (HDR) in whole plants have both been successful [120].

5.1 Selection of a target trait

The function and properties of the genes influencing the target trait, including sequencing data, transcription data, copy number in target materials, and variances compared to the reference genome, should all be completely known. Genome sequencing and gene discovery in soybeans pave the door for GE. More than 46,000 genes in the soybean genome have been predicted using a soybean reference genome assembly and Williams82 DNA sequences [121]. Hundreds of accessions of *Glycine max* and related species have recently been sequenced for new reference genomes, including a high-quality reference genome of a wild soybean W05 and a popular Chinese farmed soybean Zhonghuang 13 (ZH13) that was recently assembled [122–124]. Moreover, the soybean reference genome assemblies have been used to characterize hundreds of regulatory noncoding RNA loci, such as microRNA (miRNA) and phased small interfering RNA (phasiRNA) loci [125]. Comparative genomics can be used to examine all of the sequencing information in order to find potentially beneficial genes. More than 70% of these genes have been duplicated and survive as numerous copies as soybean is a paleopolyploid and the two duplication events occurred 59 and 13 million years ago, respectively. It is difficult to find genes that are linked to crucial agronomical properties including yield, protein, oil, and biotic and abiotic stress tolerance, which makes soybean breeding projects challenging [126–129]. As a result, finding the genes that govern significant agronomic qualities is crucial for trait selection in soybean genome editing. The fundamental problem for soybean improvement researchers has been a lack of understanding of gene activities and contributions to agronomically important target phenotypes. GE in soybean has concentrated on features with a clear genetic background, such as GmFAD2 for oleic oil content, based on existing knowledge.

5.2 Challenges and prospective for GE and related product development in soybean

In the last four decades, popular transgenic technology has been used to introduce foreign genes into crops, such as soybean, for desired qualities, and it has proven to be a viable option to expanding genetic resources. The random incorporation of transgenes in the genome, on the other hand, has triggered public outrage and rigorous government restriction, drastically increasing the cost and time required to generate a new variety. Instead of going through repeated back crosses to transfer a natural mutation in a traditional breeding method, GE technology allows crop breeders to integrate a desired feature into an elite background in a precise and predictable manner. Traditional mutagenesis breeding introduces mutations that are indistinguishable from those induced by GE. The largest constraint for GE application in soybean, such as other crops [130], is a lack of GE candidate target genes due to poor foundational research, as described above. Technical issues such as the inability to precisely mutate any target site, the lack of ways to deliver genome-editing reagents into soybean cells, the low efficiency of selecting desired events and regenerating intact plants with targeted mutation, and off-site targeting are among the remaining bottlenecks. Through the use of newly developed GE technologies and a soybean regeneration system, several attempts have been made to reduce the restrictions and enhance the efficiency of recovering GE events. Additional challenges for GE product development include transgenic GE events, intellectual property restrictions, and government control of GE. Before GE can play a significant role in soybean improvement, these challenges must be addressed.

6. Success stories of CRISPR/Cas9 mediated in soybean (*G. max* L.)

The CRISPR-Cas9 method has been used to successfully mutate the genes GmFT2a, FAD2-2, and GmSPL9 in soybean modifying flowering time, seed oil profile, and plant architecture, respectively. This success implies that employing the CRISPR-Cas9 technology to improve soybean agronomical qualities is possible.

Targeted mutants of E1 gene controlling soybean flowering were generated. Two new types of mutations were discovered: 11 bp and 40 bp deletions in the E1 coding area, respectively, and frameshift mutations that resulted in premature translation termination codons and shortened E1 proteins, causing early blooming under long day conditions. Furthermore, by predicting and analyzing the probable off-target areas of E1 targets, no off-target effects were found. The shortened E1 protein disinhibited GmFT2a/5a, and boosting GmFT2a/5a gene expression led in evident early flowering in two new mutants with significantly decreased E1 gene expression [131].

7. Conclusions

Thanks to the CRISPR/Cas9-based gene-editing method, researchers may now change crop-specific traits more accurately and effectively. The CRISPR/Cas9 system has become the most frequently used and versatile technology in crop breeding and functional genomics. Its unrivaled ability to manipulate genes contributed in the development of a number of crop varieties with beneficial agronomic traits. Most crop-improvement gene-editing research, on the other hand, is still in the early stages of uncovering genomic function and regulatory pathways. Gene-edited crops are still a long way from becoming commercially available. In addition, gene-editing approaches have yet to meet all of the requirements for changing plant genomes. Because several quality-related variables are governed by many QTLs and altering individual genes may not generate significant phenotypic change, further development will be critical for the use of CRISPR/Cas in plants. It might be possible to create a CRISPR/Cas-mediated chromosomal rearrangement technology that works well. Furthermore, delivering CRISPR cargoes remains a significant challenge. As a result, it would be advantageous to design new carrier materials. Aside from that, public concerns and the government's rigorous regulatory policies on gene-editing technologies are another roadblock to plant breeding progress. Despite the remaining hurdles, gene-editing technology is expected to become more frequently used in the future and will undoubtedly play a significant part in crop quality enhancement like in soybean.

Author details

Summra Siddique
University of Agriculture, Faisalabad, Pakistan

*Address all correspondence to: samrasiddique33@yahoo.com

References

[1] Foley JA, Ramankutty N, Brauman KA, Cassidy ES, Gerber JS, Johnston M, et al. Solutions for a cultivated planet. Nature. 2011;**478**:337-342

[2] Hsu PD, Lander ES, Zhang F. Development and applications of CRISPR-Cas9 for genome engineering. Cell. 2014;**157**:1262-1278

[3] Zhang D, Li Z, Li J. Targeted gene manipulation in plants using the CRISPR/Cas technology. Journal of Genetics and Genomics. 2016;**43**:251-262

[4] Slavin JL, Lloyd B. Health benefits of fruits and vegetables. Advances in Nutrition. 2012;**3**:506-516

[5] Wenefrida I, Utomo HS, Linscombe SD. Mutational breeding and genetic engineering in the development of high grain protein content. Journal of Agricultural and Food Chemistry. 2013;**61**:11702-11710

[6] Lusser M, Parisi C, Plan D, Rodriguez-Cerezo E. Deployment of new biotechnologies in plant breeding. Nature Biotechnology. 2012;**30**:231-239

[7] Ramesh P, Mallikarjuna G, Sameena S, Kumar A, Gurulakshmi K, Reddy BV, et al. Advancements in molecular marker technologies and their applications in diversity studies. Journal of Biosciences. 2020;**45**:1-15

[8] Chaudhary J, Alisha A, Bhatt V, Chandanshive S, Kumar N, Mir Z, et al. Mutation breeding in tomato: Advances, applicability and challenges. Plants. 2019;**8**:128

[9] Parry MA, Madgwick PJ, Bayon C, Tearall K, Hernandez-Lopez A, Baudo M, et al. Mutation discovery for crop improvement. Journal of Experimental Botany. 2009;**60**:2817-2825

[10] Gaj T, Gersbach CA, Barbas CF. ZFN, TALEN, and CRISPR/Cas-based methods for genome engineering. Trends in Biotechnology. 2013;**31**:397-405

[11] Cohen-Tannoudji M, Robine S, Choulika A, Pinto D, El Marjou F, Babinet C, et al. I-SceI-induced gene replacement at a natural locus in embryonic stem cells. Molecular and Cellular Biology. 1998;**18**:1444-1448

[12] Bibikova M, Golic M, Golic KG, Carroll D. Targeted chromosomal cleavage and mutagenesis in Drosophila using zinc-finger nucleases. Genetics. 2002;**161**:1169-1175

[13] Christian M, Cermak T, Doyle EL, Schmidt C, Zhang F, Hummel A, et al. Targeting DNA double-strand breaks with TAL effector nucleases. Genetics. 2010;**186**:757-761

[14] Cong L, Ran FA, Cox D, Lin S, Barretto R, Habib N, et al. Multiplex genome engineering using CRISPR/Cas systems. Science. 2013;**339**:819-823

[15] Joung JK, Sander JD. TALENs: A widely applicable technology for targeted genome editing. Nature Reviews. Molecular Cell Biology. 2013;**14**:49-55

[16] Shan Q, Wang Y, Chen K, Liang Z, Li J, Zhang Y, et al. Rapid and efficient gene modification in rice and Brachypodium using TALENs. Molecular Plant. 2013;**6**:1365-1368

[17] Wang Y, Cheng X, Shan Q, Zhang Y, Liu J, Gao C, et al. Simultaneous editing of three homoeoalleles in hexaploid bread wheat confers heritable

resistance to powdery mildew. Nature Biotechnology. 2014;**32**:947-951

[18] Čermák T, Baltes NJ, Čegan R, Zhang Y, Voytas DF. High-frequency, precise modification of the tomato genome. Genome Biology. 2015;**16**:1-15

[19] Ishino Y, Shinagawa H, Makino K, Amemura M, Nakata A. Nucleotide sequence of the iap gene, responsible for alkaline phosphatase isozyme conversion in Escherichia coli, and identification of the gene product. Journal of Bacteriology. 1987;**169**:5429-5433

[20] Lino CA, Harper JC, Carney JP, Timlin JA. Delivering CRISPR: A review of the challenges and approaches. Drug Delivery. 2018;**25**:1234-1257

[21] Gao C. The future of CRISPR technologies in agriculture. Nature Reviews Molecular Cell Biology. 2018;**19**(5):275-276

[22] Li C, Nguyen V, Liu J, Fu W, Chen C, Yu K, et al. Mutagenesis of seed storage protein genes in soybean using CRISPR/Cas9. BMC Research Notes. 2019;**12**(1):1-7

[23] Jinek M, Chylinski K, Fonfara I, Hauer M, Doudna JA, Charpentier EA. programmable dual-RNA–guided DNA endonuclease in adaptive bacterial immunity. Science. 2012;**337**:816-821

[24] Liu M, Rehman S, Tang X, Gu K, Fan Q, Chen D, et al. Methodologies for improving HDR efficiency. Frontiers in Genetics. 2019;**9**:691

[25] Ricroch A, Clairand P, Harwood W. Use of CRISPR systems in plant genome editing: Toward new opportunities in agriculture. Emerging Topics in Life Science. 2017;**1**:169-182

[26] Shen L, Wang C, Fu Y, Wang J, Liu Q, Zhang X, et al. QTL editing confers opposing yield performance in different rice varieties. Journal of Integrative Plant Biology. 2018;**60**:89-93

[27] Yuyu C, Aike Z, Pao X, Xiaoxia W, Yongrun C, Beifang W, et al. Effects of GS3 and GL3.1 for grain size editing by CRISPR/Cas9 in rice. Rice Science. 2020;**27**:405-413

[28] Xu R, Yang Y, Qin R, Li H, Qiu C, Li L, et al. Rapid improvement of grain weight via highly efficient CRISPR/Cas9-mediated multiplex genome editing in rice. Journal of Genetics and Genomics. 2016;**43**:529

[29] Wang W, Pan Q, Tian B, He F, Chen Y, Bai G, et al. Gene editing of the wheat homologs of TONNEAU1-recruiting motif encoding gene affects grain shape and weight in wheat. The Plant Journal. 2019;**100**:251-264

[30] Zsögön A, Čermák T, Naves ER, Notini MM, Edel KH, Weinl S, et al. De novo domestication of wild tomato using genome editing. Nature Biotechnology. 2018;**36**:1211-1216

[31] Rodríguez-Leal D, Lemmon ZH, Man J, Bartlett ME, Lippman ZB. Engineering quantitative trait variation for crop improvement by genome editing. Cell. 2017;**171**:470-480

[32] Yuste Lisbona FJ, Fernández-Lozano A, Pineda B, Bretones S, Ortíz-Atienza A, García-Sogo B, et al. ENO regulates tomato fruit size through the floral meristem development network. Proceedings of the National Academy of Sciences. 2020;**117**:8187-8195

[33] Liu J, Van Eck J, Cong B, Tanksley SD. A new class of regulatory genes underlying the cause of pear-shaped tomato fruit. Proceedings of the National Academy of Sciences. 2002;**99**:13302-13306

[34] Lin T, Zhu G, Zhang J, Xu X, Yu Q, Zheng Z, et al. Genomic analyses provide insights into the history of tomato breeding. Nature Genetics. 2014;**46**:1220-1226

[35] Tanksley SD. The genetic, developmental, and molecular bases of fruit size and shape variation in tomato. The Plant Cell. 2004;**16**:S181-S189

[36] Frary A, Nesbitt TC, Grandillo S, Knaap E, Cong B, Liu J, et al. A quantitative trait locus key to the evolution of tomato fruit size. Science. 2000;**289**:85-88

[37] Munos S, Ranc N, Botton E, Berard A, Rolland S, Duffe P, et al. Increase in tomato locule number is controlled by two single-nucleotide polymorphisms located near WUSCHEL. Plant Physiology. 2011;**156**:2244-2254

[38] Velasco C, Wan X, Salgado-Montejo A, Woods A, Oñate GA, Mu B, et al. The context of colour–flavour associations in crisps packaging: A cross-cultural study comparing Chinese, Colombian and British consumers. Food Quality and Preference. 2014;**38**:49-57

[39] Yang T, Deng L, Zhao W, Zhang R, Jiang H, Ye Z, et al. Rapid breeding of pink-fruited tomato hybrids using the CRISPR/Cas9 system. Journal of Genetics and Genomics. 2019;**46**:505

[40] Filler H, Bessudo SM, Levy CAA. Targeted recombination between homologous chromosomes for precise breeding in tomato. Nature Communications. 2017;**8**:15605

[41] Chattopadhyay T, Hazra P, Akhtar S, Maurya D, Mukherjee A, Roy S. Skin color, carotenogenesis and chlorophyll degradation mutant alleles: Genetic orchestration behind the fruit color variation in tomato. Plant Cell Reports. 2021;**40**(5):767-782

[42] Vu TV, Sivankalyani V, Kim EJ, Doan DTH, Tran MT, Kim J, et al. Highly efficient homology-directed repair using CRISPR/Cpf1-geminiviral replicon in tomato. Plant Biotechnology Journal. 2020;**18**:2133-2143

[43] Xu ZS, Yang QQ, Feng K, Xiong AS. Changing carrot color: Insertions in DcMYB7 alter the regulation of anthocyanin biosynthesis and modification. Plant Physiology. 2019;**181**:195-207

[44] Nishihara M, Higuchi A, Watanabe A, Tasaki K. Application of the CRISPR/Cas9 system for modification of flower color in Torenia fournieri. BMC Plant Biology. 2018;**18**:1-9

[45] Yu J, Tu L, Subburaj S, Bae S, Lee GJ. Simultaneous targeting of duplicated genes in Petunia protoplasts for flower color modification via CRISPR-Cas9 ribonucleoproteins. Plant Cell Reports. 2020;**40**(6):1037-1045

[46] Wang R, Lammers M, Tikunov Y, Bovy AG, Angenent GC, de Maagd RA. The rin, nor and Cnr spontaneous mutations inhibit tomato fruit ripening in additive and epistatic manners. Plant Science. 2020;**294**:110436

[47] Kopeliovitch E, Mizrahi Y, Rabinowitch HD, Kedar N. Effect of the fruit-ripening mutant genes rin and nor on the flavor of tomato fruit. Journal American Society for Horticultural Science. 1982;**107**:361-364

[48] Casals J, Cebolla-Cornejo J, Roselló S, Beltrán J, Casañas F, Nuez F. Long-term postharvest aroma evolution of tomatoes with the alcobaça (alc) mutation. European Food Research and Technology. 2011;**233**:331-342

[49] Yu QH, Wang B, Li N, Tang Y, Yang S, Yang T, et al. CRISPR/

Cas9-induced targeted mutagenesis and gene replacement to generate long-shelf life tomato lines. Scientific Reports. 2017;7:1-9

[50] Wang D, Yeats TH, Uluisik S, Rose JK, Seymour GB. Fruit softening: Revisiting the role of pectin. Trends in Plant Science. 2018;**23**:302-310

[51] Marín-Rodríguez MC, Orchard J, Seymour GB. Pectate lyases, cell wall degradation and fruit softening. Journal of Experimental Botany. 2002;**53**:2115-2119

[52] Yang L, Huang W, Xiong F, Xian Z, Su D, Ren M, et al. Silencing of SlPL, which encodes a pectate lyase in tomato, confers enhanced fruit firmness, prolonged shelf-life and reduced susceptibility to grey mould. Plant Biotechnology Journal. 2017;**15**:1544-1555

[53] Uluisik S, Chapman NH, Smith R, Poole M, Adams G, Gillis RB, et al. Genetic improvement of tomato by targeted control of fruit softening. Nature Biotechnology. 2016;**34**:950-952

[54] Wang D, Samsulrizal NH, Yan C, Allcock NS, Craigon J, Blanco-Ulate B, et al. Characterization of CRISPR mutants targeting genes modulating pectin degradation in ripening tomato. Plant Physiology. 2019;**179**:544-557

[55] Elitzur T, Yakir E, Quansah L, Zhangjun F, Vrebalov J, Khayat E, et al. Banana MaMADS transcription factors are necessary for fruit ripening and molecular tools to promote shelf-life and food security. Plant Physiology. 2016;**171**:380-391

[56] Hu C, Sheng O, Deng G, He W, Dong T, Yang Q, et al. CRISPR/Cas9-mediated genome editing of MaACO1 (aminocyclopropane-1-carboxylate oxidase1) promotes the shelf life of banana fruit. Plant Biotechnology Journal. 2020;**19**(4):654-656

[57] Pérez L, Soto E, Farré G, Juanos J, Villorbina G, Bassie L, et al. CRISPR/Cas9 mutations in the rice Waxy/GBSSI gene induce allele-specific and zygosity-dependent feedback effects on endosperm starch biosynthesis. Plant Cell Reports. 2019;**38**:417-433

[58] Zhang J, Zhang H, Botella JR, Zhu JK. Generation of new glutinous rice by CRISPR/Cas9-targeted mutagenesis of the Waxy gene in elite rice varieties. Journal of Integrative Plant Biology. 2018;**60**:369-375

[59] Xu Y, Lin Q, Li X, Wang F, Chen Z, Wang J, et al. Fine-tuning the amylose content of rice by precise base editing of the Wx gene. Plant Biotechnology Journal. 2021;**19**:11-13

[60] Gao H, Gadlage MJ, Lafitte HR, Lenderts B, Yang M, Schroder M, et al. Superior field performance of waxy corn engineered using CRISPR–Cas9. Nature Biotechnology. 2020;**38**:579-581

[61] Yang Y, Guo M, Sun S, Zou Y, Yin S, Liu Y, et al. Natural variation of OsGluA2 is involved in grain protein content regulation in rice. Nature Communications. 2019;**10**:1949

[62] Peng B, Kong H, Li Y, Wang L, Zhong M, Sun L, et al. OsAAP6 functions as an important regulator of grain protein content and nutritional quality in rice. Nature Communications. 2014;**5**:1-12

[63] Wang S, Yang Y, Guo M, Zhong C, Yan C, Sun S. Targeted mutagenesis of amino acid transporter genes for rice quality improvement using the CRISPR/Cas9 system. The Crop Journal. 2020;**8**:457-464

[64] Cruz ND, Khush G. Rice grain quality evaluation procedures. Aromatic Rices. 2000;**3**:15-28

[65] Buttery RG, Ling LC, Juliano BO, Turnbaugh JG. Cooked rice aroma and 2-acetyl-1-pyrroline. Journal of Agricultural and Food Chemistry. 1983;**31**:823-826

[66] Adams A, De Kimpe N. Chemistry of 2-acetyl-1-pyrroline, 6-acetyl-1, 2, 3, 4-tetrahydropyridine, 2-acetyl-2-thiazoline, and 5-acetyl-2, 3-dihydro-4 H-thiazine: Extraordinary Maillard flavor compounds. Chemical Reviews. 2006;**106**:2299-2319

[67] Lorieux M, Petrov M, Huang N, Guiderdoni E, Ghesquière A. Aroma in rice: Genetic analysis of a quantitative trait. Theoretical and Applied Genetics. 1996;**93**:1145-1151

[68] Singh A, Singh PK, Singh R, Pandit A, Mahato AK, Gupta DK, et al. SNP haplotypes of the BADH1 gene and their association with aroma in rice (*Oryza sativa* L.). Molecular Breeding New Strategies Plant Improvement. 2010;**26**:325-338

[69] Wakte KV, Kad TD, Zanan RL, Nadaf AB. Mechanism of 2-acetyl-1-pyrroline biosynthesis in Bassia latifolia Roxb. flowers. Physiology and Molecular Biology of Plants. 2011;**17**:231-237

[70] Niu X, Tang W, Huang W, Ren G, Wang Q, Luo D, et al. RNAi-directed downregulation of OsBADH2 results in aroma (2-acetyl-1-pyrroline) production in rice (*Oryza sativa* L.). BMC Plant Biology. 2008;**8**:100

[71] Shan Q, Zhang Y, Chen K, Zhang K, Gao C. Creation of fragrant rice by targeted knockout of the OsBADH2 gene using TALEN technology. Plant Biotechnology Journal. 2015;**13**:791-800

[72] Ashokkumar S, Jaganathan D, Ramanathan V, Rahman H, Palaniswamy R, Kambale R, et al. Creation of novel alleles of fragrance gene OsBADH2 in rice through CRISPR/Cas9 mediated gene editing. PLoS One. 2020;**15**:e0237018

[73] Stein J, Sachdev HPS, Qaim M. Potential impact and cost-effectiveness of Golden Rice. Nature Biotechnology. 2006;**24**:1200-1201

[74] Dong OX, Yu S, Jain R, Zhang N, Duong PQ, Butler C, et al. Marker-free carotenoid-enriched rice generated through targeted gene insertion using CRISPR-Cas9. Nature Communications. 2020;**11**:1-10

[75] Kaur N, Alok A, Kumar P, Kaur N, Awasthi P, Chaturvedi S, et al. CRISPR/Cas9 directed editing of lycopene epsilon-cyclase modulates metabolic flux for β-carotene biosynthesis in banana fruit. Metabolic Engineering. 2020;**59**:76-86

[76] Li X, Wang Y, Chen S, Tian H, Fu D, Zhu B, et al. Lycopene is enriched in tomato fruit by CRISPR/Cas9-mediated multiplex genome editing. Frontiers in Plant Science. 2018;**9**:559

[77] Nuss P. Anxiety disorders and GABA neurotransmission: A disturbance of modulation. Neuropsychiatric Disease and Treatment. 2015;**11**:165

[78] Nonaka S, Arai C, Takayama M, Matsukura C, Ezura H. Efficient increase of γ-aminobutyric acid (GABA) content in tomato fruits by targeted mutagenesis. Scientific Reports. 2017;**7**:7057

[79] Akama K, Akter N, Endo H, Kanesaki M, Endo M, Toki S. An in vivo targeted deletion of the calmodulin-binding domain from rice glutamate decarboxylase 3 (OsGAD3) increases γ-aminobutyric acid content in grains. Rice. 2020;**13**:20

[80] Gramazio P, Takayama M, Ezura H. Challenges and prospects of new plant breeding techniques for GABA improvement in crops: Tomato as an example. Frontiers in Plant Science. 2020;**11**:1382

[81] Li R, Li X, Fu D, Zhu B, Tian H, Luo Y, et al. Multiplexed CRISPR/Cas9-mediated metabolic engineering of γ-aminobutyric acid levels in *Solanum lycopersicum*. Plant Biotechnology Journal. 2018;**16**:415-427

[82] Che J, Yamaji N, Ma JF. Role of a vacuolar iron transporter OsVIT2 in the distribution of iron to rice grains. The New Phytologist. 2021;**230**:1049-1062

[83] Sun SK, Xu XJ, Tang Z, Tang Z, Huang XY, Wirtz M, et al. A molecular switch in sulfur metabolism to reduce arsenic and enrich selenium in rice grain. Nature Communications. 2021;**12**:1392

[84] Briggs MA, Petersen KS, Kris-Etherton PM. Saturated fatty acids and cardiovascular disease: Replacements for saturated fat to reduce cardiovascular risk. Healthcare. 2017;**5**:29

[85] Iqbal MP. Trans fatty acids—A risk factor for cardiovascular disease. Pakistan Journal of Medicine Science. 2014;**30**:194-197

[86] Li R, Li R, Li X, Fu D, Zhu B, Tian H, et al. Multiplexed CRISPR/Cas9-mediated metabolic engineering of -aminobutyric acid levels in Solanum lycopersicum. Plant Biotechnology Journal. 2018;**16**:415-427

[87] Ramirez-Tortosa MC, Granados S, Quiles JL. Chemical Composition, Types and Characteristics of Olive Oil. Oxford, UK: CABI Publishing; 2006

[88] Do PT, Nguyen CX, Bui HT, Tran LT, Stacey G, Gillman JD, et al. Demonstration of highly efficient dual gRNA CRISPR/Cas9 editing of the homeologous GmFAD2-1A and GmFAD2-1B genes to yield a high oleic, low linoleic and α-linolenic acid phenotype in soybean. BMC Plant Biology. 2019;**19**:1-14

[89] Jiang WZ, Henry IM, Lynagh PG, Comai L, Cahoon EB, Weeks DP. Significant enhancement of fatty acid composition in seeds of the allohexaploid, *Camelina sativa*, using CRISPR/Cas9 gene editing. Plant Biotechnology Journal. 2017;**15**:648-657

[90] Okuzaki A, Ogawa T, Koizuka C, Kaneko K, Inaba M, Imamura J, et al. CRISPR/Cas9-mediated genome editing of the fatty acid desaturase 2 gene in Brassica napus. Plant Physiology and Biochemistry. 2018;**131**:63-69

[91] Calyxt I. First Commercial Sale of Calyxt High Oleic Soybean Oil on the U.S. Market. 2019. Available from: https://calyxt.com/first-commercial-sale-of-calyxt-high-oleic-soybean-oil-on-the-u-s-market/ [Accessed: April 1, 2021]

[92] Oatway L, Vasanthan T, Helm JH. Phytic acid. Food Review International. 2001;**17**:419-431

[93] Sun Y, Thompson M, Lin G, Butler H, Gao Z, Thornburgh S, et al. Inositol 1, 3, 4, 5, 6-pentakisphosphate 2-kinase from maize: Molecular and biochemical characterization. Plant Physiology. 2007;**144**:1278-1291

[94] Sashidhar N, Harloff HJ, Potgieter L, Jung C. Gene editing of three BnITPK genes in tetraploid oilseed rape leads to significant reduction of phytic acid in seeds. Plant Biotechnology Journal. 2020;**18**:2241-2250

[95] Hischenhuber C, Crevel R, Jarry B, Mäki M, Moneret-Vautrin D,

Romano A, et al. Safe amounts of gluten for patients with wheat allergy or coeliac disease. Alimentary Pharmacology & Therapeutics. 2006;**23**:559-575

[96] Sanchez-Leon S, Gil-Humanes J, Ozuna CV, Gimenez MJ, Sousa C, Voytas DF, et al. Low-gluten, nontransgenic wheat engineered with CRISPR/Cas9. Plant Biotechnology Journal. 2018;**16**:902-910

[97] Bertin G, Averbeck D. Cadmium: Cellular effects, modifications of biomolecules, modulation of DNA repair and genotoxic consequences (a review). Biochimie. 2006;**88**:1549-1559

[98] Hu Y, Cheng H, Tao S. The challenges and solutions for cadmium-contaminated rice in China: A critical review. Environment International. 2016;**92**:515-532

[99] Tang L, Mao B, Li Y, Lv Q, Zhang L, Chen C, et al. Knockout of OsNramp5 using the CRISPR/Cas9 system produces low Cd-accumulating indica rice without compromising yield. Scientific Reports. 2017;**7**:14438

[100] Waltz E. Gene-edited CRISPR mushroom escapes US regulation. Nature News. 2016;**532**:293

[101] Lema MA. Regulatory aspects of gene editing in Argentina. Transgenic Research. 2019:**28**:147-150

[102] International Statement on Agricultural Applications of Precision Biotechnology. 2018. Available from: https://docs.wto.org/dol2fe/Pages/FE_Search/FE_S_S009-DP.aspx?language=E&CatalogueIdList=250406,249838,249823,249748,249641,249507,249371,249321,249324,249267&CurrentCatalogueIdIndex=7&FullTextHash=&HasEnglishRecord=True&HasFrenchRecord=True&HasSpanishRecord=True [Accessed: April 1, 2021]

[103] Hamada H, Linghu Q, Nagira Y, Miki R, Taoka N, Imai R. An in planta biolistic method for stable wheat transformation. Scientific Reports. 2017;**7**:11443

[104] Liu Q, Chen B, Wang Q, Shi X, Xiao Z, Lin J, et al. Carbon nanotubes as molecular transporters for walled plant cells. Nano Letters. 2009;**9**:1007-1010

[105] Zhao X, Meng Z, Wang Y, Chen W, Sun C, Cui B, et al. Pollen magnetofection for genetic modification with magnetic nanoparticles as gene carriers. Nature Plants. 2017;**3**:956-964

[106] Vejlupkova Z, Warman C, Sharma R, Scheller HV, Mortimer JC, Fowler JE. No evidence for transient transformation via pollen magnetofection in several monocot species. Nature Plants. 2020:**6**:1323-1324

[107] Lew TTS, Park M, Wang Y, Gordiichuk P, Yeap WC, Mohd Rais SK, et al. Nanocarriers for transgene expression in pollen as a plant biotechnology tool. ACS Materials Letters. 2020;**2**:1057-1066

[108] Demirer GS, Zhang H, Goh NS, González-Grandío E, Landry MP. Carbon nanotube–mediated DNA delivery without transgene integration in intact plants. Nature Protocols. 2019;**14**:2954-2971

[109] Grünewald J, Zhou R, Garcia SP, Iyer S, Lareau CA, Aryee MJ, et al. Transcriptome-wide off-target RNA editing induced by CRISPR-guided DNA base editors. Nature. 2019;**569**:433-437

[110] Fu Y, Foden JA, Khayter C, Maeder ML, Reyon D, Joung JK, et al. High-frequency off-target mutagenesis induced by CRISPR-Cas nucleases in human cells. Nature Biotechnology. 2013;**31**:822-826

[111] Li J, Manghwar H, Sun L, Wang P, Wang G, Sheng H, et al. Whole genome sequencing reveals rare off-target mutations and considerable inherent genetic or/and somaclonal variations in CRISPR/Cas9-edited cotton plants. Plant Biotechnology Journal. 2019;**17**:858-868

[112] Mao Y, Botella JR, Liu Y, Zhu JK. Gene editing in plants: Progress and challenges. National Science Review. 2019;**6**:421-437

[113] Hahn F, Nekrasov V. CRISPR/Cas precision: Do we need to worry about off-targeting in plants? Plant Cell Reports. 2019;**38**:437-441

[114] Slaymaker IM, Gao L, Zetsche B, Scott DA, Yan WX, Zhang F. Rationally engineered Cas9 nucleases with improved specificity. Science. 2016;**351**:84-88

[115] Kleinstiver BP, Pattanayak V, Prew MS, Tsai SQ, Nguyen NT, Zheng Z, et al. High-fidelity CRISPR–Cas9 nucleases with no detectable genome-wide off-target effects. Nature. 2016;**529**:490-495

[116] Vakulskas CA, Behlke MA. Evaluation and reduction of CRISPR off-target cleavage events. Nucleic Acid Therapeutics. 2019;**29**:167-174

[117] Haun W, Coffman A, Clasen BM, Demorest ZL, Lowy A, Ray E, et al. Improved soybean oil quality by targeted mutagenesis of the fatty acid desaturase 2 gene family. Plant Biotechnology Journal. 2014;**12**:934-940. DOI: 10.1111/pbi.12201

[118] Jacobs TB, LaFayette PR, Schmitz RJ, Parrott WA. Targeted genome modifications in soybean with CRISPR/Cas9. BMC Biotechnology. 2015;**15**:16. DOI: 10.1186/s12896-015-0131-2

[119] Curtin SJ, Zhang F, Sander JD, Haun WJ, Starker C, Baltes NJ, et al. Targeted mutagenesis of duplicated genes in soybean with zinc-finger nucleases. Plant Physiology. 2011;**156**:466-473. DOI: 10.1104/pp.111.172981

[120] Sander JD, Dahlborg EJ, Goodwin MJ, Cade L, Zhang F, Cifuentes D, et al. Selection-free zinc-finger-nuclease engineering by context-dependent assembly (CoDA). Nature Methods. 2011;**8**:67-69. DOI: 10.1038/nmeth.1542

[121] Kim H, Kim ST, Ryu J, Kang BC, Kim JS, Kim SG. CRISPR/Cpf1-mediated DNA-free plant genome editing. Nature Communications. 2017;**8**:14406. DOI: 10.1038/ncomms14406

[122] Li Z, Liu ZB, Xin A, Moon BP, Koellhoffer JP, Huang L, et al. Cas9-guide RNA directed genome editing in soybean. Plant Physiology. 2015;**169**: 960-970. DOI: 10.1104/pp.15.00783

[123] Schmutz J, Cannon SB, Schlueter J, Ma J, Mitros T, Nelson W, et al. Genome sequence of the palaeopolyploid soybean. Nature. 2010;**463**:178-183. DOI: 10.1038/nature08670

[124] Kim MY, Lee S, Van K, Kim TH, Jeong SC, Choi IY, et al. Whole-genome sequencing and intensive analysis of the undomesticated soybean (*Glycine soja* Sieb. and Zucc.) genome. Proceedings. National Academy of Sciences. United States of America. 2010;**107**:22032-22037. DOI: 10.1073/pnas.1009526107

[125] Lam HM, Xu X, Liu X, Chen W, Yang G, Wong FL, et al. Resequencing of 31 wild and cultivated soybean genomes identifies patterns of genetic diversity and selection. Nature Genetics. 2010;**42**:1053-1059. DOI: 10.1038/ng.715

[126] Zhou Z, Jiang Y, Wang Z, Gou Z, Lyu J, Li W et al. Resequencing 302 wild and cultivated accessions identifies

genes related to domestication and improvement in soybean. Nature Biotechnology. 2015;**33**:408-414. DOI: 10.1038/nbt.3096

[127] Gao CW, Gao LZ. The complete chloroplast genome sequence of semi-wild soybean, *Glycine gracilis* (Fabales: Fabaceae). Conservation Genetics Resources. 2017;**9**:343-345. DOI: 10.1007/s12686-016-0683-z

[128] Asaf S, Khan AL, Al-Harrasi A, Kim TH, Lee IJ. The first complete mitochondrial genome of wild soybean (*Glycine soja*). Mitochondrial DNA Part B. 2018;**3**:527-528. DOI: 10.1080/23802359.2018.1467228

[129] Arikit S, Xia R, Kakrana A, Huang K, Zhai J, Yan Z, et al. An atlas of soybean small RNAs identifies phased siRNAs from hundreds of coding genes. The Plant Cell. 2014;**26**:4584-4601. DOI: 10.1105/tpc.114.131847

[130] Shoemaker RC, Schlueter J, Doyle JJ. Paleopolyploidy and gene duplication in soybean and other legumes. Current Opinion in Plant Biology. 2006;**9**: 104-109. DOI: 10.1016/j.pbi.2006.01.007

[131] Zhu Y, Wu N, Song W, Yin G, Qin Y, Yan Y, et al. Soybean (*Glycine max*) expansin gene superfamily origins: Segmental and tandem duplication events followed by divergent selection among subfamilies. BMC Plant Biology. 2014;**14**(93):22-29. DOI: 10.1186/1471-2229-14-93

Section 5

Food and Biodiesel Industry

Chapter 11

Soybean Functional Proteins and the Synthetic Biology

Lilian Hasegawa Florentino, Rayane Nunes Lima and Mayla D.C. Molinari

Abstract

Recently, soybean consumption has increased, not only because of its potential for industrial and livestock use but also due to its beneficial effects on human health in the treatment and prevention of various diseases because soy can produce a wide number of functional proteins. Despite the soybean-producing high, elevated, nutritive and functional proteins, it also produces allergenic proteins, harmful secondary metabolites, and carcinogenic elements. So, recombinant protein systems that mimic the structures and functions of the natural proteins supply a single tunable and valuable source of advanced materials. But the availability of the technology to produce synthetic functional proteins is still limited. Therefore, Synthetic Biology is a powerful and promising science field for the development of new devices and systems able to tackle the challenges that exist in conventional studies on the development of functional protein systems. Thus, representing a new disruptive frontier that will allow better use of soybean functional proteins, both for animal and human food and for the pharmaceutical and chemistry industry.

Keywords: soybean, synthetic biology, bioengineering, functional proteins, proteome

1. Introduction

Synthetic Biology allows for more sophisticated and complex engineering than the old genetic modification techniques. Involving scientific subjects and non-biological engineering, including information technology, bioinformatics, and nanotechnology, synthetic biology strives to alter the organism's genes on a scale much bigger – their genome! – rewriting your genetic code, all chemical instructions needed to project, assemble, and operate a living organism. To invent new ways of life via biochemistry, created in the computer and made from off-the-shelf chemicals, will not just revolutionize biology, but it will also profoundly influence the definition of life, including what it means to be human [1]. The creation of new life forms could be a little bit threatening, although this is just a possible approach of Synthetic Biology that follows all the ethical precepts of modern science and that just has advanced in the field of microbiology, through research involving the development of the minimal genomes of a living organism, like, for example, the minimal genome of the bacteria *Mycoplasma mycoides* (JCV-Syn3.0) [2] and the project

Sc2.0 – Synthetic yeast Genome Project. Another approach less threatening and much more promising is the possibility of modifying existing organisms for "more useful" and economically applicable purposes, such as developing biofabrics to produce medicines or biofuels [3].

One of the oldest leguminous consumed by humankind, Soybean occupies an important place in the world food industry, offering oil and protein source consumption and bran-rich in proteins for animal feed. Generally composed of ~35–40% protein, ~20% lipids, ~9% dietary fibers, and ~ 8.5% moisture based on the dry weight of mature raw seed [4]. Soybean as a protein biofactory/bioreactor for various industrial purposes (cosmetic, pharmaceutical, biofuel, food) is deeply studied due to several aspects inherent to its easy cultivation in a greenhouse and its rich genetic variability [5, 6]. Two possibilities are explored for protein production in soybean, one involves the production of cisgenic proteins and the other consists of transgenic protein production. A typical example of the potential to produce bioactive proteins using soybean is in the therapeutic market. This market moves over 100 billion dollars worldwide and grows 20% annually. However, it is a market where production depends mainly on microorganisms and animals to sustain itself. Obtaining proteins from these organisms presents low productivity and high production cost, in the case of animals which involves debatable ethical issues [7]. However, with the advancement of scientific advances in biotechnology, bioinformatics, and omics technologies, the soybean has been shown a more sustainable, safe, and cheaper alternative to producing bioactive proteins when compared to production by organisms already used [7, 8].

Soybean-based agriculture faces several productivity and global sustainability challenges, including emerging threats from climate change and diseases, forcing the rapid adoption of short-to-long-term genetic innovation methods. Thus, the field of Synthetic Biology (SB or SynBio) is prepared to offer several technological solutions to rapidly improve the development of new soybean cultivars through genetic circuits, biosensors, metabolic engineering, and genome editing techniques. SynBio is a field of scientific research that integrates principles from mathematics, physics, engineering, and chemistry and applies genetic tools to develop bottom-up and top-down strategies to design technological products for industry, medicine, and agriculture (**Figure 1**). Thus, regarding the sustainable and technical management of soybeans (*Glycine max*), SynBio can usefully increase biomass production and harvest performance and dramatically transform existing genetic modification techniques. Furthermore, the omics tools have made remarkable progress [9–11] because there are few technical difficulties in obtaining complete soybean genomic sequences. However, the factor to be overcome is understanding the complex functioning and organization of the genome itself [12–18]. Thus, the combination of current genomic prediction, design, and synthesis techniques and the recently proposed genome editing methodologies could allow the rapid development of new bioreactor chassis for protein production [19–23].

Protein is one of the most important components in an essential diet for the survival of organisms; they supply adequate amounts of amino acids to the body. The availability of amino acids from food depends on various factors such as the source of the protein, prior protein processing treatments, interactions with other components of nutrition, their digestibility, absorption, and utilization in the organism [24]. Soybean is the most important source of low-cost proteins, producing more protein and oil per unit than any other leguminous crop, and it's the most consumed legume worldwide due to its functionality and nutritional value [25]. According to the USDA

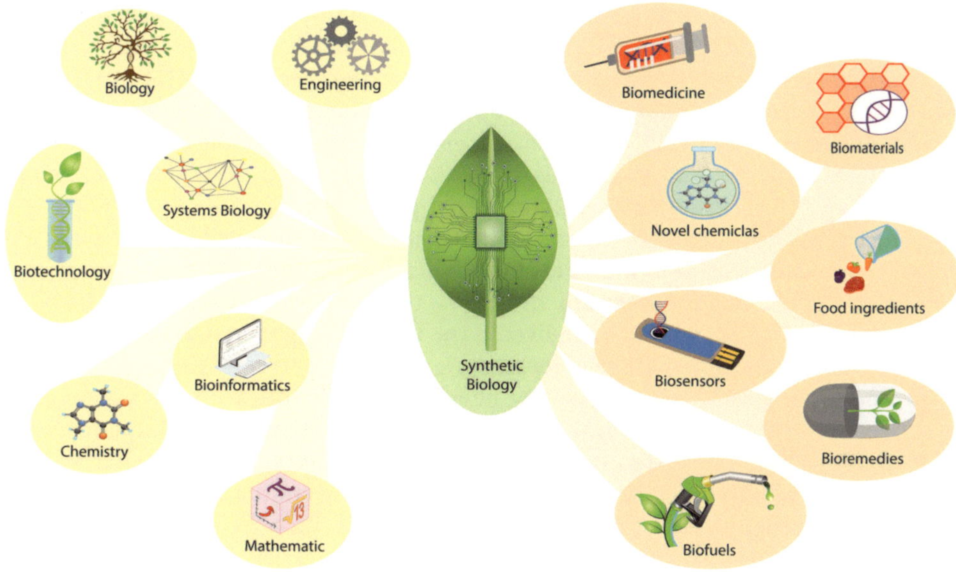

Figure 1.
Schematic representation of the concepts and applications of synthetic biology.

nutritional database, soybean seeds consist of about 36.5% protein, 19.9% lipids, 30% carbohydrates, and 9.3% dietary fiber. Moreover, soybean is the largest source of protein used in livestock, 98% of soybean meal is used for animal feed (poultry, hogs, and cattle mostly), and only 1% is used to produce food consumed by the human population. Recently, the consumption of soybean proteins has increased due to some beneficial effects of their ingestion for human health in treating and preventing various diseases, like cardiovascular diseases and various forms of cancer. Soybean-based agriculture faces several productivity and global sustainability challenges, forcing the rapid adoption of short-to-long-term genetic innovation methods. Thus, the research field of Synthetic Biology is prepared to offer several technological solutions to rapidly improve the development of new soybean cultivars through genetic circuits, biosensors, metabolic engineering, and genome editing techniques.

2. Synthetic biology applied to soybean crops

Plant Synthetic Biology has a wide range of applications in agriculture and the pharmaceutical and energy industry. In agriculture, genetic engineering can be applied to develop new cultivars that are resistant to herbicides, bugs, illness, and drought and can be used to alter the nutritional profile of a cultivar of interest. In the energy and pharmaceutical industry, SynBio allows the production of plant biofabrics for different compounds like remedies, vaccines, biofuel, etc. The major defiance of the Synthetic Biology implementation in agriculture is the time and the extensive outgoing involved in the propagation, transformation, and screening of the superior plants. Although there is an impulse in the plant biotechnology, following the development of new technologies like genome editing based on CRISPR/Cas9 [26], speed breeding [27], key genomes sequencing [28–30], and the SynBio growth as a scientific field [31], the challenge goes on. For example, the huge size of the plant genomes and their polyploidy (wheat, for example, has a hexaploidy genome >15Gb [29]) has so far limited the efficiency of the site-specific genetic manipulations. Besides that, plants usually have fewer direct homolog recombination mechanisms (HDR) than the microbial [32]. However, some works have shown the homolog recombination mechanism mediated by the CRISPR/Cas9 for obtaining genetically modified plants [33]. Thus, despite all prominent challenges for SynBio in Agriculture/Plant Biotechnology, new approaches have arisen to overcome the old problems.

Historically, the speed of productivity increase through classical breeding was not enough to meet the world's demand for food. This lack has required genetic improvement through biotechnology, thus by the 1980s, the development of molecular and plant transformation technologies delivered the first bioengineered genes into plant genomes. Limited yields due to climate stress, changes in pests and pathogens, heat waves, and other weather extremes – the new world reality due to global warming – forces biotechnology and molecular biology to evolve in a new disruptive and fast technology that allows the creation of a new productive and functional crop. Rapid crop improvement must influence naturally grown traits and transformative engineering driven by mechanistic understanding to produce the resilient production systems needed to secure future crops [34]– this will be the new Green Revolution. Currently, the most adopted genetically modified traits are herbicide and insect resistance in crops with large markets, such as soybean [35–37], canola [38] cotton [39], and corn [40, 41].

The challenges of modern agriculture are not restricted to the increment in production to attend to the huge world population growth. Therefore, actually, agriculture faces industry adaptations to digital and genetic technologies, carbon constraints, environmental and animal welfare legislation, the growing focus on "food as medicine" and their ethical production, risks associated with globalization and climate change, a global shift in diet and more discriminating customers in search of a wealthier world [30]. SynBio can overcome these obstacles, like the industry is a pioneer in the use of technological innovations, it is also the biggest beneficiary of advances in SynBio, but in recent years, primary sectors such as agriculture have benefited from this technological evolution.

According to the United States Department of Agriculture, most commercial releases of bioengineered soybeans aim to provide herbicide tolerance, biotic and abiotic factors, improved oil quality, improved yield, and growth (USDA, 2021; CTNBio, 2021). However, many other traits need to be explored, such as superior nutritional contents and the capacity of cultivars to act as biofabrics of industrial products. Primary industries such as agriculture, fisheries, and forestry have benefited directly from advances in genetic research. It is estimated that about half of the 1–3% annual increase in productivity in crops and livestock to date has been driven by improved genetics, with genetic gain rates predicted to more than double with the implementation of emerging molecular technologies [42].

Soybean [*Glycine max* (L.) Merr.], the most consumed legume in the world, originated and domesticated in North-Eastern Asian regions, especially China and Korea, its consumption has been disseminated worldwide since it arrived in American colonies in 1765 [31]. Growing demand for a nutritious, quality, low-cost, low-environment impact, source of protein to feed the growing human population turned soybean into one of the most important global agricultural commodities [43]. Because it is one of the major protein sources, such as food, for animal nutrition, including humans, livestock, pets, and fish, thus soybean seeds of commercial crops contain about 40% protein and about 20% oil [44]. In addition to being a source of protein,

	Target trait	Function target gene and technology	Ref.
Yield			
Plant Architecture	Alters plant height and internode length	CRISPR/Cas9 multiplex knockout of GmLHY / (GmLHY1a, GmLHY1b, GmLHY2a, GmLHY2b)	[58]
	Shorter plastochron length	CRISPR/Cas9 multiplex knockout of GmSPL9/ (GmSPL9a, GmSPL9b, GmSPL9c, GmSPL9d)	[59]
	Control flowering time and plant height	CRISPR/Cas9 multiplex knockout of GmAP1/ (GmAP1a, GmAP1b, GmAP1c, GmAP1d)	[60]
Photoperiod	Expand the regional adaptability	CRISPR/Cas9 multiplex knockout of GmFT2a e GmFT5a	[61]
	phenotypic diversity associated with important agronomic traits	BE base editor of GMFT2a and GmFT4	[62]
	Affect photoperiodic flowering	CRISPR/Cas9 knockout of GmPRR37	[63]
Nutrition and Quality			
Storage Protein		CRISPR/Cas9 knockout:	
	Altered expression pattern of three storage proteins	Glyma.20 g148400	[64]
		Glyma.03 g163500	
		Glyma.19 g164900	
Seed Oil	Increase of oleic acid rate and decrease linoleic acid content	TALENs targeted mutagenesis of GmFAD2- 1A and GmFAD2-1B	[60]
	High oleic, low linoleic and α-linolenic acid	CRISPR/Cas9 multiplex knockout of GmFAD2-1A and GmFAD2-1B	[65]
	Increase in oleic acid, a decrease of linoleic acid, and a higher protein rate	CRISPR/Cas9 multiplex knockout of GmFAD2-1A and GmFAD2-2A	[66]
	Increase of oleic acid rate and decrease linoleic acid content	TALENs directly delivery of mutation in GmFAD3A in fad2-1a fad2-1b soybean plants.	[67]
	High oleic and low linoleic acid rate	CRISPR/Cas9 mediated targeted disruption GmFAD2-2	[68]
Bean Flavor-free	Reduction of beany flavor	CRISPR/Cas9 knockout GmLox1, GmLox2, and GmLox3	[69]
Abiotic Stress Tolerance			
Herbicide Resistance	Chlorsulfuron-resistant	CRISPR/Cas9 knock-in (HDR) of GmAls	[70]
	Fertile transgenic soybean with herbicide tolerance	ZFNs knock-in (NHEJ): aad-1 (2,4D tolerance maker)	[71]
		dgt-28 (glyphosate tolerance maker) and dsm-2 (glufosinate tolerance maker) at GmFAD2-1a locus	
Nitrogen Fixation			
Root Nodulation	gmric1/gmric2 double mutants with increased nodule numbers and gmrdn1-1/1-2/1-3 triple mutant lines with decreased nodulation	CRISPR/Cas9 multiplex knockout of GmRIC1 (Glyma.13 g292300) and GmRIC2 (Glyma.06 g284100), GmRDN1-1 (Glyma.02g279600), GmRDN1-2 (Glyma.14 g035100) and GmRDN1-3 (Glyma.20 g040500)	[72]

Table 1.
Roll some examples of engineered soybean crops with improved traits by synthetic biology. Based on [57].

soy has recently been used to produce biofuels. Currently, the United States, Brazil, and Argentina together produce more than 80% of the world's soybean crop. On the other hand, China is the largest soybean importer in the world, consuming 30% of the world's soybean production [45]. Thus, the global soy market is governed by two major producers (the United States and Brazil, respectively), which produce around 68% of the world's crop, and a major consumer (China) [46].

The agriculture sector has a heavy history of fast improving new transformative techniques innovations, for example, in 2005–2006 the worldwide soybean production was 220.809 million tons, already in 2021–2022 the production reached 385.524 million tons (data from https://www.sopa.org/statistics/world-soybean-production), this increment can be attributed to the development of new disruptive genetic technologies. In the past, mutagenesis is presented as an alternative for classical plant breeding to increase genetic variation in soybean germplasm. Random mutagens techniques were normally used to introduce changes in genes aleatorily, including radiation (such as X-ray), fast neutrons, and gamma rays, chemicals (such as EMS (ethyl methanesulfonate)), and biological mutagenesis (such as T-DNA insertion and transposons) [47]. Although it is hereditary and stable, random mutagenesis demand intensive, specific, time-consuming, and expensive techniques to identify the intended phenotypes in the mutants [48], and in the most cases, it is impossible to locate and obtains the specific allele to determinate function due to the imprecision of the random mutation [43].

In this way, to solve this challenge, in the last 20 years, with the advent of the SynBio, new biotechnologies based on site-directed nucleases (SNDs) or site-specific nucleases (SSNs), such as Zinc Finger Nucleases (ZFNs) [49, 50]. Transcription Activator-Like Effector Nucleases (TALENs) [51, 52] or the Clustered Regularly Interspaced Short Palindromic Repeat (CRISPR) [53, 54] has been developed for generation of site-specific mutagenesis and, as the multiple SDN platforms are a very useful tool, they have been integrated into the new and actual plant breeding programs[55, 56]. Thus, with the rise of the SynBio tools and techniques, there was an exponential acceleration in the speed, quality, and several launches of new commercially interesting crops and, on the other hand, a great reduction in costs and time spent. These new approaches also increased the scope and size of genetic variability available for crop improvement, allowing the creation of diverse new engineered soybean crops for a wider range of traits (**Table 1**).

3. Soybean as a protein source

According to the *Poverty and Shared Prosperity Report*, after the pandemic, the estimative world percentage of people in extreme poverty (characterized by a daily income up to U$1,90) will reach about 9.1% to 9.4% of the world population. This estimate is alarming because, before the pandemic, it was estimated that poverty would fall to 7.9% in 2020. Unfortunately, people in poverty have a low caloric intake and nutritional deficiencies, especially regarding access to micronutrients and essential amino acids [73]. Food proteins, from animal or vegetable sources, supply the essential amino acids important for the construction and maintenance of the basic body structures therefore, they are fundamental for the right physical and mental development of children. Vegetable proteins represent a low-cost source of nutrients and energy, but, in many cases, they are poor in highly digestible essential amino acids. Experts predict a large increase in the world population. Consequently, the demand for plant proteins will increase in the same proportion due to the low cost and lower environmental impact of their production [74].

For centuries, humankind has cultivated seed crops as a protein source, legumes and cereals, as the principal cultivated crops. These cultures currently provide more than 70% of the protein for human consumption. Nonetheless, legumes accumulate higher quantities of protein in contrast to cereal, and among the legume crops, soybean is the one with the highest percentage of proteins [75]. Soybean seeds are a rich source of high-quality digestible proteins and contain all the essential amino acids found in animal proteins, without cholesterol and with a low level of saturated fatty acids [76]. Most soybean seed components, including proteins and peptides, isoflavones, saponins, and protease inhibitors, have been shown to have biological activity [77, 78].

The principal advantages of soybean as a protein source are (1) good balance in amino acids composition, containing all the essential amino acids; (2) presence of components physiologically beneficial to human health, which are shown to lower the cholesterol and reduce the risk of hyperlipidemia and cardiovascular diseases; (3) excellent processing ability, as emulsification, gelling, water- and oil- holding capacity; and (4) excellent nutritional and functional properties of their proteins, for example, solubility, emulsifying, film-forming and foaming properties[79, 80]. Their composition varies according to the variety, location, and conditions of its planting, such as climate and farming practices [76] and besides that, stimuli, like genetic modification, can also modify protein profile, expression, and accumulation rates [81].

Soybean proteins can be classified into four main groups: albumins, globulins, prolamins, and glutelins [82]. Based on solubility patterns, soybean proteins can be classified into two categories, albumins (water-soluble) and globulins (salt solution-soluble), which represent the primary protein type [83]. When separated with ultracentrifugation, two major storage proteins can be identified, glycinin (11S) and β-conglycinin (βCG, 7S), corresponding, respectively, to ~40% and ~ 50% of soy proteins amount [81] corresponding to the largest mass of the soy seed. The sedimentation coefficient values are a more precise pattern for identifying soybean proteins, where larger S (Svedberg units) numbers correspond to a larger protein. By ultracentrifugation under appropriate buffer conditions (0.5 ionic strength and pH 7.6), soybean proteins can be separated into four main groups, 2S, 7S, 11S, and 15S fractions [84]. While the 2S fraction (20% of total proteins) contains the most albumins, such as 2S globulin, cytochrome C, Kunitz trypsin inhibitor, and Bowman-Birk trypsin inhibitor, both the inhibitors are associated with delayed growth in children [84, 85]. The globulins are mainly present in 7S, 11S, and 15S fractions of soybean proteins. Like the 2S fraction, the 7S fraction (40% of total protein) is also highly heterogeneous, containing β-conglycinin, α-amylase, lipoxygenase, and hemagglutinin [86], and the 11S fraction (30% of total protein) consists of only the glycinin, which is the major protein in soybean seeds. The β-conglycinin and the glycinin correspond to the major component of storage proteins. The minor component, typically about 10% of the total proteins, the 15S fraction, maybe be a polymer (possibly a dimer of glycinin) [84].

3.1 Soybean-based meat

The plant-based meat industry is focused on developing burgers, patties, mince, and sausages. Besides that, at the moment, the production of the primary meat cuts, such as steak, is not the primary search worry, due to their structure composition complexity, many groups have progressed in the research of synthetic beef. Between the commercially available plant-based meat, the Beyond Burger (BB) and the Impossible Burger (IB) can be highlighted. The main ingredients of this burger can vary, but usually, it contains soy protein, wheat gluten, egg protein, or milk proteins. In the specific case of BB, non-genetically engineered ingredients such as beetroot are incorporated to give red color to meat analogs and promote the feeling of a "bleeding" meat when cooked [87, 88].

At the beginning of the 21st century, meat analogs entered the mainstream due to the demand for healthy foods, and the worry about the sustainability implications of the consumer's diet continued to increase. Also called meat substitute, meat alternatives, fake or mock meat, and imitation meat, the meat analog is, for definition, the product of replacing the main ingredient with meat [89]. The search for sustainable, healthy, and tasty meat analogs started in the early 1960s [90].

Usually, the meat analogs found in the markets are advanced plant-based meat. The texture and taste are like the conventional meat, which uses plant-derived ingredients that have attributed almost exactly to animal-derived meat and can be indistinguishable from their animal-based equivalents. The biggest challenge for food producers is developing acceptable quality meat analogs because their characteristics depend on the ingredients used. Plant-based meat needs to be unfolded, cross-linked, and realigned to form microscopic and macroscopic fibers [89]. Different techniques are applied to plant-based meat proteins to improve their "meat qualities" to texture, processes such as extrusion, spinning, and simple shear flow have been used [91]. To solidify the structure, following the previous treatment, heating, cooling, drying, or coagulation can be applied [92].

Among the vegetables used to produce plant-based meat, the soybean stands out due to the presence of the leghemoglobin protein that mimics animal myoglobin. The soybean-based meat was the first kind of plant-based meat; in the early 1960s, traditionally, soybean proteins were used as ingredients for food analogs such as tofu and tempeh (fermented soybean cake). These products were made basically by simple processing/fermentation techniques and have been highly consumed in southeast Asia countries for centuries since 965 BCE [93]. In addition to these traditional Asian products, in the mid to late 20th century, the Texturized Vegetable Protein (TVP) was introduced as meat alternative, obtained from the extruded defatted soybean meal soybean proteins concentrates or wheat gluten, made most from soybeans [94, 95].

Meanwhile, Soybean Leghemoglobin (SLH) is used in IB, whose function is not only to provide red-colored liquid mimicking the 'bleeding' of minced meat but can also impart a meat flavor profile in plant-based products of meat [96]. SLH is a close structural ortholog of animal myoglobin that plays a crucial role in the consumption of animal-based meat because, during the cooking, this especially abundant heme protein unfolds and exposes the heme cofactor, responsible for the catalyzes the transformation of the amino acids, nucleotides, vitamins, and sugars naturally present in animal muscle tissue, into a mainly specific and diverse set of flavor and aroma compounds, which combination creates the distinctive and unmistakable meat flavor [96]. SLH, in your turn, acts the parallel role of unfolding under cooking, releasing their heme cofactor to catalyze the transformation of the same ubiquitous biomolecules, isolated from plant sources, into a wide range of compounds that mimic the unique meat flavor and aroma [97].

Besides all these advantages, in many cases, the plant-based meat can present insufficient rates of essential amino acids and trace elements, which can become more challenging to produce plant-based products that perfectly mimic the meat's nutritional values, such as the meat flavor and aroma [89]. The SynBio rises like an efficient approach that will allow the perfect plant-based meat production. By engineering the plant's existing compounds, like SLH, into more "animal-like" compounds and by creating and introducing artificial and synthetic compounds that can improve the meat quality. This development can be achieved by protein and metabolic engineering with the aim to produce the needed ingredients to create a synthetic mimic of plant-based meat.

3.2 Proteomics studies in soybean

Proteomics is a useful tool for examining changes in protein profile generated by the response to various external or internal stimuli such as salt concentration, drought, desiccation, cold, heat, mineral toxicity, mineral deficiency, mutations, and gene introduction or silencing. In addition, proteomics can analyze differences in nutrition-relevant food proteomes, such as identifying marks for the quality of processed foods [98].

It is safe to work with soybean because several materials about their genetic information are available in the literature. A high-quality soybean genome (Wm82) is currently available in databases and is used as a reference for several studies involving omics [12]. Genomes of wild and cultivated soybeans are also available. Recently, the genome and proteome of a highly productive tropical Brazilian species (BRS 537) (EMBRAPA, 2021) are deposited in NCBI in accession GCA_012273815.2. In addition, data from several soybean proteomes under several conditions are also available for prospecting proteins. All these data provide solid information about the genetics and behavior of the crop, facilitating the identification of targets of interest with precision. Species proteomic sets are one of the richest materials for finding these targets, as they indicate that genes present in the genome are being translated into functional proteins.

The fact that soybean is naturally rich in protein content (40%) demonstrates a wide range of protein material to be explored, which is why soybean proteomes are widely studied worldwide. The search for new targets and potential uses is getting bigger every day. Currently, with large-scale proteomics, the isolation of a greater number of proteins is made possible, and much remains to be explored in this crop [99]. In addition to the genetic variability and availability of data on the culture, its importance in producing recombinant proteins for industrial purposes is also due to several other factors. How reducing production costs, easy cultivation in the greenhouse, high protein/biomass ratio, production safety, dosage accuracy allow generating of marketable formulations that may not require purification, low technology sustainability for the production line, reducing the risk of contamination, scalability, and minimal waste production. No other protein expression technology is as efficient as the soybean system [7].

The state-of-the-art studies involving proteomic analysis of soybean seeds in the last 16 years are described in **Table 2**. Since 2005, 50 articles that evaluated the set of proteins in soybean seeds were computed, these studies involve a wide range of proteomic scenarios that come helping in the screening for proteins that can play roles in metabolic pathways for the synthesis of essential amino acids, bioactive proteins, tolerance to various environmental factors, production of sustainable fuels, and others [139], besides providing increasingly solid knowledge about the behavior of the crop under different conditions and stages of development.

Objective	Tissue	Methodology	Ref.
Protein profile in conventional soybean	Mature Seed	2-DE; MALDI-TOF-MS; LC/MS-MS	[100–104]
	All seed stages	2-DE; MALDI-TOF-MS	[105]
	Mature Seed	2-DE; MALDI-TOF-MS	[106–108]
		NanoUPLC-MS	[109]
		SDS-PAGE; MALDI-TOF-MS; ESI-Q-TOF MS	[110]
		NanoHPLC-MS/MS	[111]
		SDS-PAGE; LC-MS/MS	[112]
	Seed in development	2-DE; LC-ESI-MS/MS	[113]
		2-DE; nESI-LC-MS/MS; Sec-MudPIT	[114]
		2-DE; LC-MS-MS	[115]
		MS/MS	[116]
	Seed cotyledons	2-DE; MALDI-TOF-MS	[117]
	Germinating Seeds	2-DE; MALDI-TOF/MS	[118]
		SDS-PAGE; LC-MS/MS	[119]
Protein profile in soybean with high protein content	Mature Seed	2-DE; MALDI-TOF-MS	[120–122]
		2-DE; SDS-PAGE; LC/MS-MS	[123]
		Tandem Mass Tag -TMT; LC-MS/MS	[124]
Protein profile in transgenic soybean	Mature Seed	2-DE; MALDI-TOF-MS	[125]
		SDS-PAGE; 2-D PAGE; MALDI-QTOF MS/MS	[126]
		SDS-PAGE	[127]
	Seed Cotyledons	ITRAQ; LC-MS/MS	[128]
	Germinating Seeds	LC-MS-MS	[129]
Protein profile of mutated soybean	Mature Seed	SDS-PAGE	[130]
		2-DE; MS/MS	[131]
		Tandem Mass Tag -TMT	[132]
Protein profile of Chernobyl area soybean	Mature Seed	2-DE; MS/MS	[133]
Protein profile under High temperature	Mature Seed	DIGE; MALDI-TOF MS; MS/MS	[134]
Protein profile under salt stress	Germinating Seeds	2-DE PAGE; MALDI-TOF-MS	[135]
Protein profile under high temperature and humidity stress	Seed in development	2-DE; MALDI-TOF-MS	[136]
Protein profile under biological fermentation	Meal – SBM	2-DE; LC-MS-MS	[137]
Protein profile under chilling temperature	Germinating Seeds	2-DE; MALDI-TOF/MS	[138]

Table 2.
Studies involving proteomics analysis in soybean seeds.

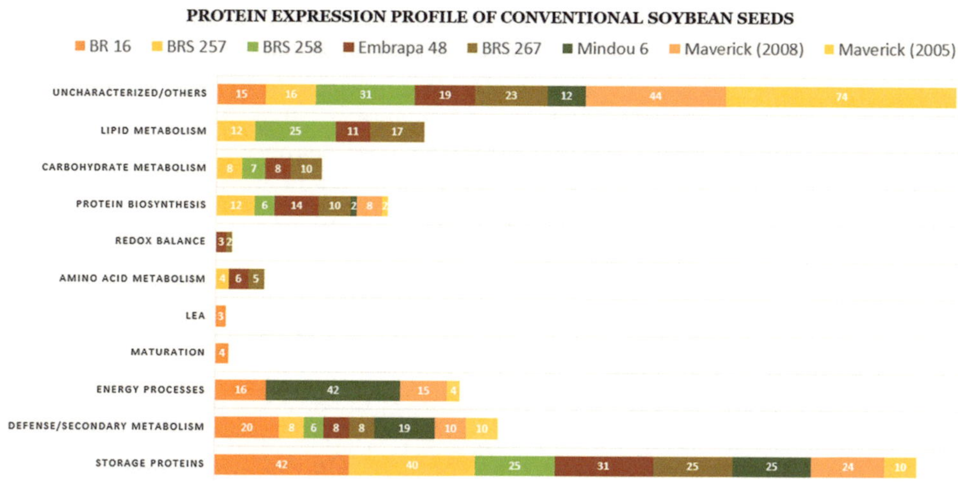

Figure 2.
Examples of conventional soybean seeds are characterized by functional annotation. **Label:** *BR16; BRS 257, BRS 258, Embrapa 48, BRS 267; Mindou 6; Maverick. Numbers represent the percentage of protein in each category.*

Proteomes for studies of nutritional factors are also widely observed in the literature, as they directly influence protein digestibility. Natarajan and team [107] investigated protein and genetic profiles of Kunitz trypsin inhibitors (KTIs) in seeds of 16 different soybean genotypes that included four groups consisting of wild soybean (*Glycine soja*), ancestor of cultivated soybean. They identified that KTI exists as multiple isoforms in soybean. The authors noted that the number and intensity of proteins between wild and cultivated genotypes varied. These data suggest that the greatest variation in protein profiles occurred between wild and cultivated soybean genotypes rather than between genotypes in the same group. However, genetic variation of genes related to KTI1, KTI2, and KTI3 was detected within and between groups (**Figure 2**).

A proteomic study on developing soybean seeds (*G. max* var. Mindou 6) showed 48 differentially expressed proteins [109]. Among these proteins, 25% were related to protein destination and storage, 42% to energy and metabolism, 15% to disease/defense, 6% to transporters, 4% to secondary metabolism, 4% to transcription, 2% to the synthesis of proteins and 2% for cell growth/division. It was observed that with the maturity of the seeds, the number of proteins varied, some decreased, and others increased their concentrations. The sucrose-binding protein (SBP) 2 precursors, which can contribute to improving the digestibility, nutritional value, and food quality of seeds, were increased with maturity (**Figure 2**).

4. Soybean functional proteins

The function proprieties of proteins are physicochemical aspects that influence the behavior of proteins in food preparations, for example. Based on epidemiological studies, soybean consumption has been associated, for many years ago, with several potential health benefits in reducing chronic diseases such as insulin resistance/type II diabetes, cardiovascular disease, obesity, certain types of cancer, and immune disorders [76]. Nowadays, it is already proven that soybeans are a rich source of phytochemicals, and many of these compounds have important benefits to human and animal health. Among these phytochemicals, phytoestrogens, mainly isoflavones (genistein and daidzein) and lignans, usually get more attention [140]. Nonetheless, in recent years, those physiological functions have been attributed to soybean proteins intact or more commonly bioactive and functional peptides derivate from soybean processing. These bioactive peptides are small protein fragments produced by enzymatic hydrolysis, fermentation, food processing, and gastrointestinal digestion of larger soybean proteins [141], showing multiple beneficial metabolic effects [78, 142].

The soybean seeds contain ~38% protein, ~18% oil, ~30% carbohydrates, ~14% moisture, ash, and secondary metabolites, are a considerable source of vitamins (A, thiamin, riboflavin, pyridoxine, and folic acid) and minerals (Fe, Zn, Mg, K, Ca, Mn, and Se), phytoestrogens and fibers, as well as a widely important source of protein [143]. More important than quantity is the quality of the proteins found in soybean, all eight essential amino acids, which are necessary to human nutrition, but are not produced by the human body, are found in soybean. While the sulfur-containing amino acids (methionine and cysteine) are a limiting factor with a chemical score of 47, compared to 100, as ideal protein for human nutrition [144], soybean proteins are an extraordinary source of lysine.

The two major storage proteins, glycinin (11S) and β-conglycinin (βCG, 7S), are considered naturally bio-inactive, but different ratios of βCG and glycinin may have other nutritional and physiological effects. However, many bioactive peptides are inert while still constituting a larger protein but become activated when released from the original structure by gastrointestinal digestion, enzyme and food processing, or fermentation. These peptides common are 2 to 20 amino acids in length and can be absorbed by the human intestine, falling into the bloodstream, where they can exercise systemic or local physiological effects in target tissues [76]. It has shown a difference in the human intestinal absorption of 11S peptides compared to 11S globulin or amino acids mixture, being that the 11S peptides take to a significantly greater increase in venous blood amino acids concentration. This difference is more notable for aromatic and branched-chain amino acids, which could indicate that hydrolyzed soybean proteins are faster and more efficiently absorbed in the human intestine [145].

Thus, in the last decade, the focus of research on the functionality of soy-based foods has shifted from proteins to bioactive peptides. Moreover, numerous bioactive soybean peptides have been identified with widespread beneficial physiological effects, such as anti-diabetic, anti-cancer, hypotensive, anti-inflammatory, antioxidant, and lipid-lowering (hypocholesterolemic, hypotriglyceridemic, anti-obesity) (**Table 3**) [76]. Among these is the lunasin, one chemopreventive peptide that consists of 43 amino acids residues with a C-terminal of nine aspartic acid and cell adhesion motif, enabling the binding to non-acetylated H3 and H4 histones, preventing their acetylation, which gives they the anti-carcinogenic activity [160, 169].

Lectin (hemagglutinin or agglutinin), a highly specific carbohydrate-binding protein with an important role in biological recognition, can be founded in soybean seed, ~0.2–1% of total protein [171]. Trypsin and protease inhibitors encompass several proteins and peptides, such as the Bowman-Birk protease inhibitor (BBI), the Kunitz trypsin inhibitor (KTI), and lunasin. BBI is a small protein of ~10 kDa, belonging to the serine protease inhibitor family, that mightily interacts with trypsin and/or chymotrypsin and strongly inhibits their enzymatic function [172]. The soybean KTI consists of a protein of ~20 kDa, with a single polypeptide chain cross-linked by two disulfide bridges, which inhibits trypsin and, at a lesser rate, chymotrypsin [171].

Soybean Protein Source	Bioactive Peptide	Properties	Ref.
βCG	YVVNPDNDEN	Hypocholesterolemic	[146, 147]
	YVVNPDNNEN		
	LAIPVNKP	ACE inhibition	[77, 148]
	LPHF		
Glycinin	IAVPGEVA	Hypocholesterolemic Anti-diabetic	[147, 149–151]
	IAVPTGVA		[147, 149, 152, 153]
	LPYP		[146, 147, 150, 154, 155]
	VLIVP	ACE inhibition	[77]
	SPYP		
	WL		
	SFGVAE	Hypocholesterolemic	[150]
	HCQRPR	Phagocytosis stimulatory peptide	[155–157]
	QRPR		
Lunasin	SKWQHQQDSCRKQKQ GVNLTPCEKHIMEKIQ GRGDDDDDDDDD	Antioxidative Anti-inflammatory Anti-cancer Hypocholesterolemic	[77, 141, 157–159]
Bowman-Birk Inhibitor		Anti-cancer Proteinase inhibition Chemoprevention	[160–168]
Soybean Protein	YVVFK; IPPGVPYWT; PNNKFPQ; NWGLPV; TPRVF	Hypotensive	[157, 169, 170]
	WGAPSL; VAWWMY; FVVNATSN	Hypocholesterolemic	[77, 155, 157]

Table 3.
Some examples of the principal soybean bioactive peptides and their properties. Adapted from [76].

Both soybean lectins and protease inhibitors usually have been classified as antinutrients because they may lower the nutritional value of soybean. Therefore, their consumption has shown preventing effects against many diseases, such as cancer. The BBI already demonstrated an anti-inflammatory function that prevents the development of cancer and coronary diseases [171, 173].

In addition to these proteins with known functions and characteristics, several aspects of human health are attributed to soybean proteins. The proteins of many animal species show a high-fat content that can be implicated in increasing blood cholesterol, triglycerides, and Low-Density Lipoprotein (LDL-c), which stimulated the search for other protein sources [143]. Studies about the soybean protein consumption effect on subjects with hypercholesterolemia concluded that it could reduce total blood cholesterol, triglycerides, and LDL-c levels [174]. In 1999, the US Food and Drug Administration (FDA) approved the label for foods containing soybean proteins as protection against coronary heart disease. Its potential role in reducing risk factors for cardiovascular disease is one of the highest causes of death worldwide [143].

Soybean proteins can positively impact the angiotensin-converting enzyme (ACE) activity, acting as ACE inhibitor peptides that can be released enzymatically from a larger protein precursor *in vivo* during gastrointestinal digestion and *in vitro* by food processing. These peptides can reduce blood pressure by limiting the effects of ACE II in vasoconstriction and improving the vasodilatory effects of bradykinin, a potent endothelium-dependent vasodilator and mild diuretic [175, 176].

4.1 Synthetic biology applied to functional proteins

Being a wide broad domain with many new and emerging fields, SynBio can give the necessary tools to face many of the challenges of the modern world. The inherent complexity and redundancy of the plant genome represent a problem to be solved by SynBio, too, for this, it applies the most important engineering principles: decoupling, abstraction, and standardization [177]. Decoupling is the simplification of complex problems into smaller ones that can be solved individually. Abstraction divides the topology of information into hierarchical levels, allowing limited and selected data to be exchanged between levels. Standardization is used to determine and characterize orthogonal parts and standardized conditions for testing. Engineering a biological system is one method to manipulate information, process chemicals, provide food, constructing materials, and help to maintain or enhance human health and our environment [178].

Plants are the great chemist of nature, being a perfect platform for SynBio approaches. The rise of SynBio broadened the horizons of plant engineering. However, as SynBio is still dependent on existing transformation techniques, the major challenge to implementing SynBio in the production of modified interest plants is the time and expense involved in the propagation, transformation, and screening of higher plants [39]. But with, the application of the SynBio approach in other fields of functional, protein production, like food production systems, will save water resources, improve land-use efficiency, and avoid the use of pesticides and fertilizers [179]. In addition, the SynBio based functional proteins and food manufacturing systems are less affected by uncontrollable environmental factors and are easier to carry out according to high-quality standards and scale at an industrial level. By constructing cell-based food factories, foods such as plant-based meat analogs, animal-free bioengineered milk, and sugar substitutes can be created from completely renewable resources [179].

Soybean proteins have been widely used to produce many protein-based food formulations due to their excellent nutritional and functional properties. Physical modification, chemical modification, and enzymatic modification have been applied to improve the functional aspects of soybean proteins [80]. But to overcome the future food and global climate challenges, just improvement in processing techniques is not enough, it is a critical step in the development of new soybean varieties that can be able to meet both the demands of the consumer market and the producers. In many cases, the traditional plant breeding cannot attend to this demand, proving to be necessary classical and, mainly, new genetic engineering techniques to add value to the soybean crop, such as reduction of allergens and antinutrients factors along with the increase of quantity and quality proteins, oil, and carbohydrates [98].

The first option for the improved soybean crops in a "functional way" is the generation of genetically modified varieties. For example, the genetic engineering techniques, such as CRISPR-Cas9, represents a great opportunity to improve the nutritional value of soybean-based foods, for instance, by developing carotenoid-enriched functional crops and oilseeds crops with elevated levels of omega 3 fatty acid [39]. Besides the increment/silencing of the expression of target genes, soybean can also represent an important vehicle for the creation of bio fabric, to produce a wide range of bioactive compounds by the heterology expression of desirable genes or by the metabolic engineering of the plant. For example, based on classical genetic transformation techniques, soybean has already been used to produce functional human growth hormone and coagulation factor IX [180]; and anti-HIV Cyanovirin-N [181].

5. Conclusion

Soybean seeds are an excellent source of proteins, as they provide all essential amino acids and a promising source of biologically active proteins/peptides with a wide range of effects such as anti-diabetic, anti-hypertensive, anti-cancer, antioxidant, anti-inflammatory, hypolipidemic, immunostimulatory, and neuromodulatory properties. However, soybean has a low content of sulfur amino acids, and many consumers may exhibit allergenic and antinutritional reactions due to the presence of certain proteins and peptides, such as protease inhibitors. But the same inhibitors, like KTI and BBI, show anti-cancer and anti-inflammatory activity, respectively. Thus, the future of soybean-based foods is not just about the classic plant breeding and/or new processing techniques to remove undesirable characters because they may be interesting for other applications. SynBio rises as a modern solution to create a more "consumable" soybean, through protein and metabolic engineering, to remove just the exact allergenic and antinutritional factors. In addition, soybean can be a great platform to create biofabrics combined with SynBio techniques.

Author details

Lilian Hasegawa Florentino[1]*, Rayane Nunes Lima[1] and Mayla D.C. Molinari[2]

1 Embrapa Genetic Resources and Biotechnology, Brasilia, DF, Brazil

2 Arthur Bernardes Foundation, Embrapa Soybean, Londrina, PR, Brazil

*Address all correspondence to: lilian.florentino@embrapa.br

References

[1] Solomon LD. In: Solomon LD, editor. Synthetic Biology: Science, Business, and Policy [Internet]. 1st ed. New York: Routledge; 2017. pp. 01-174. Available from: https://www.taylorfrancis.com/books/9781351487245

[2] Hutchison CA, Chuang RY, Noskov VN, Assad-Garcia N, Deerinck TJ, Ellisman MH, et al. Design and synthesis of a minimal bacterial genome. Science 2016;351(6280), aad6253.

[3] Dymond J, Boeke J. The saccharomyces cerevisiae SCRaMbLE system and genome minimization. Bioengineered Bugs. 2012;**3**(3):168-171

[4] He FJ, Chen JQ. Consumption of soybean, soy foods, soy isoflavones and breast cancer incidence: Differences between Chinese women and women in Western countries and possible mechanisms. Food Science and Human Wellness. 2013;**2**(3-4):146-161

[5] Herman EM. Soybean seed proteome rebalancing. Frontiers in Plant Science. 2014;**5**:437

[6] Herman EM, Schmidt MA. Towards using biotechnology to modify soybean seeds as protein bioreactors. Recent Advancements in Gene Expression and Enabling Technologies in Crop Plants. 2015:193-212. Available from: https://arizona.pure.elsevier.com/en/publications/towards-using-biotechnology-to-modify-soybean-seeds-as-protein-bi

[7] Bost K, Piller K. Protein expression systems: Why soybean seeds? In: Soybean - Molecular Aspects of Breeding. London: IntechOpen; 2011

[8] Kumar V, Vats S, Kumawat S, Bisht A, Bhatt V, Shivaraj SM, et al. Omics advances and integrative approaches for the simultaneous improvement of seed oil and protein content in soybean (*Glycine max* L.). Critical Reviews in Plant Sciences [Internet]. 2021;**40**(5):398-421. Available from: https://scholars.ttu.edu/en/publications/omics-advances-and-integrative-approaches-for-the-simultaneous-im

[9] Yang Y, Saand MA, Huang L, Abdelaal WB, Zhang J, Wu Y, et al. Applications of multi-omics technologies for crop improvement. Frontiers in Plant Science. 2021;**3**(12):1846

[10] Wang X, Sakata K, Komatsu S. An integrated approach of proteomics and computational genetic modification effectiveness analysis to uncover the mechanisms of flood tolerance in soybeans. International Journal of Molecular Sciences. 2018;**19**(5):1301. Available from: https://www.mdpi.com/1422-0067/19/5/1301/htm

[11] Zhao B, Zhang S, Yang W, Li B, Lan C, Zhang J, et al. Multi-omic dissection of the drought resistance traits of soybean landrace LX. Plant, Cell & Environment [Internet]. 2021;**44**(5):1379-1398. Available from: https://onlinelibrary.wiley.com/doi/full/10.1111/pce.14025

[12] Schmutz J, Cannon SB, Schlueter J, Ma J, Mitros T, Nelson W, et al. Genome sequence of the palaeopolyploid soybean. Nature. 2010;**463**(7278):178-183. Available from: https://www.nature.com/articles/nature08670

[13] Severin AJ, Woody JL, Bolon YT, Joseph B, Diers BW, Farmer AD, et al. RNA-Seq Atlas of glycine max: A guide to the soybean transcriptome.

BMC Plant Biology [Internet]. 2010;**10**(1):1-16. Available from: https://bmcplantbiol.biomedcentral.com/articles/10.1186/1471-2229-10-160

[14] Jones SI, Gonzalez DO, Vodkin LO. Flux of transcript patterns during soybean seed development. BMC Genomics [Internet]. 2010;**11**(1):1-15. Available from: https://bmcgenomics.biomedcentral.com/articles/10.1186/1471-2164-11-136

[15] Jones SI, Vodkin LO. Using RNA-Seq to profile soybean seed development from fertilization to maturity. PLOS ONE [Internet]. 2013;**8**(3):e59270. Available from: https://journals.plos.org/plosone/article?id=10.1371/journal.pone.0059270

[16] Du J, Wang S, He C, Zhou B, Ruan YL, Shou H. Identification of regulatory networks and hub genes controlling soybean seed set and size using RNA sequencing analysis. Journal of Experimental Botany [Internet]. 2017;**68**(8):1955-1972. Available from: https://academic.oup.com/jxb/article/68/8/1955/2901051

[17] Qi Z, Zhang Z, Wang Z, Yu J, Qin H, Mao X, et al. Meta-analysis and transcriptome profiling reveal hub genes for soybean seed storage composition during seed development. Plant, Cell & Environment [Internet]. 2018;41(9):2109-2127. Available from: https://onlinelibrary.wiley.com/doi/full/10.1111/pce.13175

[18] Lee SH, Lee S, Lee SH, Kim HJ, Singh D, Lee CH. Integrated metabolomics and volatolomics for comparative evaluation of fermented soy products. Food. 2021, 10(11):2516. Available from: https://www.mdpi.com/2304-8158/10/11/2516/htm

[19] Liu J, Gunapati S, Mihelich NT, Stec AO, Michno JM, Stupar RM. Genome editing in soybean with CRISPR/Cas9. Methods in Molecular Biology [Internet]. 2019;**1917**:217-234. Available from: https://link.springer.com/protocol/10.1007/978-1-4939-8991-1_16

[20] Bao A, Tran LSP, Cao D. CRISPR/Cas9-based gene editing in soybean. Methods in Molecular Biology [Internet]. 2020;**2107**:349-364. Available from: https://link.springer.com/protocol/10.1007/978-1-0716-0235-5_19

[21] Wang T, Xun H, Wang W, Ding X, Tian H, Hussain S, et al. Mutation of GmAITR genes by CRISPR/Cas9 genome editing results in enhanced salinity stress tolerance in soybean. Frontiers in Plant Science. 2021;**12**:2752

[22] Nguyen CX, Dohnalkova A, Hancock CN, Kirk KR, Stacey G, Stacey MG. Critical role for uricase and xanthine dehydrogenase in soybean nitrogen fixation and nodule development. The Plant Genome [Internet]. 2021:e20171. Available from: https://onlinelibrary.wiley.com/doi/full/10.1002/tpg2.20172

[23] Ge H, Wang X, Xu J, Lin H, Zhou H, Hao T, et al. A CRISPR/Cas12a-mediated dual-mode electrochemical biosensor for polymerase chain reaction-free detection of genetically modified soybean. Analytical Chemistry [Internet]. 2021;**93**(44):14885-14891. Available from: https://pubs.acs.org/doi/full/10.1021/acs.analchem.1c04022

[24] Vagadia BH, Vanga SK, Raghavan V. Inactivation methods of soybean trypsin inhibitor – A review. In: Trends in Food Science and Technology. Vol. 64. Europe: Elsevier Ltd; 2017. pp. 115-125

[25] Song XP, Hansen MC, Potapov P, Adusei B, Pickering J, Adami M, et al. Massive soybean expansion in South America since 2000 and implications

for conservation. Nature Sustainability [Internet]. 2021;**4**(9):784-792. Available from: https://www.nature.com/articles/s41893-021-00729-z

[26] Puchta H. Using CRISPR/Cas in three dimensions: Towards synthetic plant genomes, transcriptomes and epigenomes. The Plant Journal. 2016;**87**(1):5-15

[27] Watson A, Ghosh S, Williams MJ, Cuddy WS, Simmonds J, Rey MD, et al. Speed breeding is a powerful tool to accelerate crop research and breeding. Nature Plants. 2018;**4**(1):23-29. Available from: https://www.nature.com/articles/s41477-017-0083-8

[28] Jarvis DE, Ho YS, Lightfoot DJ, Schmöckel SM, Li B, Borm TJA, et al. The genome of Chenopodium quinoa. Nature. 2017;**542**(7641):307-312. Available from: https://www.nature.com/articles/nature21370

[29] Zimin Av, Puiu D, Hall R, Kingan S, Clavijo BJ, Salzberg SL. The first near-complete assembly of the hexaploid bread wheat genome, Triticum aestivum. GigaScience [Internet]. 2017;6(11):1. Available from: /pmc/articles/PMC5691383/

[30] Zimin Av, Puiu D, Luo MC, Zhu T, Koren S, Marçais G, et al. Hybrid assembly of the large and highly repetitive genome of Aegilops tauschii, a progenitor of bread wheat, with the MaSuRCA mega-reads algorithm. Genome Research [Internet]. 2017;27(5):787-792. Available from: /pmc/articles/PMC5411773/

[31] Liu W, Stewart CN. Plant synthetic biology. Trends in Plant Science. 2015;**20**(5):309-317

[32] Gao C. The future of CRISPR technologies in agriculture. Nature Reviews Molecular Cell Biology. 2018;19(5):275-276.

[33] Sun Y, Zhang X, Wu C, He Y, Ma Y, Hou H, et al. Engineering herbicide-resistant rice plants through CRISPR/Cas9-mediated homologous recombination of acetolactate synthase. Molecular Plant. 2016;**9**:628-631

[34] Bailey-Serres J, Parker JE, Ainsworth EA, Oldroyd GED, Schroeder JI. Genetic strategies for improving crop yields. Nature. 2019;**575**(7781):109-118. Available from: https://www.nature.com/articles/s41586-019-1679-0

[35] Stewart CN, Adang MJ, All JN, Boerma HR, Cardineau G, Tucker D, et al. Genetic transformation, recovery, and characterization of fertile soybean transgenic for a synthetic Bacillus thuringiensis cryIAc gene. Plant Physiology [Internet]. 1996;**112**(1):121-129. Available from: https://pubmed.ncbi.nlm.nih.gov/8819322/

[36] Castle LA, Siehl DL, Gorton R, Patten PA, Chen YH, Bertain S, et al. Discovery and directed evolution of a glyphosate tolerance gene. Science (New York, NY) [Internet]. 2004;**304**(5674):1151-1154. Available from: https://pubmed.ncbi.nlm.nih.gov/15155947/

[37] Behrens MR, Mutlu N, Chakraborty S, Dumitru R, Wen ZJ, LaVallee BJ, et al. Dicamba resistance: Enlarging and preserving biotechnology-based weed management strategies. Science (New York, NY) [Internet]. 2007;**316**(5828):1185-1188. Available from: https://pubmed.ncbi.nlm.nih.gov/17525337/

[38] Stewart CN, Adang MJ, All JN, Raymer PL, Ramachandran S, Parrott WA. Insect control and dosage effects in transgenic canola containing a

synthetic Bacillus thuringiensis cryIAc gene. Plant Physiology [Internet]. 1996;112(1):115. Available from: /pmc/articles/PMC157930/?report=abstract

[39] Perlak FJ, Deaton RW, Armstrong TA, Fuchs RL, Sims SR, Greenplate JT, et al. Insect resistant cotton plants. Bio/Technology. 1990;8(10):939-943. Available from: https://www.nature.com/articles/nbt1090-939

[40] Koziel MG, Beland GL, Bowman C, Carozzi NB, Crenshaw R, Crossland L, et al. Field performance of elite transgenic maize plants expressing an insecticidal protein derived from Bacillus thuringiensis. Bio/Technology. 1993;11(2):194-200. Available from: https://www.nature.com/articles/nbt0293-194

[41] Jansens S, van Vliet A, Dickburt C, Buysse L, Piens C, Saey B, et al. Transgenic corn expressing a Cry9C insecticidal protein from Bacillus thuringiensis protected from European Corn Borer Damage. Crop Science [Internet]. 1997;37(5):1616-1624. Available from: https://onlinelibrary.wiley.com/doi/full/10.2135/cropsci1997.0011183X003700050035x

[42] Goold HD, Wright P, Hailstones D. Emerging opportunities for synthetic biology in agriculture. Genes [Internet]. 2018;9(7):341. Available from: /pmc/articles/PMC6071285/

[43] Xu H, Zhang L, Zhang K, Ran Y. Progresses, challenges, and prospects of genome editing in soybean (Glycine max). Frontiers in Plant Science. 2020;**11**:1593

[44] Singh RJ. Botany and cytogenetics of soybean. In: The Soybean Genome. Compendium of Plant Genomes. Cham: Springer; 2017. pp. 11-40. Available from: https://link.springer.com/chapter/10.1007/978-3-319-64198-0_2

[45] Hart C. The economic evolution of the soybean industry. In: The Soybean Genome. Compendium of Plant Genomes. Cham: Springer; 2017. pp. 1-9. Available from: https://link.springer.com/chapter/10.1007/978-3-319-64198-0_1

[46] Gale F, Valdes C, Ash M. Interdependence of China, United States, and Brazil in Soybean Trade [Internet]. USA: USDA; 2019. Available from: https://www.ers.usda.gov/webdocs/outlooks/93390/ocs-19f-01.pdf?v=226.2

[47] O'Rourke JA, Graham MA, Whitham SA. Soybean functional genomics: Bridging the genotype-to-phenotype gap. In: The Soybean Genome. Compendium of Plant Genomes. Cham: Springer; 2017. pp. 151-170. Available from: https://link.springer.com/chapter/10.1007/978-3-319-64198-0_10

[48] Shiming L, Lakhssassi N, Zhou Z, Colantonio V, Kassem MA, Meksem K. Soybean genomic libraries, TILLING, and genetic resources. In: The Soybean Genome. Compendium of Plant Genomes. 2017. pp. 131-149. Available from: https://link.springer.com/chapter/10.1007/978-3-319-64198-0_9

[49] Kim YG, Cha J, Chandrasegaran S. Hybrid restriction enzymes: Zinc finger fusions to Fok I cleavage domain. Proceedings of the National Academy of Sciences [Internet]. 1996;**93**(3):1156-1160. Available from: https://www.pnas.org/content/93/3/1156

[50] Bibikova M, Carroll D, Segal DJ, Trautman JK, Smith J, Kim Y-G, et al. Stimulation of homologous recombination through targeted cleavage by chimeric nucleases. Molecular and Cellular Biology [Internet]. 2001;**21**(1):289-297. Available from: https://journals.asm.org/doi/abs/10.1128/MCB.21.1.289-297.2001

[51] Christian M, Cermak T, Doyle EL, Schmidt C, Zhang F, Hummel A, et al. Targeting DNA double-strand breaks with TAL effector nucleases. Genetics [Internet]. 2010;**186**(2):757-761. Available from: https://academic.oup.com/genetics/article/186/2/757/6063632

[52] Miller JC, Tan S, Qiao G, Barlow KA, Wang J, Xia DF, et al. A TALE nuclease architecture for efficient genome editing. Nature Biotechnology. 2010;**29**(2):143-148. Available from: https://www.nature.com/articles/nbt.1755

[53] Doudna JA, Charpentier E. The new frontier of genome engineering with CRISPR-Cas9. Science [Internet]. 2014;**346**(6213):1258096. Available from: http://science.sciencemag.org/

[54] Ran FA, Hsu PD, Wright J, Agarwala V, Scott DA, Zhang F. Genome engineering using the CRISPR-Cas9 system. Nature Protocols. 2013;**8**(11):2281-2308

[55] Zhang Y, Malzahn AA, Sretenovic S, Qi Y. The emerging and uncultivated potential of CRISPR technology in plant science. Nature Plants. 2019;**5**(8):778-794. Available from: https://www.nature.com/articles/s41477-019-0461-5

[56] Chen K, Wang Y, Zhang R, Zhang H, Gao C. CRISPR/Cas genome editing and precision plant breeding in agriculture. Annual Reviews of Plant Biology. 2019;**70**:667-697. Available from: https://www.annualreviews.org/doi/abs/10.1146/annurev-arplant-050718-100049

[57] Adamczyk JJ, Adams LC, Hardee DD. Field efficacy and seasonal expression profiles for terminal leaves of single and double Bacillus thuringiensis toxin cotton genotypes. Journal of economic entomology [Internet]. 2001;**94**(6):1589-1593. Available from: https://pubmed.ncbi.nlm.nih.gov/11777069/

[58] Cheng Q, Dong L, Su T, Li T, Gan Z, Nan H, et al. CRISPR/Cas9-mediated targeted mutagenesis of GmLHY genes alters plant height and internode length in soybean. BMC Plant Biology [Internet]. 2019;**19**(1):1-11. Available from: https://pubmed.ncbi.nlm.nih.gov/31852439/

[59] Bao A, Chen H, Chen L, Chen S, Hao Q, Guo W, et al. CRISPR/Cas9-mediated targeted mutagenesis of GmSPL9 genes alters plant architecture in soybean. BMC Plant Biology [Internet]. 2019;**19**(1):1-12. Available from: https://pubmed.ncbi.nlm.nih.gov/30961525/

[60] Chen L, Nan H, Kong L, Yue L, Yang H, Zhao Q, et al. Soybean AP1 homologs control flowering time and plant height. Journal of Integrative Plant Biology [Internet]. 2020;**62**(12):1868-1879. Available from: https://pubmed.ncbi.nlm.nih.gov/32619080/

[61] Cai Y, Wang L, Chen L, Wu T, Liu L, Sun S, et al. Mutagenesis of GmFT2a and GmFT5a mediated by CRISPR/Cas9 contributes for expanding the regional adaptability of soybean. Plant Biotechnology Journal [Internet]. 2020;**18**(1):298-309. Available from: https://pubmed.ncbi.nlm.nih.gov/31240772/

[62] Cai Y, Chen L, Zhang Y, Yuan S, Su Q, Sun S, et al. Target base editing in soybean using a modified CRISPR/Cas9 system. Plant Biotechnology Journal [Internet]. 2020;**18**(10):1996-1998. Available from: https://onlinelibrary.wiley.com/doi/full/10.1111/pbi.13386

[63] Wang L, Sun S, Wu T, Liu L, Sun X, Cai Y, et al. Natural variation and CRISPR/Cas9-mediated mutation in GmPRR37 affect photoperiodic flowering and contribute to regional adaptation of soybean. Plant Biotechnology Journal

[Internet]. 2020;**18**(9):1869-1881. Available from: https://pubmed.ncbi.nlm.nih.gov/31981443/

[64] Li C, Nguyen V, Liu J, Fu W, Chen C, Yu K, et al. Mutagenesis of seed storage protein genes in soybean using CRISPR/Cas9. BMC Research Notes [Internet]. 2019;**12**(1):1-7. Available from: https://pubmed.ncbi.nlm.nih.gov/30917862/

[65] Do PT, Nguyen CX, Bui HT, Tran LTN, Stacey G, Gillman JD, et al. Demonstration of highly efficient dual gRNA CRISPR/Cas9 editing of the homeologous GmFAD2-1A and GmFAD2-1B genes to yield a high oleic, low linoleic and α-linolenic acid phenotype in soybean. BMC Plant Biology [Internet]. 2019;**19**(1):1-14. Available from: https://pubmed.ncbi.nlm.nih.gov/31307375/

[66] Wu N, Lu Q, Wang P, Zhang Q, Zhang J, Qu J, et al. Construction and analysis of GmFAD2-1A and GmFAD2-2A soybean fatty acid desaturase mutants based on CRISPR/Cas9 technology. International Journal of Molecular Sciences [Internet]. 2020;**21**(3):1104. Available from: https://pubmed.ncbi.nlm.nih.gov/32046096/

[67] Demorest ZL, Coffman A, Baltes NJ, Stoddard TJ, Clasen BM, Luo S, et al. Direct stacking of sequence-specific nuclease-induced mutations to produce high oleic and low linolenic soybean oil. BMC Plant Biology [Internet]. 2016;**16**(1):1-8. Available from: https://bmcplantbiol.biomedcentral.com/articles/10.1186/s12870-016-0906-1

[68] Al Amin N, Ahmad N, Wu N, Pu X, Ma T, Du Y, et al. CRISPR-Cas9 mediated targeted disruption of FAD2-2 microsomal omega-6 desaturase in soybean (Glycine max.L). BMC Biotechnology [Internet]. 2019;**19**(1):1-10. Available from: https://bmcbiotechnol.biomedcentral.com/articles/10.1186/s12896-019-0501-2

[69] Wang J, Kuang H, Zhang Z, Yang Y, Yan L, Zhang M, et al. Generation of seed lipoxygenase-free soybean using CRISPR-Cas9. The Crop Journal. 2020;**8**(3):432-439

[70] Li Z, Liu Z-B, Xing A, Moon BP, Koellhoffer JP, Huang L, et al. Cas9-guide RNA directed genome editing in soybean. Plant Physiology [Internet]. 2015;169(2):960-970. Available from: https://pubmed.ncbi.nlm.nih.gov/26294043/

[71] Bonawitz ND, Ainley WM, Itaya A, Chennareddy SR, Cicak T, Effinger K, et al. Zinc finger nuclease-mediated targeting of multiple transgenes to an endogenous soybean genomic locus via non-homologous end joining. Plant Biotechnology Journal [Internet]. 2019;17(4):750-761. Available from: https://pubmed.ncbi.nlm.nih.gov/30220095/

[72] Bai M, Yuan J, Kuang H, Gong P, Li S, Zhang Z, et al. Generation of a multiplex mutagenesis population via pooled CRISPR-Cas9 in soya bean. Plant Biotechnology Journal [Internet]. 2020;18(3):721-731. Available from: https://pubmed.ncbi.nlm.nih.gov/31452351/

[73] Soria-Hernández CG, Serna-Saldívar SO, Chuck-Hernández C. Comparison of physicochemical, functional and nutritional properties between proteins of soybean and a novel mixture of soybean-maize. Applied Sciences. 2020;10(19):6998. Available from: https://www.mdpi.com/2076-3417/10/19/6998/htm

[74] Roell MS, Zurbriggen MD. The impact of synthetic biology for future agriculture and nutrition. Current

Opinion in Biotechnology. 2020;**1**(61):102-109

[75] Krishnan HB. Engineering soybean for enhanced sulfur amino acid content. In: Crop Science. New York, NY: Crop Science Society of America; 2005. pp. 454-461

[76] Chatterjee C, Gleddie S, Xiao CW. Soybean bioactive peptides and their functional properties. Nutrients. 2018;**10**(9):1211. Available from: https://www.mdpi.com/2072-6643/10/9/1211/htm

[77] Wang W, Dia VP, Vasconez M, de Mejia EG, Nelson RL. Analysis of soybean protein-derived peptides and the effect of cultivar, environmental conditions, and processing on lunasin concentration in soybean and soy products. Journal of AOAC International. 2008;**91**(4):936-946

[78] Omoni AO, Aluko RE. Soybean foods and their benefits: Potential mechanisms of action. Nutrition Reviews [Internet]. 2005;**63**(8):272-283. Available from: https://pubmed.ncbi.nlm.nih.gov/16190314/

[79] Nishinari K, Fang Y, Guo S, Phillips GO. Soy proteins: A review on composition, aggregation and emulsification. Food Hydrocolloids. 2014;**39**:301-318

[80] Zhang QT, Tu ZC, Wang H, Huang XQ, Fan LL, Bao ZY, et al. Functional properties and structure changes of soybean protein isolate after subcritical water treatment. Journal of Food Science and Technology [Internet]. 2015;52(6):3412. Available from: /pmc/articles/PMC4444865/

[81] Natarajan S, Luthria D, Bae H, Lakshman D, Mitra A. Transgenic soybeans and soybean protein analysis: An overview. Journal of Agricultural and Food Chemistry [Internet]. 2013;**61**(48):11736-11743. Available from: https://pubs.acs.org/doi/full/10.1021/jf402148e

[82] Osborne TB. Our present knowledge of plant proteins. Science [Internet]. 1908;**28**(718):417-427. Available from: https://www.science.org/doi/abs/10.1126/science.28.718.417

[83] Liu K. Soybean: Overview. Encyclopedia of Food Grains. 2nd ed. 2016;1;1-4:228-236.

[84] Sui X, Zhang T, Jiang L. Soy protein: Molecular structure revisited and recent advances in processing technologies. Annual Review of Food Science and Technology. 2021;**12**:119-147. Available from: https://www.annualreviews.org/doi/abs/10.1146/annurev-food-062220-104405

[85] Sugawara M, Ito D, Yamamoto K, Akita M, Oguri S, Momonoki YS. Kunitz soybean trypsin inhibitor is modified at its C-terminus by novel soybean thiol protease (protease T1). Plant Production Science. 2007;**10**(3):314-321. Available from: https://www.tandfonline.com/doi/abs/10.1626/pps.10.314

[86] Liu K. Soybean improvements through plant breeding and genetic engineering. In: Soybeans. Boston (MA): Springer; 1997. pp. 478-523. Available from: https://link.springer.com/chapter/10.1007/978-1-4615-1763-4_11

[87] US20090208633A1 - Protein Composition for Meat Products or Meat Analog Products - Google Patents [Internet]. Available from: https://patents.google.com/patent/US20090208633A1/en

[88] CA2314727C - Food coloring composition - Google Patents [Internet]. Available from: https://patents.google.com/patent/CA2314727C/en

[89] Ismail I, Hwang YH, Joo ST. Meat analog as future food: A review. Journal of Animal Science and Technology [Internet]. 2020;62(2):111. Available from: /pmc/articles/PMC7142285/

[90] Sadler MJ. Meat alternatives — Market developments and health benefits. Trends in Food Science & Technology. 2004;**15**(5):250-260

[91] Manski JM, van Riemsdijk LE, van der Goot AJ, Boom RM. Importance of intrinsic properties of dense caseinate dispersions for structure formation. Biomacromolecules [Internet]. 2007;**8**(11):3540-3547. Available from: https://pubmed.ncbi.nlm.nih.gov/17929972/

[92] Kyriakopoulou K, Dekkers B, van der Goot AJ. Plant-based meat analogues. Sustainable Meat Production and Processing. 2019;**1**:103-126

[93] Annotated E. History of soy IN China and TAIWAN 1 history of soybeans and SOYFOODS IN China and TAIWAN , and IN CHINESE cookbooks , restaurants , and CHINESE work with SOYFOODS outside China (1024 BCE TO 2014). EXTENSIVELY ANNOTATED BIBLIOGRAPHY AND SOURCEBOOK Inclu. 2014:1-3015

[94] Meat the alternative: Australia's $3B opportunity [Internet]. Available from: https://apo.org.au/node/257891

[95] Kinsella JE. Texturized proteins: Fabrication, flavoring, and nutrition. CRC Critical Reviews in Food Science and Nutrition [Internet]. 1978;**10**(2):147-207. Available from: https://pubmed.ncbi.nlm.nih.gov/365461/

[96] Fraser RZ, Shitut M, Agrawal P, Mendes O, Klapholz S. Safety evaluation of soy leghemoglobin protein preparation derived from *Pichia pastoris*, intended for use as a flavor catalyst in plant-based meat. International Journal of Toxicology [Internet]. 2018;37(3):241. Available from: /pmc/articles/PMC5956568/

[97] Methods and compositions for affecting the flavor and aroma profile of consumables - Patent US-9700067-B2 - PubChem [Internet]. Available from: https://pubchem.ncbi.nlm.nih.gov/patent/US9700067

[98] Natarajan SS. Analysis of soybean seed proteins using proteomics. Journal of Data Mining in Genomics & Proteomics. 2014;**05**(01):2153-0602

[99] Dastmalchi M, Dhaubhadel S. Proteomic insights into synthesis of isoflavonoids in soybean seeds. PROTEOMICS [Internet]. 2015;**15**(10):1646-1657. Available from: https://onlinelibrary.wiley.com/doi/full/10.1002/pmic.201400444

[100] Natarajan S, Xu C, Caperna TJ, Garrett WM. Comparison of protein solubilization methods suitable for proteomic analysis of soybean seed proteins. Analytical Biochemistry [Internet]. 2005;**342**(2):214-220. Available from: https://pubmed.ncbi.nlm.nih.gov/15953580/

[101] Natarajan SS, Xu C, Garrett WM, Lakshman D, Bae H. Assessment of the natural variation of low abundant metabolic proteins in soybean seeds using proteomics. Journal of Plant Biochemistry and Biotechnology. 2012;**21**(1):30-37

[102] Maria John KM, Khan F, Luthria DL, Garrett W, Natarajan S. Proteomic analysis of anti-nutritional factors (ANF's) in soybean seeds as affected by environmental and genetic factors. Food Chemistry [Internet]. 2017;**218**:321-329. Available from: https://pubmed.ncbi.nlm.nih.gov/27719916/

[103] Natarajan S, Xu C, Bae H, Caperna TJ, Garrett WM. Proteomic analysis of allergen and antinutritional proteins in wild and cultivated soybean seeds. Journal of Plant Biochemistry and Biotechnology. 2006;**15**(2):103-108

[104] Natarajan SS, Xu C, Bae H, Caperna TJ, Garrett WM. Characterization of storage proteins in wild (Glycine soja) and cultivated (Glycine max) soybean seeds using proteomic analysis. Journal of Agricultural and Food Chemistry [Internet]. 2006;**54**(8):3114-3120. Available from: https://pubmed.ncbi.nlm.nih.gov/16608239/

[105] Xu C, Caperna TJ, Garrett WM, Cregan P, Bae H, Luthria DL, et al. Proteomic analysis of the distribution of the major seed allergens in wild, landrace, ancestral, and modern soybean genotypes. Journal of the Science of Food and Agriculture. 2007;**87**(13):2511-2518

[106] Koo SC, Bae DW, Seo JS, Park KM, Choi MS, Kim SH, et al. Proteomic analysis of seed storage proteins in low allergenic soybean accession. Journal of Applied Biological Chemistry [Internet]. 2011;54(3):332-339. Available from: https://applbiolchem.springeropen.com/articles/10.3839/jksabc.2011.053

[107] Natarajan S, Xu C, Bae H, Bailey BA. Proteomic and genomic characterization of Kunitz trypsin inhibitors in wild and cultivated soybean genotypes. Journal of Plant Physiology [Internet]. 2007;**164**(6):756-763. Available from: https://pubmed.ncbi.nlm.nih.gov/16884824/

[108] Natarajan S, Xu C, Bae H, Bailey BA, Cregan P, Caperna TJ, et al. Proteomic and genetic analysis of glycinin subunits of sixteen soybean genotypes. Plant Physiology and Biochemistry : PPB [Internet]. 2007;**45**(6-7):436-444. Available from: https://pubmed.ncbi.nlm.nih.gov/17524657/

[109] Murad AM, Rech EL. NanoUPLC-MSE proteomic data assessment of soybean seeds using the Uniprot database. BMC Biotechnology. 2012;**12**:1-17

[110] Gomes LS, Senna R, Sandim V, Silva-Neto MAC, Perales JEA, Zingali RB, et al. Four conventional soybean [Glycine max (L.) Merrill] seeds exhibit different protein profiles as revealed by proteomic analysis. Journal of Agricultural and Food Chemistry [Internet]. 2014;**62**(6):1283-1293. Available from: https://pubs.acs.org/doi/abs/10.1021/jf404351g

[111] Capriotti AL, Caruso G, Cavaliere C, Samperi R, Stampachiacchiere S, Zenezini Chiozzi R, et al. Protein profile of mature soybean seeds and prepared soybean milk. Journal of Agricultural and Food Chemistry [Internet]. 2014;**62**(40):9893-9899. Available from: https://pubs.acs.org/doi/abs/10.1021/jf5034152

[112] Riascos JJ, Weissinger SM, Weissinger AK, Kulis M, Burks AW, Pons L. The seed biotinylated protein of soybean (Glycine max): A boiling-resistant new allergen (Gly m 7) with the capacity to induce IgE-mediated allergic responses. Journal of Agricultural and Food Chemistry [Internet]. 2016;**64**(19):3890-3900. Available from: https://pubmed.ncbi.nlm.nih.gov/27108990/

[113] Zhang YM, Hu R, Li H, Zhu H, Zhao JM, Guo N, et al. Proteomics changes in filling seeds of vegetable soybean. HortScience. 2016;**51**(11):1397-1401

[114] Agrawal GK, Hajduch M, Graham K, Thelen JJ. In-depth investigation of the soybean seed-filling proteome

and comparison with a parallel study of rapeseed. Plant Physiology. 2008;**148**(1):504-518

[115] Hajduch M, Ganapathy A, Stein JW, Thelen JJ. A systematic proteomic study of seed filling in soybean. Establishment of high-resolution two-dimensional reference maps, expression profiles, and an interactive proteome database. Plant physiology [Internet]. 2005;**137**(4):1397-1419. Available from: https://pubmed.ncbi.nlm.nih.gov/15824287/

[116] Meyer LJ, Gao J, Xu D, Thelen JJ. Phosphoproteomic analysis of seed maturation in Arabidopsis, rapeseed, and soybean. Plant Physiology [Internet]. 2012;**159**(1):517-528. Available from: https://academic.oup.com/plphys/article/159/1/517/6109308

[117] Cho S-W, Kwon S-J, Roy SK, Kim H-S, Lee C-W, Woo SH. A systematic proteome study of seed storage proteins from two soybean genotypes. KOREAN JOURNAL OF CROP SCIENCE [Internet]. 2014;**59**(3):359-363. Available from: http://dx.doi.org/10.7740/kjcs.2014.59.3.359

[118] Kim HT, Choi UK, Ryu HS, Lee SJ, Kwon OS. Mobilization of storage proteins in soybean seed (Glycine max L.) during germination and seedling growth. Biochimica et biophysica acta [Internet]. 2011;**1814**(9):1178-1187. Available from: https://pubmed.ncbi.nlm.nih.gov/21616178/

[119] Han C, Yin X, He D, Yang P. Analysis of proteome profile in germinating soybean seed, and its comparison with rice showing the styles of reserves mobilization in different crops. PloS One [Internet]. 2013;**8**(2):e56947. Available from: https://pubmed.ncbi.nlm.nih.gov/23460823/

[120] Krishnan HB, Nelson RL. Proteomic analysis of high protein soybean (Glycine max) accessions demonstrates the contribution of novel glycinin subunits. Journal of Agricultural and Food Chemistry [Internet]. 2011;59(6):2432-2439. Available from: https://pubs.acs.org/doi/full/10.1021/jf104330n

[121] Min CW, Gupta R, Kim SW, Lee SE, Kim YC, Bae DW, et al. Comparative biochemical and proteomic analyses of soybean seed cultivars differing in protein and oil content. Journal of Agricultural and Food Chemistry [Internet]. 2015;**63**(32):7134-7142. Available from: https://pubs.acs.org/doi/full/10.1021/acs.jafc.5b03196

[122] Xu XP, Liu H, Tian L, Dong XB, Shen SH, Qu LQ. Integrated and comparative proteomics of high-oil and high-protein soybean seeds. Food Chemistry [Internet]. 2015;**172**:105-116. Available from: https://pubmed.ncbi.nlm.nih.gov/25442530/

[123] Yang A, Yu X, Zheng A, James AT. Rebalance between 7S and 11S globulins in soybean seeds of differing protein content and 11SA4. Food Chemistry [Internet]. 2016;**210**:148-155. Available from: https://pubmed.ncbi.nlm.nih.gov/27211633/

[124] Min CW, Gupta R, van Truong N, Bae JW, Ko JM, Lee BW, et al. A TMT-based quantitative proteomic analysis provides insights into the protein changes in the seeds of high- and low- protein content soybean cultivars. Journal of Plant Biotechnology [Internet]. 2020;**47**(3):209-217. Available from: https://doi.org/10.5010/JPB.2020.47.3.209

[125] Brandão AR, Barbosa HS, Arruda MAZ. Image analysis of two-dimensional gel electrophoresis for comparative proteomics of transgenic and non-transgenic soybean seeds. Journal of Proteomics. 2010;**73**(8):1433-1440

[126] Barbosa HS, Arruda SCC, Azevedo RA, MAZ A. New insights on proteomics of transgenic soybean seeds: Evaluation of differential expressions of enzymes and proteins. Analytical and Bioanalytical Chemistry [Internet]. 2012;**402**(1):299-314. Available from: https://pubmed.ncbi.nlm.nih.gov/21947011/

[127] Yamada T, Mori Y, Yasue K, Maruyama N, Kitamura K, Abe J. Knockdown of the 7S globulin subunits shifts distribution of nitrogen sources to the residual protein fraction in transgenic soybean seeds. Plant Cell Reports [Internet]. 2014;**33**(12):1963-1976. Available from: https://link.springer.com/article/10.1007/s00299-014-1671-y

[128] Liu W, Xu W, Li L, Dong M, Wan Y, He X, et al. iTRAQ-based quantitative tissue proteomic analysis of differentially expressed proteins (DEPs) in non-transgenic and transgenic soybean seeds. Scientific Reports. 2018;**8**(1):1-10. Available from: https://www.nature.com/articles/s41598-018-35996-y

[129] Yu X, Jin H, Fu X, Yang Q, Yuan F. Quantitative proteomic analyses of two soybean low phytic acid mutants to identify the genes associated with seed field emergence. BMC Plant Biology [Internet]. 2019;**19**(1):1-14. Available from: https://bmcplantbiol.biomedcentral.com/articles/10.1186/s12870-019-2201-4

[130] Lee KJ, Kim JB, Kim SH, Ha BK, Lee BM, Kang SY, et al. Alteration of seed storage protein composition in soybean [Glycine max (L.) Merrill] mutant lines induced by γ-irradiation mutagenesis. Journal of Agricultural and Food Chemistry [Internet]. 2011;**59**(23):12405-12410. Available from: https://pubs.acs.org/doi/abs/10.1021/jf202809j

[131] Schmidt MA, Barbazuk WB, Sandford M, May G, Song Z, Zhou W, et al. Silencing of soybean seed storage proteins results in a rebalanced protein composition preserving seed protein content without major collateral changes in the metabolome and transcriptome. Plant Physiology [Internet]. 2011;**156**(1):330-345. Available from: https://academic.oup.com/plphys/article/156/1/330/6111500

[132] Natarajan S, Islam N, Krishnan HB. Proteomic profiling of fast neutron-induced soybean mutant unveiled pathways associated with increased seed protein content. Journal of Proteome Research [Internet]. 2020;**19**(10):3936-3944. Available from: https://pubs.acs.org/doi/abs/10.1021/acs.jproteome.0c00160

[133] Danchenko M, Skultety L, Rashydov NM, Berezhna V, Mátel L, Salaj T, et al. Proteomic analysis of mature soybean seeds from the Chernobyl area suggests plant adaptation to the contaminated environment. Journal of Proteome Research [Internet]. 2009;**8**(6):2915-2922. Available from: https://pubs.acs.org/doi/abs/10.1021/pr900034u

[134] Ren C, Bilyeu KD, Beuselinck PR. Composition, vigor, and proteome of mature soybean seeds developed under high temperature. Crop Science [Internet]. 2009;**49**(3):1010-1022. Available from: https://onlinelibrary.wiley.com/doi/full/10.2135/cropsci2008.05.0247

[135] Xu XY, Fan R, Zheng R, Li CM, Yu DY. Proteomic analysis of seed germination under salt stress in soybeans. Journal of Zhejiang University SCIENCE B [Internet]. 2011;**12**(7):507-517. Available from: https://link.springer.com/article/10.1631/jzus.B1100061

[136] Wang L, Ma H, Song L, Shu Y, Gu W. Comparative proteomics analysis reveals the mechanism of pre-harvest seed deterioration of soybean under high temperature and humidity stress. Journal of Proteomics. 2012;**75**(7):2109-2127

[137] Seo SH, Cho SJ. Changes in allergenic and antinutritional protein profiles of soybean meal during solid-state fermentation with Bacillus subtilis. LWT - Food Science and Technology [Internet]. 2016;**70**:208-212. Available from: http://dx.doi.org/10.1016/j.lwt.2016.02.035

[138] Cheng L, Gao X, Li S, Shi M, Javeed H, Jing X, et al. Proteomic analysis of soybean [Glycine max (L.) Meer.] seeds during imbibition at chilling temperature. Molecular Breeding [Internet]. 2010;**26**(1):1-17. Available from: https://link.springer.com/article/10.1007/s11032-009-9371-y

[139] Hashiguchi A, Komatsu S. Proteomics of soybean plants. Proteomics in Food Science: From Farm to Fork. 2017:89-105

[140] Salgado JM, Donado-Pestana CM. Soy as a functional food. In: Soybean and Nutrition [Internet]. London: IntechOpen; 2011. Available from: https://www.intechopen.com/chapters/19973

[141] Lule VK, Garg S, Pophaly SD, Hitesh TSK. Potential health benefits of lunasin: A multifaceted soy-derived bioactive peptide. Journal of Food Science [Internet]. 2015;**80**(3):R485-R494. Available from: https://onlinelibrary.wiley.com/doi/full/10.1111/1750-3841.12786

[142] Seber LE, Barnett BW, McConnell EJ, Hume SD, Cai J, Boles K, et al. Scalable purification and characterization of the anticancer lunasin peptide from soybean. PLoS One. 2012;7(4):e35409

[143] Barbalho SM, Flávia Farinazzi-Machado M v. Soybean: Food or Remedy? Soybean and Nutrition [Internet]. London: IntechOpen; 2011. Available from: https://www.intechopen.com/chapters/19992

[144] Burssens S, Pertry I, Ngudi DD, Kuo Y-H, Montagu M v, Lambein F. Soya, human nutrition and health. In: Soybean and Nutrition [Internet]. London: IntechOpen; 2011. Available from: https://www.intechopen.com/chapters/19979

[145] Maebuchi M, Samoto M, Kohno M, Ito R, Koikeda T, Hirotsuka M, et al. Improvement in the intestinal absorption of soy protein by enzymatic digestion to oligopeptide in healthy adult men. Food Science and Technology Research. 2007;**13**(1):45-53

[146] Lammi C, Zanoni C, Arnoldi A, Vistoli G. Two peptides from soy β-conglycinin induce a hypocholesterolemic effect in HepG2 cells by a statin-like mechanism: Comparative in vitro and in silico modeling studies. Journal of Agricultural and Food Chemistry [Internet]. 2015;**63**(36):7945-7951. Available from: https://pubmed.ncbi.nlm.nih.gov/26310992/

[147] Lammi C, Zanoni C, Arnoldi A. IAVPGEVA, IAVPTGVA, and LPYP, three peptides from soy glycinin, modulate cholesterol metabolism in HepG2 cells through the activation of the LDLR-SREBP2 pathway. Journal of Functional Foods. 2015;**14**:469-478

[148] Kuba M, Tana C, Tawata S, Yasuda M. Production of angiotensin I-converting enzyme inhibitory peptides from soybean protein with Monascus

purpureus acid proteinase. Process Biochemistry. 2005;**40**(6):2191-2196

[149] Pak V, Koo MS, Kasymova TD, Kwon DY. Isolation and identification of peptides from soy 11s-globulin with hypocholesterolemic activity. Chemistry of Natural Compounds. 2005;41(6):710-714. Available from: https://link.springer.com/article/10.1007/s10600-006-0017-6

[150] Pak V, Koo M, Kwon DY, Yun L. Design of a highly potent inhibitory peptide acting as a competitive inhibitor of HMG-CoA reductase. Amino Acids [Internet]. 2012;**43**(5):2015-2025. Available from: https://link.springer.com/article/10.1007/s00726-012-1276-0

[151] Erdmann K, Cheung BWY, Schröder H. The possible roles of food-derived bioactive peptides in reducing the risk of cardiovascular disease. The Journal of Nutritional Biochemistry. 2008;**19**(10):643-654

[152] Lammi C, Zanoni C, Arnoldi A, Vistoli G. Peptides derived from soy and lupin protein as dipeptidyl-peptidase IV inhibitors: In vitro biochemical screening and in silico molecular modeling study. Journal of Agricultural and Food Chemistry [Internet]. 2016;**64**(51):9601-9606. Available from: https://pubs.acs.org/doi/full/10.1021/acs.jafc.6b04041

[153] Pak V, Koo M, Lee N, Kim MS, Kwon DY. Structure—Activity relationships of the peptide Ile-Ala-Val-Pro and its derivatives revealed using the semi-empirical am1 method. Chemistry of Natural Compounds. 2005;**41**(4):454-460. Available from: https://link.springer.com/article/10.1007/s10600-005-0176-x

[154] Yoshikawa M, Fujita H, Matoba N, Takenaka Y, Yamamoto T, Yamauchi R, et al. Bioactive peptides derived from food proteins preventing lifestyle-related diseases. BioFactors [Internet]. 2000;**12**(1-4):143-146. Available from: https://onlinelibrary.wiley.com/doi/full/10.1002/biof.5520120122

[155] Yoshikawa M. Bioactive peptides derived from natural proteins with respect to diversity of their receptors and physiological effects. Peptides. 2015;**72**:208-225

[156] Yoshikawa M, Kishi K, Takahashi M, Watanabe A, Miyamura T, Yamazaki M, et al. Immunostimulating peptide derived from soybean protein. Annals of the New York Academy of Sciences [Internet]. 1993;**685**(1):375-376. Available from: https://onlinelibrary.wiley.com/doi/full/10.1111/j.1749-6632.1993.tb35892.x

[157] Singh BP, Vij S, Hati S. Functional significance of bioactive peptides derived from soybean. Peptides. 2014;**54**:171-179

[158] Pabona JMP, Dave B, Su Y, Montales MTE, de Lumen BO, de Mejia EG, et al. The soybean peptide lunasin promotes apoptosis of mammary epithelial cells via induction of tumor suppressor PTEN: Similarities and distinct actions from soy isoflavone genistein. Genes and Nutrition [Internet]. 2013;**8**(1):79-90. Available from: https://link.springer.com/articles/10.1007/s12263-012-0307-5

[159] de Mejia E, de Lumen BO. Soybean bioactive peptides: A new horizon in preventing chronic diseases. Sexuality, Reproduction and Menopause. 2006;**4**(2):91-95

[160] Jae HP, Hyung JJ, de Lumen BO. In vitro digestibility of the cancer-preventive soy peptides lunasin and BBI. Journal of Agricultural and Food Chemistry [Internet]. 2007;**55**(26):10703-10706. Available

from: https://pubs.acs.org/doi/full/10.1021/jf072107c

[161] Losso JN. The biochemical and functional food properties of the Bowman-Birk inhibitor. Critical Reviews in Food Science and Nutrition. 2008;**48**(1):94-118. Available from: https://www.tandfonline.com/doi/abs/10.1080/10408390601177589

[162] Baturay NZ, Roque H. In vitro reduction of peroxidation in UVC-irradiated cell cultures by concurrent exposure with Bowman-Birk protease inhibitor. Teratogenesis, Carcinogenesis, and Mutagenesis [Internet]. 1991;**11**(4):195-202. Available from: https://onlinelibrary.wiley.com/doi/full/10.1002/tcm.1770110404

[163] Meyskens FLJ. Development of difluoromethyl-ornithine and Bowman-Birk inhibitor as chemopreventive agents by assessment of relevant biomarker modulation: Some lessons learned. IARC Scientific Publications. 2001;**154**:49-55

[164] Kennedy AR, Radner BS, Nagasawa H. Protease inhibitors reduce the frequency of spontaneous chromosome abnormalities in cells from patients with Bloom syndrome. Proceedings of the National Academy of Sciences [Internet]. 1984;**81**(6):1827-1830. Available from: https://www.pnas.org/content/81/6/1827

[165] Kennedy AR, Steven WX. Effects of the Bowman-Birk inhibitor on growth, invasion, and clonogenic survival of human prostate epithelial cells and prostate cancer cells*. The Prostate [Internet]. 2002;**50**(2):125-133. Available from: https://onlinelibrary.wiley.com/doi/full/10.1002/pros.10041

[166] Bruce Malkowicz S, McKenna WG, Vaughn DJ, Wan XS, Propert KJ, Rockwell K, et al. Effects of Bowman–Birk inhibitor concentrate (BBIC) in patients with benign prostatic hyperplasia*†. The Prostate [Internet]. 2001;**48**(1):16-28. Available from: https://onlinelibrary.wiley.com/doi/full/10.1002/pros.1077

[167] Armstrong WB, Kennedy AR, Wan XS, Atiba J, McLaren CE, Meyskens FLJ. Single-dose administration of Bowman-Birk inhibitor concentrate in patients with oral leukoplakia. Cancer epidemiology, biomarkers & prevention : a publication of the American Association for Cancer Research, cosponsored by the American Society of Preventive Oncology. 2000;**9**(1):43-47

[168] Armstrong WB, Kennedy AR, Wan XS, Taylor TH, Nguyen QA, Jensen J, et al. Clinical modulation of oral leukoplakia and protease activity by Bowman-Birk inhibitor concentrate in a phase IIa chemoprevention trial. Clinical Cancer Research : An Official Journal of the American Association for Cancer Research. 2000;**6**(12):4684-4691

[169] Wang W, Gonzalez De Mejia E. A new frontier in soy bioactive peptides that may prevent age-related chronic diseases. Comprehensive Reviews in Food Science and Food Safety [Internet]. 2005;**4**(4):63-78. Available from: https://onlinelibrary.wiley.com/doi/full/10.1111/j.1541-4337.2005.tb00075.x

[170] Kim SE, Kim HH, Kim JY, Kang YI, Woo HJ, Lee HJ. Anticancer activity of hydrophobic peptides from soy proteins. BioFactors [Internet]. 2000;**12**(1-4):151-155. Available from: https://onlinelibrary.wiley.com/doi/full/10.1002/biof.5520120124

[171] Villares A, García-Lafuente A, Palacios I, Lozano M, Moro C, Guillamón E. Soy and soy-based foods: Role in health and nutrition. In: Soybean

and Nutrition [Internet]. London: IntechOpen; 2011. Available from: https://www.intechopen.com/chapters/19986

[172] Odani S, Ikenaka T. Studies on soybean trypsin inhibitors. IV. Complete amino acid sequence and the anti-proteinase sites of Bowman-Birk soybean proteinase inhibitor. Journal of Biochemistry [Internet]. 1972;**71**(5):839-848. Available from: https://pubmed.ncbi.nlm.nih.gov/4672481/

[173] Dia VP, Berhow MA, de Mejia EG. Bowman-Birk inhibitor and genistein among soy compounds that synergistically inhibit nitric oxide and prostaglandin E2 pathways in lipopolysaccharide-induced macrophages. Journal of Agricultural and Food Chemistry [Internet]. 2008;**56**(24):11707-11717. Available from: https://pubmed.ncbi.nlm.nih.gov/19053380/

[174] Maki KC, Butteiger DN, Rains TM, Lawless A, Reeves MS, Schasteen C, et al. Effects of soy protein on lipoprotein lipids and fecal bile acid excretion in men and women with moderate hypercholesterolemia. Journal of Clinical Lipidology [Internet]. 2010;**4**(6):531-542. Available from: https://pubmed.ncbi.nlm.nih.gov/21122701/

[175] Martin M, Deussen A. Effects of natural peptides from food proteins on angiotensin converting enzyme activity and hypertension. Critical Reviews in Food Science and Nutrition. 2019;**59**(8):1264-1283

[176] de Leo F, Panarese S, Gallerani R, Ceci L. Angiotensin converting enzyme (ACE) inhibitory peptides: Production and implementation of functional food. Current Pharmaceutical Design [Internet]. 2009;**15**(31):3622-3643. Available from: https://pubmed.ncbi.nlm.nih.gov/19925416/

[177] Slusarczyk AL, Lin A, Weiss R. Foundations for the design and implementation of synthetic genetic circuits. Nature Reviews Genetics. 2012;**13**(6):406-420. Available from: https://www.nature.com/articles/nrg3227

[178] Endy D. Foundations for engineering biology. Nature. 2005;**438**(7067):449-453. Available from: https://www.nature.com/articles/nature04342

[179] Stephens N, di Silvio L, Dunsford I, Ellis M, Glencross A, Sexton A. Bringing cultured meat to market: Technical, socio-political, and regulatory challenges in cellular agriculture. Trends in Food Science & Technology. 2018;**1**(78):155-166

[180] da Cunha NB, Murad A, Vianna G, Rech E. Recombinant biosynthesis of functional human growth hormone and coagulation factor IX in transgenic soybean seeds. BMC Proceedings. 2014;**8**(4):1-2. Available from: https://link.springer.com/articles/10.1186/1753-6561-8-S4-P112

[181] Murad A, Cunha N, Lacorte C, Coelho M, Vianna G, Rech E. Expression, purification and analysis of the anti-HIV cyanovirin-N produced in transgenic soybeans seeds. BMC Proceedings. 2014;**8**(4):1-2. Available from: https://bmcproc.biomedcentral.com/articles/10.1186/1753-6561-8-S4-P105

Chapter 12

Innovative Application of Soy Protein Isolate and Combined Crosslinking Technologies to Enhance the Structure of Gluten-Free Rice Noodles

Moses Ojukwu and Azhar Mat Easa

Abstract

Soy protein isolate (SPI) may serve as a health-enhancing functional ingredient in many food products due to the content of isoflavones. However, the high protein content may also be exploited as a structure modifier in gluten-free noodles. We applied Soy protein isolate to improve rice flour noodles' structure, textural, and cooking properties by combining cross-linking and cold gelation of soy protein isolate using microbial transglutaminase and glucono-δ-lactone, respectively. The simultaneous cross-linking yielded noodles with improved structure and textural properties, mainly due to a more robust microstructure resulting from an increase in intermolecular protein cross-linking promoted by microbial transglutaminase and glucono-δ-lactone. However, the structurally enhanced noodles showed longer cooking time and reduced cooking yield upon drying. This was solved by employing pre-drying steaming treatments for 5 or 10 min to yield noodles with shorter cooking times, lower cooking losses, and improved cooking yield. We have also developed an alternative process technology using superheated steam (SHS) technology. The superheated steam technology made it possible to open up the structurally enhanced air-dried noodles by promoting faster gelatinization, as evidenced by reduced enthalpy, increased cooking yield, and sustained crystallinity of the starch granules noodle matrix.

Keywords: soy protein isolates, noodles, texture, cross-linking, cooking properties

1. Introduction

Soy protein contains all essential amino acids, and its consumption has been beneficial to human physiological needs as it lowers cholesterol and reduces the risk of coronary heart and cardiovascular diseases [1, 2]. Soy protein has been used in food systems for its excellent gelling, water-holding capacity, sensory and functional properties [3]. Soybeans have been an excellent source of proteins in Asia and are consumed in various forms, such as soymilk, soy curd (tofu), and fermented soy paste (miso).

Soy protein isolates (SPI) are produced by removing the oil content of soy proteins. When the pH of aqueous soy proteins is reduced to pH 4.5–4.8, they are separated into whey fractions and storage globulins. SPI comprises different proteins, and the essential components are grouped into 2S,7S,11S, and 15S, representing their sedimentation coefficients when subjected to a centrifugal field, with the β-conglycinin (7S) and glycinin (11S) being the majority. While the former lacks disulfide bonds, glycine has a more compact structure equilibrated by disulfide bonds and thus possesses lower gelling, emulsifying, and foaming capacity when compared with β-conglycinin [1]. Because of this, SPI has been used to affect the textural and structural properties of noodles [2].

The consumer acceptability of noodles is dependent on the overall texture of rice noodles [3]. The texture of noodles is simply the surface mouthfeel and the resistance to chewing and the textural properties of noodles are influenced by various factors, such as the properties of the type of flour, such as amylose/amylopectin ratio of starch, the protein, and lipid composition, the processing conditions, and the thermal properties of the flour [4].

The Texture Profile Analyzer (TPA) proposed by Szczesniak [5] can be used to measure the textural properties of solid food materials, including rice noodles [6]. The TPA method is a 2-time compression type test from which parameters, such as hardness, chewiness, adhesiveness, cohesiveness, fracturability, springiness, gumminess, chewiness, of the noodles can be obtained [7]. It is the most straightforward and frequent technique to relate instrumental measurement with sensory evaluation.

Hardness is the force required to cause a pre-determined deformation. It also measures the resistance of noodles to compressions which is the maximum force of the first compression [6]. The amount of energy required to break down the noodles while chewing before swallowing is expressed in chewiness. Cohesiveness demonstrates how well the noodles withstand the second deformation and influence chewiness by indicating the extent of the structure breakdown throughout chewiness [7]. While the degree of adherence of the noodle on the probe after the first compression is the adhesiveness of the noodle [5], the noodles' ability to return to their original shape after compression is called the noodle's springiness.

In noodles' research, cooking parameters indicate the impact of the noodles' structure on sustainability [8]. The cooking quality of noodles can be evaluated by measuring the length of cooking time, the cooking loss, and the cooking yield. The cooking quality of noodles is critical to noodles' sensory and textural properties [8]. Starch gelatinization, protein coagulation, and other structural changes occur in proteins during cooking. These account for how long it takes to get the noodles properly cooked, their water retention capacity, and their ability to withstand the cooking processes and maintain their structure.

Consumers show a preference for noodles with short cooking time, less cooking loss, and high cooking yield [8, 9]. Noodles are usually cooked in boiling water. The optimum cooking time is the time, it takes for the white core in the center of a noodle strand to disappear [10]. Rice noodles cook faster than wheat-based noodles. This is because the primary raw material for rice noodles, rice flour, does not contain gluten. Robust protein networks by gluten limit water ingress into the noodles, thus elongating the starch gelatinization, a precursor for the noodles to be cooked [8, 11]. The cooking time of rice noodles is influenced by the rice starch properties and other additives, such as hydrocolloids and starches. Rice noodles prepared by blending rice flour with other starches have a longer cooking time due to alteration in the starch's gelatinization temperatures, which ultimately expanded the water retention capacity of the noodles [12].

The cooking loss of rice noodles shows the number of substances lost from the noodles during cooking. It is indicative of the structural integrity of the noodles. Rice noodles with high cooking loss are undesirable because they become sticky due to increased leaching of amylose and starch recrystallization. According to the Chinese standard for starch-based noodles, noodles with a maximum of 10% cooking loss are acceptable [13, 14]. Rice noodles prepared from rice flour with high amylose content have less cooking loss. Also, the inclusion of protein isolates and protein cross-linking impart some form of structural integrity on the noodles, minimizing the cooking loss [15].

Cooking yield can be used to estimate the water retention capacity of the noodles. During cooking, noodles absorbed water and increased weight [9, 16]. The differences in the cooking yield of rice noodles with noodles made from wheat and other flours are due to differences in amylose concentrations, swelling power, and the pasting properties of starches from different sources [12].

2. Gelation of soy proteins

The ability of SPI to form gel is an essential functional property for its use in noodle structure modification [17, 18]. Soy proteins are susceptible to denaturation when thermally treated. Heat exposes the hydrophobic groups in proteins, increasing hydrophobicity, decreasing net charges, thus promoting protein networks [19]. In the heat-induced process, aggregation, network formation, and the unfolding of the hydrophobic parts are interwoven (**Figure 1**). Cold gelation could be achieved by introducing an acidulant such as glucono- δ -lactone that releases gluconic acid slowly into the system to promote gelation of thermally aggregated protein molecules.

The gelation and coagulation of SPI have involved hydrophobic and covalent cross-links [20]. SPI gels can be enhanced using microbial transglutaminase (MTG), which is a well-known food-grade enzyme that utilizes the acyltransferase process to join the γ-carboxamide (acyl donor) of a glutamine residue to the γ-amine (acyl acceptor) of lysine residues along protein chains. The cross-linking action of MTG on SPI in noodles has been shown by Ojukwu et al. [18]. Upon examination using sodium

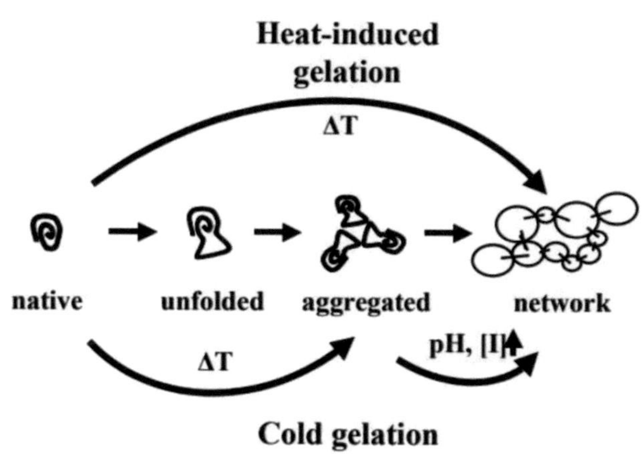

Figure 1.
Conversion of native globular protein into a protein network according to the heat-induced or cold gelation procedure (ΔT = change in temperature, [I] = iso-electric point).

dodecyl sulfate-polyacrylamide gel electrophoresis (SDS-PAGE), it was evident that protein networks formed after incubating for 30 min.

3. Innovating functional (health-enhancing) noodles

In the beginning, we have innovated health-enhancing noodles by employing banana peel or banana pulp flours [21]. It was quite possible to use banana peel flour to replace that of wheat partially. Banana peels which make up almost 40% of the fruit can control starch hydrolysis in noodles. In this study, cooked noodles made from *Cavendish* banana peel flour were characterized for physicochemical properties, including elasticity, pH, tensile strength, and color. Banana peel noodles exhibited higher elasticity and lower glycemic index than the control (wheat noodles), while the tensile strength was similar. The addition of banana pulp or banana peel to the formulation lowers the predicted glycemic index of the cooked noodles. The effect is more evident in the banana peel formulation because of its higher dietary fiber and resistant starch contents. Overall, the substitution of banana peel or banana pulp can alter the physicochemical properties of the cooked noodles [22].

Yellow alkaline noodles prepared with green banana pulp flour effectively control starch hydrolysis, which can lower the ingestion rate and absorption of carbohydrates in noodles [21]. The green banana pulp can also be used as an effective flour substitute. Despite the potential of the banana peel or banana pulp flours as functional ingredients in noodles, partial substitution of the wheat flour at higher levels was not successful due to lack of protein network to impart structure and texture. We then sought plant protein isolates that could be cross-linked at relatively low application levels to enhance the noodles' network. Consequently, it would also impart good cooking, handling, and eating quality. We evaluated soy protein isolate and achieved various functionalities to improve the quality characteristics of the noodles.

3.1 Structural modification of wheat flour noodles using SPI

Noodles are structurally altered for various purposes, such as to make them able to withstand processing and cooking and release glucose more slowly during digestion. Alteration of structure in canned noodles can lower the cooking loss during the retort process. For instance, it was shown that cross-linking agents, such as MTG and ribose could prevent damage to canned SPI-yellow alkaline noodles' during thermal processing [2]. Thus, partially substituted yellow alkaline noodles treated with MTG and ribose separately and in combination generated an enhanced structure that can withstand thermal processing.

3.2 Use of SPI in coating yellow alkaline noodles

Capsaicin is an antiobesity agent and an active component of chili powder. However, chili can cause pain, therefore attempts to affect the release of chili in the mouth were also tried using cross-linking agents, SPI, and layering technology.

For the capsaicin to be effectively delivered, the compound should enter the body in a considerable quantity. However, a higher dose can cause a burning sensation in the mouth. Noodles, one of the widely eaten products, can be used for safe delivery inside the body. It is suggested that chili powder can be added to noodles by sandwiching the dough between layers of wheat dough. The addition of resistant starch flour or SPI and microbial transglutaminase to the sandwich layers can lower the glycemic index [23]. The capsaicin retaining ability of noodles formulated with SPI at the core was higher than that prepared using resistant starch flour. MTG was used to enhance the elasticity, while SPI was used as a texture enhancer. A combination of MTG and SPI had increased the network structure of "sandwiched noodles" that aids in capsaicin retention. Compared to other formulations, layered noodles prepared with 1.5 g MTG and 5 g SPI per 100 g mixture exhibited the best capsaicin retention and tensile parameters [24].

4. Enhancing the textural, structural, and cooking properties of fresh rice noodles using SPI and combined cross-linking technologies

Glucono- δ -lactone (GDL) is a cyclic ester. It contains a lactone group and gradually dissolves in water to form gluconic acid, causing a reduction in pH in food systems (**Figure 2**). It is widely used as an acidulant in yogurts, sausages, and different dessert mixes. GDL is used as a coagulant in the production of *tofu*, a soymilk curd [25, 26]. The introduction of GDL into protein alters the pH toward the isoelectric point and subsequently gels the proteins through electrostatic repulsion, salt-bridging, and noncovalent interactions, such as hydrogen bonding and van der Waals forces [27, 28]. The aggregation of protein in GDL-induced acidification of soymilk proteins occurred at pH 5.9 [29].

The combined control of the acidification and gelling of proteins by heat treatment creates a "cold gelation" [20, 30]. In the cold gelation process, aggregates and the formation of a protein network are separated in time. Cold gelation is made of two steps. First, at neutral pH well above the isoelectric point, low concentrations of native globular proteins are thermally treated, causing the native proteins to unfold and form disulfide cross-linked aggregates. The soluble aggregates formed were made possible due to net surface charge proteins and repulsive forces, which prevent random aggregation. Depending on the denaturation conditions, a stable dispersion of aggregates is obtained after cooling to room temperature. Secondly, an alteration in the quality of the solvent causes gelation. Typically, cold-set gels by acidification are stronger than salt-induced cold-set gels for the same protein concentration [31].

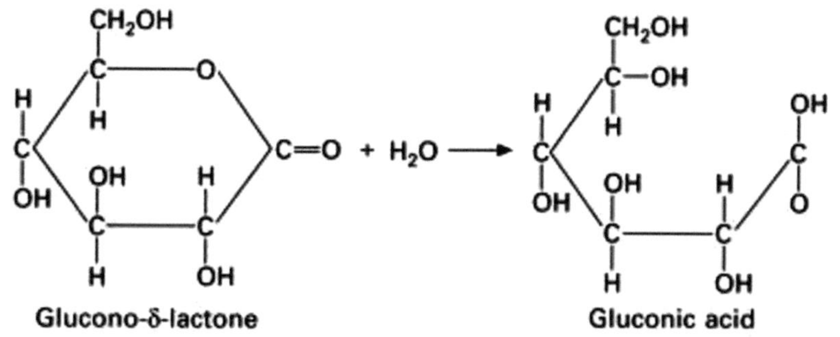

Figure 2.
Hydrolysis of GDL.

SPI and MTG have been used to improve noodles' texture and mechanical properties, but a mixture of SPI, MTG, and GDL has never been tested to improve the texture and other properties in rice noodles. Rice flour-SPI noodles containing MTG and GDL were tested for various parameters to develop a gluten-free alternative to wheat flour noodles with comparable or better characteristics. Rice noodles were prepared to contain 5% SPI, were cross-linked with MTG (1% w/w rice flour), and cold-gelled using GDL (1% w/w rice flour). This restructured fresh rice noodle termed RNS-COM (SPI-rice noodle formed via combined cross-linking technology) showed heavier protein bands than a typical rice protein band, indicating the polymerization of proteins into bigger matrixes. The cooking time for RNS-COM was significantly lower than commercially available yellow alkaline noodles (YAN). No significant difference was found between the percentage cooking loss between the two. RNS-COM had better textural parameters, including hardness and chewiness, while compactness and tensile strength were comparable (**Figure 3**). Overall, the RNS-COM noodles performed better than regular rice noodles and can potentially be used in place of YAN [18]. The addition of SPI in rice noodles made them have better textural properties than the control (RN). Increased SPI cross-linking by MTG and GDL in RNS-COM gave a compact and robust protein matrix embedded with the starch granules within the rice noodles (data not shown).

SPI, MTG, and GDL are additives commonly used in the food industry to enhance foods' texture and physiochemical properties. Adding these additives to rice noodles makes them more palatable and may mimic those of wheat flour noodles, such as YAN. It is crucial to optimize the values of such additives to make the product marketable. Optimization is required for factors, including hardness, springiness, chewiness, tensile strength, and cooking time. It is suggested that an addition of SPI, 68.32 (g/kg of rice flour), MTG, 5.06 g/kg of rice flour), and GDL, 5.0 (g/kg of rice flour) improved the various parameters. The hardness of the final product with optimized values of SPI, MTG, and GDL was hardness (53.19 N), springiness (0.76), chewiness (20.28 J), tensile strength (60.35 kPa), and cooking time (5.15 min) [32].

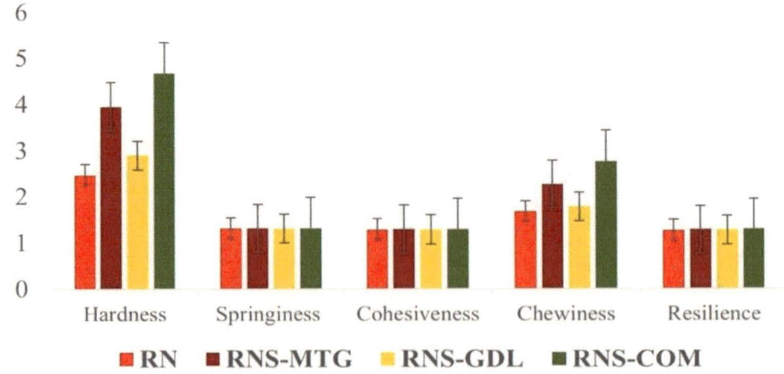

Figure 3.
Standardized textural properties of rice noodles (RN = rice noodles with 100% flour, RNS-MTG = rice noodles with 5% SPI, crosslinked with 1% MTG, RNS-GDL = rice noodles with 5% SPI, gelled with 1% GDL, RNS-COM = rice noodles with 5% SPI, crosslinked with 1% MTG and gelled with 1% GDL).

5. Improving the textural, structural, and cooking properties of air-dried rice noodles using SPI and combined cross-linking technologies

Despite having improved texture, taste, and flavor, fresh rice noodles deteriorate after preparation and are susceptible to spoilage due to their high moisture levels. Furthermore, fresh noodles are prone to discoloration, which is unappealing [33]. As a result, reducing the moisture content will extend the shelf life of the noodles while also preserving their texture and aroma, potentially increasing their market value [34]. Because of their less porous structure, air-dried noodles have been confirmed to shrink during processing, have poor rehydration characteristics, develop a tough texture, and have long cooking times [35, 36]. This problem could be solved by carrying out a steaming process before drying.

Therefore, RNS-COM was steamed before drying to yield air-dried RNS-COM (**Figure 4**). Findings revealed that RNS-COM was chewier and springier than air-dried RN because of the improved swelling of starch molecules in the noodle framework. Furthermore, the protein cross-links formed by MTG and GDL reinforced covalent networks in the noodles, increasing the chewiness. The springier the noodles, the denser the cross-links between amylose and other network forming molecules in a noodle structure [37, 38]. The improved springiness of the RNS-COM could be attributed to increased polymerization prompted by the additional proteins and cross-linking agents. GDL may have prompted some cold protein gelation and better overall interactions between starch molecules, resulting in stronger and harder RNS-COM with enhanced starch retrogradation on the noodles' surface [39].

Figure 4.
Air-dried RNS-COM (RNS-COM = rice noodles with 5% SPI, crosslinked with 1% MTG and gelled with 1% GDL)

Initially, the structurally optimized air-dried RNS-COM suffered a longer cooking time and a reduced yield; however, this was resolved by using pre-drying steaming treatments for 5 or 10 min. The subsequent air-dried RNS-COM had shorter cooking times, lower cooking losses, and higher yields. The microstructures of the steamed and dried combinedly-treated RNS-COM differed noticeably, but the relative crystalline structure of starch was preserved just after the steaming and drying treatments [40].

6. Our latest approach

Superheated steam (SHS) drying of food materials is a technological innovation with inherent benefits over hot air-drying processes, such as shorter drying rates due to higher temperatures, relatively low energy demand, little or emission of harmful pollutants in the atmosphere, and the absence of oxidation process owing to the unavailability of oxygen [41]. SHS enhanced the structurally improved dried RNS-COM (RNS-COM-SHS), cooking, and textural properties. The SHS, generated with a superheated steam oven (31L SHARP Healsio AX1700VMR) set at 120°C was able to unlock the framework of RNS-COM-SHS, leading to faster gelatinization of starch granules and a 1-minute decrease in optimum cooking time. RNS-COM-SHS had a compact structure with starch granules tightly integrated with the proteins (**Figure 5**). More cavities and less tight areas in RNS-COM-SHS would offer less restriction to moisture penetration into the noodle structure, lowering the cooking time and improving cooking yield.

Superheated steam processing had no adverse effects on the essential textural properties of the noodles, but it resulted in noodles that took less time to cook due to faster starch gelatinization, as evidenced by lower enthalpy. As a result, superheated steam processing of RNS improves the cooking and textural properties of dried rice-flour soy protein isolate noodles.

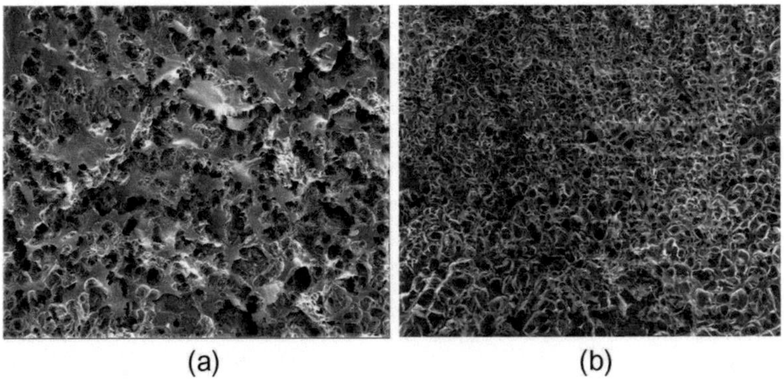

Figure 5.
Structural morphology of the noodles. (a) RN-SHS, (b) RNS-COM-SHS.

7. Conclusion

Fresh rice flour noodles with enhanced textural and mechanical properties were prepared by incorporating soy protein isolate, MTG, and GDL. Improvements in specific properties can be attributed to enhanced cross-linking of proteins due to MTG and GDL-induced cold gelation at reduced pH values.

Upon air-drying, the structurally enhanced RNS-COM showed longer cooking time and reduced cooking yield that could be solved by employing pre-drying steaming treatments. The robust network in air-dried RNS-COM can be opened by superheated steam, which allows for faster gelatinization, evidenced by reduced enthalpy, increased cooking yield, and sustained crystallinity of the starch granules in the noodle matrix.

Therefore, the superheated steam processing of RNS improves the cooking characteristics and textural qualities of dried rice-flour soy protein isolate noodles.

Acknowledgements

This research was funded by the Ministry of Higher Education, Fundamental Research Grant Scheme (FRGS/1/2019/STG01/USM/01/1). The authors would like to acknowledge Universiti Sains Malaysia (USM) for the Fellowship awarded to Moses Ojukwu.

Conflict of interest

The authors declare no conflict of interest.

Author details

Moses Ojukwu[1,2] and Azhar Mat Easa[1*]

1 Food Technology Division, School of Industrial Technology, Universiti Sains Malaysia, Minden, Penang, Malaysia

2 Department of Food Science and Technology, Federal University of Technology, Owerri, Imo State, Nigeria

*Address all correspondence to: azhar@usm.my

References

[1] Adachi M, Kanamori J, Masuda T, Yagasaki K, Kitamura K, Mikami B, et al. Crystal structure of soybean 11S globulin: Glycinin A3B4 homohexamer. Proceedings of the National Academy of Sciences. 2003;**100**(12):7395-7400

[2] Yeoh SY, Alkarkhi AF, Easa AM. Effect of cross-linking agents on physicochemical, textural properties and microstructure of canned soy protein isolate-yellow alkaline noodles prepared by retort processing. Journal of Food Processing and Preservation. 2014;**38**(3):1187-1197

[3] Lubowa M, Yeoh SY, Varastegan B, Easa AM. Effect of pre-gelatinised high-amylose maize starch combined with Ca^{2+}-induced setting of alginate on the physicochemical and sensory properties of rice flour noodles. International Journal of Food Science and Technology. 2021;**56**(2):1021-1029

[4] Loubes MA, Flores SK, Tolaba MP. Effect of formulation on rice noodle quality: Selection of functional ingredients and optimization by mixture design. LWT-Food Science and Technology. 2016;**69**:280-286

[5] Szczesniak AS. Texture is a sensory property. Journal of Food Quality and Preference. 2002;**13**(4):215-225

[6] Wee MSM, Goh AT, Stieger M, Forde CG. Correlation of instrumental texture properties from textural profile analysis (TPA) with eating behaviours and macronutrient composition for a wide range of solid foods. Food and Function. 2018;**9**(10):5301-5312

[7] Wangtueai S, Phimolsiripol Y, Vichasilp C, Regenstein JM, Schoenlechner R. Optimization of gluten-free functional noodles formulation enriched with fish gelatin hydrolysates. LWT. 2020;**133**:109977

[8] Karim R, Sultan MT. Yellow Alkaline Noodles: Processing Technology and Quality Improvement. NewYork: Springer; 2014

[9] Ahmed I, Qazi IM, Li Z, Ullah J. Rice noodles: Materials, processing and quality evaluation. Proceedings of the Pakistan Academy of Sciences, B. Life and Environmental Sciences. 2016;**53**(3):215-238

[10] Hatcher DW, Dexter JE, Anderson MJ, Bellido GG, Fu BX. Effect of blending durum wheat flour with hard white wheat flour on the quality of yellow alkaline noodles. Food Chemistry. 2009;**113**(4):980-988

[11] Tan TC, Easa AM. Comparative study of cooking quality, microstructure, and textural and sensory properties between fresh wheat noodles prepared using sodium chloride and salt substitutes. LWT-Food Science and Technology. 2018;**97**:396-403

[12] Qazi IM, Rakshit SK, Tran T, Ullah J, Khan MZ. Effect of blending selected tropical starches with rice flour on the cooking quality and texture of rice based noodles. Sarhad Journal of Agriculture. 2014;**30**(2):257-264

[13] Raungrusmee S, Shrestha S, Sadiq MB, Anal AK. Influence of resistant starch, xanthan gum, inulin and defatted rice bran on the physicochemical, functional and sensory properties of low glycemic gluten-free noodles. LWT-Food Science and Technology. 2020;**126**:109279

[14] Tan ZGL, Tan B. Starch noodles: History, classification, materials, processing, structure, nutrition, quality evaluating and improving. Food Research International. 2009;**42**(5-6):551-576

[15] Kim JI, Kee SL, Yoo SH. Quality improvement of rice noodle restructured with rice protein isolate and transglutaminase. Food Chemistry. 2014;**145**:409-416

[16] Lu Z, Collado LS. Rice noodles. In: Rice. Minnesota: Elsevier; 2019. pp. 557-588

[17] Nishinari K, Fang Y, Guo S, Phillips G. Soy proteins: A review on composition, aggregation and emulsification. Food Hydrocolloids. 2014;**39**:301-318

[18] Ojukwu JST, Easa AM. Cooking, textural, and mechanical properties of rice flour-soy protein isolate noodles prepared using combined treatments of microbial transglutaminase and glucono-δ-lactone. Journal of Food Science. 2020;**85**(9):2270-2277

[19] Chiti F, Dobson CM. Amyloid formation by globular proteins under native conditions. Nature Chemical Biology. 2009;**5**(1):15-22

[20] Tang H, Wu HP, Yu L, Li ZC, Yang XQ. Coagulation and gelation of soy protein isolates induced by microbial transglutaminase. Journal of Food Biochemistry. 2006;**30**(1):35-55

[21] Saifullah R, Abbas F, Yeoh S, Easa AM. Utilization of green banana flour as a functional ingredient in yellow noodle. International Food Research Journal. 2009;**16**(3):373-379

[22] Ramli S, Alkarkhi AF, Shin Yong Y, Min-Tze L, Easa AM. Effect of banana pulp and peel flour on physicochemical properties and in vitro starch digestibility of yellow alkaline noodles. International Journal of Food Sciences. 2009;**60** (Suppl. 4):326-340

[23] Li L-Y, Tan T-C, Liong M-T, Easa AM. Preparation and evaluation of chili powder-enriched layered noodles. LWT-Food Science and Technology. 2014;**59**(1):566-571

[24] Li LY, Easa AM, Liong MT, Tan TC, Foo WT. The use of microbial transglutaminase and soy protein isolate to enhance retention of capsaicin in capsaicin-enriched layered noodles. Food Hydrocolloids. 2013;**30**(2):495-503

[25] Ringgenberg E, Alexander M, Corredig M. Effect of concentration and incubation temperature on the acid induced aggregation of soymilk. Food Hydrocolloids. 2013;**30**(1):463-469

[26] Guo ST, Ono T. The role of composition and content of protein particles in soymilk on tofu curding by glucono-δ-lactone or calcium sulfate. Journal of Food Science. 2005;**70**(4):C258-C262

[27] Hsia S-Y, Hsiao Y-H, Li W-T, Hsieh J-F. Aggregation of soy protein-isoflavone complexes and gel formation induced by glucono-δ-lactone in soymilk. Scientific Reports. 2016;**6**:35718

[28] Dybowska B, Fujio Y. Optical analysis of glucono-δ-lactone induced soy protein gelation. Journal of Food Engineering. 1998;**36**(1):123-133

[29] Campbell LJ, Gu X, Dewar SJ, Euston SR. Effects of heat treatment and glucono-δ-lactone-induced acidification on characteristics of soy protein isolate. Food Hydrocolloids. 2009;**23**(2):344-351

[30] de Faria JT, Minim VPR, Minim LA. Evaluating the effect

of protein composition on gelation and viscoelastic characteristics of acid-induced whey protein gels. Food Hydrocolloids. 2013;**32**(1):64-71

[31] Alting AC, Hamer RJ, de Kruif CG, Paques M, Visschers RW. Number of thiol groups rather than the size of the aggregates determines the hardness of cold set whey protein gels. Food Hydrocolloids. 2003;**17**(4):469-479

[32] Ojukwu M, Ofoedu C, Seow EK, Easa AM. Optimization of soy protein isolate, microbial transglutaminase and glucono-δ-lactone in gluten-free rice noodles. Journal of the Science of Food and Agriculture. 2021;**101**(9):3732-3741

[33] Li K-X, Zhu Q-J, Sun T, Amza X-NG, Zhou H-M. Quality characteristics, structural changes, and storage stability of semi-dried noodles induced by moderate dehydration: Understanding the quality changes in semi-dried noodles. Food Chemistry. 2016;**194**:797-804

[34] Ismail M, Law C, Hii C. Transparency phenomena of flat-rice noodles (kuew teow) at drying at soaking variation. International Food Research Journal. 2016;**23**:195-202

[35] Oh I, Lee S. Rheological, microstructural, and tomographical studies on the rehydration improvement of hot air-dried noodles with oleogel. Journal of Food Engineering. 2020;**268**:109750

[36] Gulia N, Dhaka V, Khatkar B. Instant noodles: Processing, quality, and nutritional aspects. Critical Reviews in Food Science Nutrition. 2014;**54**(10):1386-1399

[37] Yi C, Zhu H, Bao J, Quan K, Yang R. The texture of fresh rice noodles as affected by the physicochemical properties and starch fine structure of aged paddy. LWT. 2020;**130**:109610

[38] Zhou Z, Xiong JC, Xiong H. Effect of cross-linked waxy maize starch on the quality of non-fried instant noodles. Starch-Stärke. 2015;**67**(11-12):1035-1043

[39] Low YK, Esah EM, Cheng LH. The impact of glucono-delta lactone (GDL) on rice flour pasting properties and GDL's dipping effects on the quality of rice noodles. Journal of Food Processing and Preservation. 2021;**45**(3):e14944

[40] Ojukwu M, Ahaotu N, Easa AM. Evaluation of air-dried soy protein isolate-rice noodles prepared via combined treatment with microbial transglutaminase and glucono-δ-lactone. International Journal of Food Science and Technology. 2022;**57**(2):1186-1196. DOI: 10.1111/ijfs.15491

[41] Sehrawat R, Nema PK, Kaur BP. Effect of superheated steam drying on properties of foodstuffs and kinetic modeling. Innovative Food Science Emerging Technologies. 2016;**34**:285-301

Chapter 13

Biodiesel Production Using Reactive Distillation Column Based on Intensification Processes

Ana Kiesse Zeleme and António André Chivanga Barros

Abstract

Environment concerns related to the use of fossil fuels are reflected in proposals for new conversion technologies to produce biofuels from biomass. The biofuels produced in this context have the same characteristics as petroleum derivatives, however, with reduced greenhouse gas emissions and with no sulfur in their molecular structures. In this context, a reactive distillation (RD) column was designed, constructed, installed, and operated using process intensification principles. It was applied in the production of biodiesel, using residual frying oil as the raw material, by the transesterification reaction, in a continuous regime. The process started with alcohol in excess in the reboiler, located in the bottom of the RD, which was heated through the combustion of liquefied petroleum gas (LPG) to produce ethanol vapor, which was recirculated in the column until stabilization. In this stage, the reagents were inserted into the feed tanks. Thus, the tank valves were opened for each reactant. The reaction products were recovered during the experiment from the bottom of the column and they were distilled to obtain two phases, biodiesel and glycerol. The results obtained from this study show that the use of an RD column can produce biodiesel in a continuous regime.

Keywords: reactive distillation, transesterification, biodiesel, column, process intensification

1. Introduction

Interest in fuel production from renewable sources has increased significantly in the recent years due to political issues and the economic instability of the oil market. This approach also brings environmental benefits associated with the lower levels of gaseous emissions from the combustion of fuels from non-renewable sources [1–3].

Thus, many procedures have been proposed for the production of biofuels, particularly biodiesel, which is a non-petroleum fuel produced by applying transesterification to break down the triglyceride present in fatty acids [3, 4]. In this reaction, the chemical structure of triglycerides in oil is broken down *via* the exchange of alkyl groups between an ester and alcohol, with the alcohol being used as a reactant.

Biodiesel is a promising alternative to conventional diesel fuel due to its renewable nature and the associated reduction in the emission of particles and greenhouse gases [1, 4, 5]. Homogeneous catalysis can be applied in the transesterification reactions involved in biodiesel production to obtain high yields in a relatively short time. However, since the catalysts cannot be reused and must be neutralized after the reaction, biodiesel does not compete favorably with fossil fuels [2]. This, together with environmental concerns, has led to the search for heterogeneous catalysts, which are eco-friendly and effective [2–7].

The use of a heterogeneous catalyst in the chemical reaction involved in biodiesel production can also reduce the time and cost of the process. In addition, it could be reused, resulting in a decrease in the level of impurities in the reaction products, and would allow continuous operation in a fixed bed reactor [4, 6].

The conversion efficiency is related to the catalytic activity of the heterogeneous catalyst used. In general, basic heterogeneous catalysts are more active than acidic heterogeneous catalysts, requiring shorter reaction times and relatively low reaction temperatures. Therefore, there are many advantages to performing the transesterification of triglycerides using a basic heterogeneous catalyst [2].

The synthesis of biodiesel with a heterogeneous catalyst can enhance the production and an economically feasible process can be achieved, mainly due to the lower costs related to the reuse of the catalyst and reduced levels of impurities in the products and byproducts of the reaction [6].

The methodology developed in this study facilitates a step change in the conceptual design process, offering a tool for the study of reactive distillation (RD) columns, with distillation, extractive distillation, esterification, and transesterification being performed in the same equipment, to allow biodiesel production in a continuous process.

The RD process is based on the process intensification concept, taking advantage of the synergy that occurs when combining the catalysis reaction and separation in a single unit, which allows the concurrent production and removal of products. This improves productivity and selectivity, reduces energy use, eliminates the need for solvents, and leads to highly efficient systems with improved environmental sustainability [8].

The use of RD as a reactive separation technology has attracted considerable attention given its high levels of efficiency [9] and the advantages include energy and capital savings, increased reaction conversion, high selectivity, and the use of the reaction heat [10]. Therefore, RD processes have been previously examined for application in esterification and etherification reactions [11–14].

A study developed by Kousoulidou et al. [15], based on the tests on the engine and vehicle, concluded that blending of biodiesel in petroleum diesel at 10% (v/v) changes fuel properties to a degree that can affect combustion characteristics, such as the start of ignition and heat-release rate. For the authors, the limited evidence on the effects of such blends on the combustion and emissions of diesel engines was not originally designed to operate on biodiesel blends.

The authors made a number of experiments with 10% (v/v) (B10) biodiesel fuel of palm oil origin performed on a light duty. The measurements included in-cylinder pressure, pollutants emissions, and fuel consumption. Combustion effects were limited but changes in the start of ignition and heat-release rate could be identified. Emission effects included both higher and lower smoke and NOx, depending on the operation point. Others studies developed by others authors cited in the literature show the possibility to increase the biodiesel perceptual in the blend with diesel that the results show the possibility to increase this perceptual without the modification of engine characteristics.

The complexity involved in improving the performance of this process is associated with the multi-feed positions and selection of the reactants, catalysts, and solvents.

In this study, the transesterification reaction was carried out with homogeneous catalysts, in a continuous distillation column reactor, and the operational performance was evaluated in terms of biodiesel production quality. Experiments were also conducted using an RD column, where the reactant was supplied at the feed position.

The process intensification concept was adopted, using an RD column for the transesterification reaction, to recover the reactant (ethanol) used and to break the azeotrope point between ethanol and water, using the extractive distillation being performed in the top section of this equipment, as shown in **Figure 1**.

In this case, a homogeneous catalyst was used in the RD column. The same operational conditions were applied in all cases studied, including the residence time, the mass of reactant, and molar mass used in the reaction. The results obtained show good reaction efficiency, which was improved by applying the process intensification principles. This concept can be applied to other synthesis processes carried out in a continuous regime.

2. Experimental

2.1 Material

The use of residual frying oil for bio-fuel production is associated with advantageous economic and environmental factors. In general, when food is processed, the vegetal oil or raw fats are exposed to environments with high temperatures, where

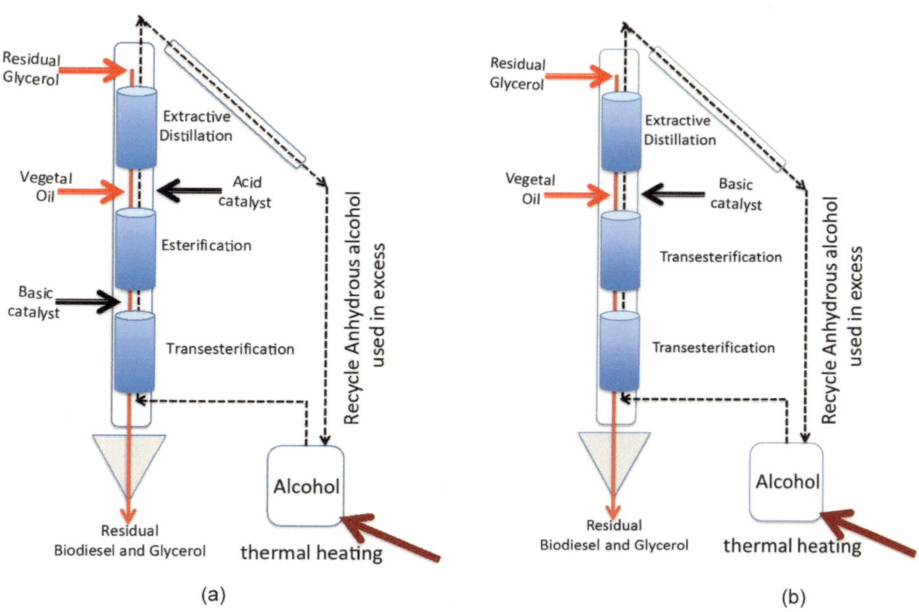

Figure 1.
Reactive distillation column proposed herein where (a) processing of raw material with high acid index and (b) processing of raw material with a reduced acid index.

the thermal degradation of oil occurs. The residual frying oil produced can be used for biofuel production due to its potential for energy production through a chemical reaction. For this study, the residual frying oil (soybean oil), with an acid value of 0.2 mgKOH/g and average molecular weight of 884.65 g/mole, was used in all experiments.

2.2 Reactive distillation column

An RD column was designed, constructed and installed (**Figure 1**). It was composed of three modules, which incorporate internally packing to increase the interfacial area for vapor-liquid contact, mass transfer and chemical reaction. The module located in the intermediate part of the column (**Figure 2a**) receives the limiting reactant (oil) and the excess reactant (ethyl alcohol) with dissolved sulfuric acid, used as a catalyst in the esterification reaction, based on the conversion of free fatty acids into biodiesel. The flow rate of the reagents is associated with the stoichiometric balance. In this module, the mixtures are fed vertically, downward, through the action of gravity, and interact strongly with the ascendant vapor of the alcohol used in excess coming from the reboiler, installed at the base of the equipment.

After the esterification reaction, the downward mixture flows to the transesterification module (**Figure 2b**) where feeding with the basic catalyst occurs. In this case, the feed catalyst must neutralize the acid catalyst from the esterification and the remaining mass acts as a catalyst for the transesterification reaction in the bottom module. As previously mentioned, the transesterification reaction produces biodiesel and glycerol, which are collected in a decanting funnel installed at the bottom of the column. In the decanting funnel, the separation of the products into two phases occurs, that is, a light phase of the crude biodiesel and a heavy phase of crude glycerol (**Figure 3**).

In the esterification reaction, with the dissolution of the catalysts in excess reagent, water is produced. This reduces the level of purity of the alcohol in the

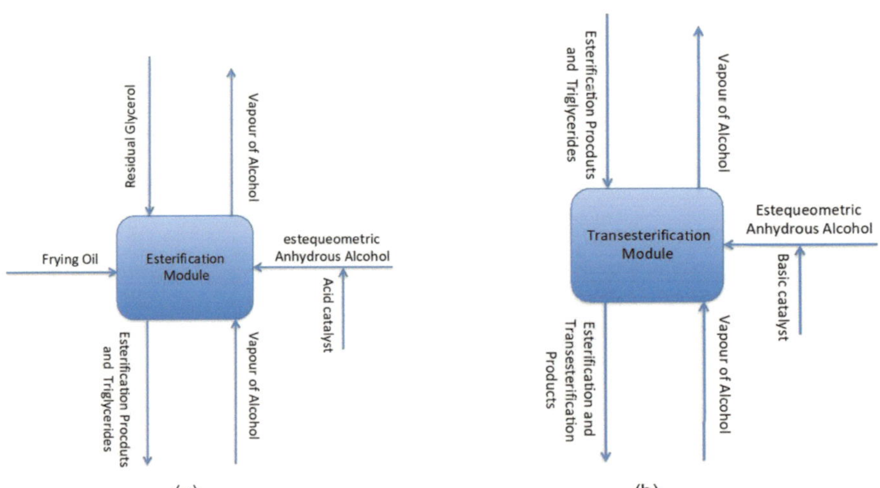

Figure 2.
Modules that composed of the reactive distillation column: (a) Esterification module and (b) Transesterification module.

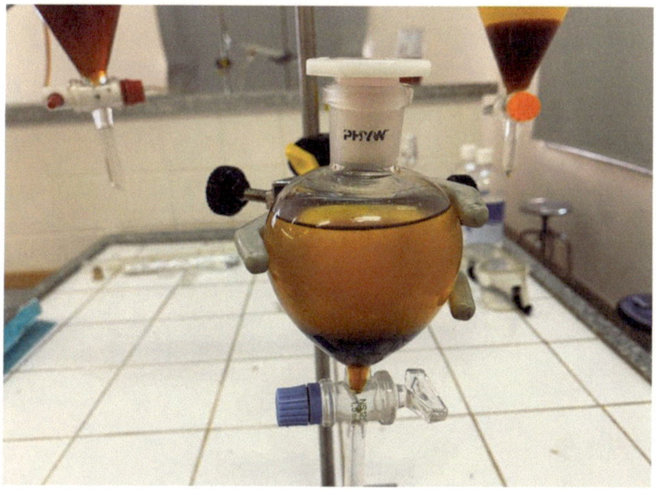

Figure 3.
Decanting funnel installed at the bottom of the RD column.

vapor phase and thus the capacity of the chemical reaction. For this, a module for extractive distillation (**Figure 4a**) is therefore required at the top of the column to break the azeotrope point between ethanol and water and to produce the anhydrous ethyl alcohol.

In this process, the residual glycerol recovered at the bottom of the RD column is used as a solvent in the extractive distillation, being fed at the top of this equipment and flowing down through the action of gravity. Along its path, the solvent interacts with the ethanol vapor and the interaction between the flows modifies the thermodynamic characteristics of the ethanol, breaking the azeotrope point and increasing the purity of this compound (**Figure 4a**).

The top products are condensed and flow downward to the reboiler where they are heated to produce the vapor needed for the RD column (**Figure 4b**). Based on this principle, the alcohol vapor in excess operates in the recirculation circuit, being involved in the esterification and transesterification reactions and the extractive distillation, and in the last step, the crude glycerol is used as a solvent. On the other hand, the steady-state characteristic of this process guarantees continuous feeding and removal of the products at the bottom of this equipment, with the recirculation of ethanol, used in this study as the reactant in excess.

Figures 2 and **4** are compressed into **Figure 1**, which represents the RD column, with the three modules described herein. There are four feed flows in the RD column. The operations involved in this process occur in the top condenser, the bottom decanting funnel, and the reboiler, used to produce the vapor flow, and vapor is used to allow effective chemical reactions to produce biodiesel and glycerol.

2.3 Experimental procedures

2.3.1 Reactive distillation column operation

To carry out the biodiesel production experiments in the RD column, the procedures involved the following steps:

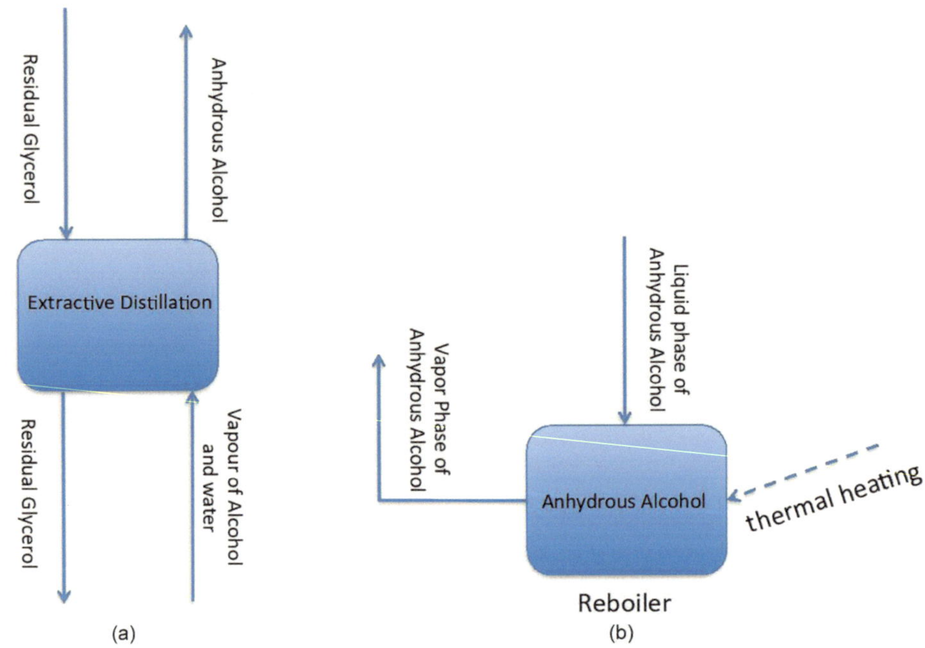

Figure 4.
Description of streams associated with chemical reaction modules: a) Module of Extractive Distilation used to break the azeotrope point of mixture and; b) Reboiler used to vaporization of alcohol.

a. The limiting reactant was characterized to determine the acidity index. Under these conditions, the operational sequence can be defined with or without the esterification reaction. If the acidity index is lower than that established in the literature, only the transesterification reaction is used (**Figure 1b**);

b. After establishing the quality of the raw material, the stoichiometric equation is used to determine the masses of reagents needed for the chemical reaction. The calculated mass is then converted into flow rates;

c. The appropriate masses of limiting and excess reactants and the catalyst are weighed out. In this case, the catalyst mass was determined from a previous transesterification reaction study performed in a laboratory batch-scale reactor. Under the study conditions, 1.2% catalyst mass in relation to the oil mass was defined;

d. In the next step, the mass of the limiting reactant was heated to the reaction temperature, in this case until reaching the boiling point of ethanol. The mass of alcohol in excess was heated and used to dissolve of catalyst mass;

e. The previously heated masses were inserted into the respective feed tanks. The use of heated mass minimizes the thermal barrier due to the temperature difference between the ascendant vapor and the feed flow;

f. The anhydrous alcohol, previously held in the reboiler tank, produces the vapor that flows into the RD column, based on the recirculation process described above. Thus, the valve of the reactant tank is open, and the flow rate is based on the values established in the stoichiometric balance. Since the acidity index value was below that established in the literature, esterification and transesterification modules were used in the transesterification reaction, as described in **Figure 1b** and;

g. The feed flow rates and the operation temperature were adequately controlled, and the products were removed progressively until the end of the reaction.

2.3.2 Measuring the process performance

When the reaction was concluded, the products were purified using a batch distillation column to recover the excess alcohol. For this, phase separation into a light phase containing biodiesel and a heavy phase of glycerol occurred. The biodiesel recovered was used to determine the yield, in terms of biodiesel mass-produced in relation to the frying oil mass initially used in the process. The biodiesel was then neutralized and dried by evaporation to ensure appropriate quality for use in internal combustion to produce electricity.

2.3.3 Batch biodiesel production

Before the experiments were performed in the RD column, tests were performed in a batch reactor using the reagents under the same operational conditions, with the following methodology:

a. The acidity index of the frying oil was determined. As this parameter is below the limit established in the literature, only the transesterification reaction was carried out;

b. The stoichiometric balance was determined and the masses of reactants for the transesterification reaction were weighed out. For the excess reactant, an amount six times more than the stoichiometric mass was used;

c. The reagent masses previously weighed out were heated until the reaction temperature and the basic catalyst was dissolved in the heated alcohol;

d. The mass of alcohol with the catalyst was inserted in the reactor with the previously heated oil, starting the transesterification reaction, which proceeded for 1 h;

e. At the end of the reaction, the mixture was placed into a distillation flask to recover the alcohol in excess. The remaining mass was then placed in the decanting funnel for the phase separation into biodiesel (light phase) and glycerol (heavy phase).

2.3.4 Characterization of reaction products

The characterization of the biodiesel was carried out through the analysis of the ester content, kinematic viscosity, acid index, and specific mass. The kinematic viscosity was obtained using an Ostwald viscometer based on the ABNT NBR 10441 methodology and, to determine the acidity index, titration with potassium hydroxide was carried out based on the ABNT NBR 14448 method. The specific mass was determined at room temperature (20± 2°C) using a 25-mL pycnometer calibrated with distilled water. These procedures were also used in a study by Barros et al. [1], who experimented using a batch reactor.

3. Results and discussion

3.1 Batch experiments

The experiments in the RD column were carried out based on reference data obtained in a batch reactor, mainly related to the following parameters: (a) reaction time; (b) reagents and catalyst mass; (c) agitation degree; and (d) mass ratio. The transesterification data from the batch reactor were used to close the mass balances and to determine the conversion rates in terms of the biodiesel mass-produced. In this study, frying oil (92.352 g), anhydrous ethanol (88.9 g), and KOH catalyst (1.108 g) were used. The equivalent catalyst mass was 1.2% of the frying oil mass. The crude biodiesel produced was characterized to determine the quality of the biofuel and the results were as follows: density 0.8785 g/mL; kinematic viscosity 1.920 mm^2/s; and acidity index 3.1668 mgOH/g. These parameters deviated from the specifications established by the ANP [16], but the biofuel needs to be purified before the final evaluation.

The operational parameters of the batch reactor were used to carry out the experiments in the RD distillation column, with the configuration shown in **Figure 1b**. Only two modules were used for the transesterification and the purification of the ascending and recycled alcohol vapor by extractive distillation (performed in the top module). The separation of the products from the RD column in the decanting funnel is shown in **Figure 2**, where biodiesel is the light phase and glycerol is the heavy phase.

3.2 Reactive distillation column operation

The experiments were performed in an RD column with the transesterification reaction carried out using two modules. Extractive distillation was conducted in the top module of the unit. During the experiments, the performance efficiency was monitored, with the ethanol vapor flowing into the column and passing through each module until stabilization (the steady-state) was reached. When the reagents are inside in the tank, the valves of the feed tank is open, there is an interaction between liquid and vapor phases, ensuring the turbulence necessary for effective mass and heat transfer and the chemical reaction to occur. The feed flow was previously heated to minimize the thermal barriers in the column, characterized by the condensation of the vapor flow, due to interaction between the phases. If this phenomenon occurs, there will be a decrease in the vapor flow rate toward the top of the equipment, which can limit the operational quality of the process. During the operation of the RD column, the bottom products are progressively recovery and after distillation, they are placed in the decanter tank where two phases are formed: a light phase (biodiesel) and a heavy phase (glycerol).

To obtain the results 1240.00 g of frying oil, 208.79 g of anhydrous ethanol (99.9%) and 14.888 g of catalyst (equivalent to 1.2% of the frying oil) were used. Based on the performance of the experiments over 1 h, the following feed flow rates were defined: (a) frying oil 20.67 g/min and (b) anhydrous ethanol with dissolved catalyst 3.728 g/min. Using these operational conditions, with the RD column in the steady state, the measured flow rate of the ethanol vapor in excess was 16.42 g/min, corresponding to 4.4 times less than the stoichiometric flow rate. During the experiments, the products were collected from the bottom of the RD column, as shown in **Figure 1** (**a** and **b**).

3.3 Characterization of products

The physico-chemical characteristics of the residual biodiesel from the batch reactor and the RD column produced based on process intensification principles, were determined and the results were compared with international standards. The values obtained for the density, acidity index, refractive index, and kinematic viscosity can be seen in **Table 1**.

The international standard values for the biodiesel parameters were extracted from the Brazilian standards NBR 7148 and NBR 14065 for density at 20°C, NBR 14448 for the acidity index, and NBR 10441 for kinematic viscosity at 40°C. A comparison between the data shows significant deviations, mainly for the acidity index and kinematic viscosity. In this study, purification of the product is required for the biodiesel to meet international standards. In general, the purification process is based on extraction with an acid solution followed by evaporation to dry the biodiesel [17].

In this context, for better qualification of the crude biodiesel, liquid-liquid extraction was carried out with an acid water solution, and the masses used were previously defined based on a titration procedure. An acid water solution was employed in association with the basic catalyst used in the transesterification reaction. In this procedure, the acid mass required for the titration was determined, and the acid solution was prepared, heated up to 60°C, and then mixed with biodiesel under agitation. After this step, the mixture was transferred to a decanting funnel and left to stabilize for approximately 40 min. The phases were separated, and the biodiesel phase was transferred to a distillation flask to eliminate traces of water by drying.

When the purification step was finished, the biodiesel was characterized using the procedures described in the methodology section of this chapter, and the results are given in **Table 2**.

Table 2 shows the characteristics of the biodiesel samples obtained from the RD column and batch reactor. The acidity index for the former adheres to the established international standard, but this is not the case for the biodiesel from the batch reactor. For both cases studied, the kinematic viscosity was below the value given in the

Reference	Processed mass (g)	Density 20°C (kg/m³)	Refraction index at 20°C	Acidity index (mgKOH/g)	Kinematic viscosity at 40°C (mm²/s)	Yield (%)
Batch reactor	92.352	878.5	1.453	3.1668	1.920	92.306
RD column	1240.0	876.7	1.454	2.0255	1.812	15.441
International standard	—	850.0–900.0	1.45	≤0.5	3.0–6.0	—

Table 1.
Physico-chemical parameters for biodiesel obtained in this study.

Reference	Processed mass (g)	Density 20°C (kg/m³)	Refraction index at 20°C	Acidity index (mgKOH/g)	Kinematic viscosity at 40°C (mm²/s)	Yield (%)
Batch reactor	92.352	876.387	1.456	1.1077	1.885	92.306
RD column	1240.0	880.682	1.455	0.5027	1.651	15.441
International standard	—	850.0–900.0	1.450	≤0.5	3.0–6.0	—

Table 2.
Physico-chemical data for purified biodiesel.

standard, probably due to the quality of the raw material used in this study. However, the refraction index and density of both samples lie within the ranges set by the international standard.

The results obtained in this study are similar to those reported by Colombo et al. [2, 5], Pitt et al. [6], and Barros et al. [1], who investigated biodiesel production by esterification and transesterification, in recirculation reactors and a batch reactor, using a raw material with a reduced acidity index.

A limitation of this study is related to the rate of conversion of the raw material into biodiesel (approximately 15.441%). This could be due to the reduced residence time in the RD column and/or the turbulence resulting from the interaction between the two phases involved in the process. Thus, further studies need to be conducted to determine the optimum feed reactant flow rate to maximize the residence time and increase the degree of conversion. Alternatively, the length or number of modules in the RD column could be increased to increase the reaction rate.

However, the turbulence phenomenon involves strong interaction between the liquid and vapor phases and this should be further investigated to better understand the relevance in terms of conversion rates. In addition, the use of other types of internal packing may improve the yield or the degree of conversion of the raw material to biodiesel.

4. Conclusions

The results reported herein can be used to conclude that:

a. The RD column developed and used in this study is an apparatus with high operational flexibility to carry out esterification, transesterification, and conventional and extractive distillation, under different conditions and evaluate the parameters that affect the conversion rate;

b. The batch reactor can be used to support continuous processes, characterized by the use of RD columns, with high operational flexibility;

c. The reduced residence time in RD columns limits the application of this technical process, which requires optimization to increase the conversion rate and obtain products with high purity;

d. The RD column can be characterized as innovative, mainly when the reactant in excess is used as a circulating vapor, which is purified in the top module by extractive distillation and recycled to the bottom tank and;

e. The RD column promotes a continuous process with the potential to increase the production flow rate to meet industrial demand.

Acknowledgements

The authors wish to thank Prof. Dr. A. A. Chivanga Barros for his dedication to developing this project, mainly for spending particular money to finance this study.

Conflict of interest

The authors declare no conflict of interest.

Author details

Ana Kiesse Zeleme and António André Chivanga Barros*
Department of Engineering and Technology (DET), Instituto Superior Politecnico de Tecnologias e Ciências (ISPTEC), Talatona, Luanda, Angola

*Address all correspondence to: chivanga.barros@isptec.co.ao

References

[1] Barros AAC, Wust E, Meier HF. Study of the technical-scientific feasibility of biodiesel production from fatty residues. Engenharia Sanitaria e Ambiental. 2008;**13**(3):255-262

[2] Colombo K, Ender L, Santos MM, Barros AAC. Production of biodiesel from soybean oil and methanol, catalyzed by calcium oxide in a recycle reactor. South African Journal of Chemical Engineering. 2019;**28**:19-25

[3] Chiarello L, Porto T, Barros AAC, Simionatto E, Botton V, Wiggers V. Bosting an oil refinery into a biorefinery. Angolan Mineral, Oil & Gas Journal. 2020;**1**:1-5

[4] Colombo K, Ender L, Barros AAC. The study of biodiesel production using CaO as a heterogeneous catalytic reaction. Egyptian Journal of Petroleum. 2017;**26**(2):341-349

[5] Colombo K. Biodiesel production via transesterification with heterogeneous catalysis using CaO as a catalyst [master's thesis]. Brazil: Blumenau Regional University; 2013

[6] Pitt FD, Domingos AM, Chivanga Barros AA. Purification of residual glycerol recovered from biodiesel production. South African Journal of Chemical Engineering. 2019;**29**:42-51

[7] Semwal S, Arora AK, Badoni RP, Tulli DK. Biodiesel production using heterogeneous catalysts. Bioresource Technology. 2011;**102**:2151-2161

[8] Kiss AA. Novel catalytic reactive distillation processes for a sustainable chemical industry. Topics in Catalysis. 2019;**62**:1132-1148. DOI: 10.1007/s11244-018-1052-9

[9] Stankiewicz AI, Moulijn JA. Process intensification: Transforming chemical engineering. Chemical Engineering Progress. 2000;**96**:22-34

[10] Cárdenas-Guerra JC, LópezArenas T, Lobo-Oehmichen R, Pérez-Cisneros ESA. Reactive distillation process for deep hydrodesulfurization of diesel: Multiplicity and operation aspects. Computers and Chemical Engineering. 2010;**34**:196-209

[11] Subawalla H, Fair JR. Design guidelines for solid-catalyzed reactive distillation systems. Industrial and Engineering Chemistry Research. 1999;**38**:3696-3709

[12] Taylor R, Krishna R. Modeling reactive distillation. Chemical Engineering Science. 2000;**55**: 5183-5229

[13] Huss RS, Chen M, Malone MF, Doherty MF. Reactive distillation for methyl acetate production. Computers and Chemical Engineering. 2003;**27**:1855-1866

[14] Huang K, Wang SJ. Design and control of a methyl tertiary butyl ether (MTBE) decomposition reactive distillation column. Industrial & Engineering Chemistry Research. 2007;**46**:2508-2519

[15] Marina K, Georgios F, Leonidas N, Zissis S. Biodiesel blend effects on common-rail diesel combustion and emissions. Fuel. 2010;**89**(11):3442-3449

[16] Agência Nacional do Petróleo, Gás Natural e Biocombustíveis (ANP). Biodiesel [Internet]. Brazil; 2020.

Available from: http://www.anp.gov.br/biocombustiveis/biodiesel [Cited: 18 May 2020]

[17] Dib FH. Production of biodiesel from recycled waste oil [master's thesis]. Ilha Solteira (SP), Brazil; 2010

Chapter 14

Routes to Aggregate Value to Soybean Products

Fernando Luiz Pellegrini Pessoa,
Ewerton Emmanuel da Silva Calixto,
Hugo Gomes D'amatto Villardi, Ana Lucia Barbosa de Souza
and Bruna Aparecida Souza Machado

Abstract

This chapter presents routes to aggregate value to soybean oil products by sustainable and economical sources for biofuels. The traditional production routes, such as pyrolysis, allows, by mixing oils and plastics, to generate bio-oils with high burning power. One example of an alternative route is single-step interesterification, where the methyl acetate reacts with the triglycerides in the oil, forming fatty acid methyl esters and triacetin as a by-product. This is a great advantage of this route, as in addition to its commercial value being greater than that of glycerol, it can be mixed with biofuel without changing its characteristics. The main objective is present routes that may reduce cost in general, in addition to generating co-products that allow an increase in the process added value.

Keywords: soybeans products, aggregated values, triglycerides, triacetin, methyl esters

1. Introduction

Originated from Asia, soy is the largest oilseed in volume, production and international trade. In terms of food, soy is the main source of protein and the second most consumed oil in the world, only behind palm [1]. Furthermore, it has an importance that involves the development of a productive complex including the processing of the bean and its main products: crude oil and bran; also, its use for animal food edible oil and fuel (see **Figure 1**).

The soybean industrial processing can be summarized in two steps: (1) the crude oil production with soybean residue; (2) Crude oil refining to obtain other products (e.g.: refined oils, margarine, hydrogenated fat).

The volume of soy sold is extremely important for the economy, especially for the main world producers, such as Brazil. In order to have a positive trade balance, Brazil needs to export tons of primary products, which have low unit values to offset the import of value-added products, such as electronic appliances and machines. In addition, processing soybeans increases the job offer, further heating up the economy. Therefore, adding value to the soy production process is very important for a country's economy.

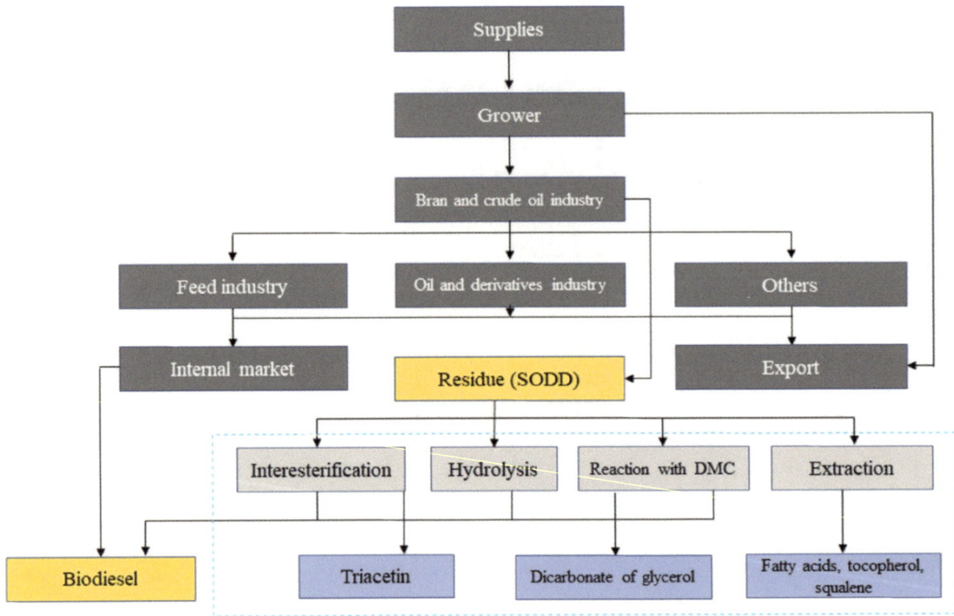

Figure 1.
Soy production chain.

Based on this, this chapter aims to discuss and present routes for the processing of soy, especially the Deodorized Soybean Oil Distillate (DSOD) and viable options to produce biodiesel to add value to the soy chain.

2. Oil refining

Refining consists of a set of processes that aim to transform crude oils into edible oils. The refining process aims to improve the appearance, odor and taste of the oil, which occurs with the removal of certain components from the crude oil. There are two types of refining, chemical and physical. These names are related to the process of removing unwanted fatty acids in the oil. The main steps in the refining processes are: degumming (hydration), neutralization (deacidification), whitening (clarification) and deodorization.

The removal of traces of components responsible for undesirable odors and tastes occurs in the step called deodorization. According to Sangoi and Almeida [2], the reduction of free fatty acids contained in oils promotes the removal of flavors and odors, at a range of 0.01% to 0.03%, eliminating the undesirable characteristics of the oil. This step generates a residue known as a distillate from soybean oil deodorization. According to Ma and Hanna [3], this substance is concentrated in tocopherols (vitamin E), sterols and fatty acids. DSOD is obtained from the precipitation of an aqueous phase during deodorization, and the compounds of interest are insoluble in water [4].

According to Aranda and Morlock [5], to produce a ton of biodiesel 0.91 tons of fatty acids are needed. According to Fontana [6], 0.1% of DSOD are generated in the refining of soybean oil, which would provide a quantity of 6440 tons of fatty acids, considering a production of 9.2 million tons of oil, according to Abiove [7] and an average percentage of 75% of fatty acids in DSOD. These values show the potential of the residue, increased by the logistical facility, as it is found in oil refining industries, unlike residual oils, which have a high logistical cost and variable production.

3. Sludge extraction

The most interesting components present in the sludge are fatty acids, however some works have evaluated the potential of other components such as sterols, tocopherols and squalene.

There are studies in the literature that evaluated the extraction of components from the deodorized distillate, such as sterols using supercritical extraction. The results show a recovery of 76% with a purity of 60%, both by weight. These substances are important because they act in the reduction of cholesterol in the blood, being of great interest to the pharmaceutical industry [8]. Similar works using enzymes are also presented and recovery of up to 87.7% of sterols [9].

4. Biodiesel

Biodiesel is currently an important biofuel of global interest, as it is considered a sustainable energy source of renewable origin [10]. As the most used raw material for production is refined soy oil, the production of fuel has the potential to close the carbon cycle [11]. In addition, the use of biodiesel generates less pollutants than fossil diesel, mainly CO_2, CO and SO_2, although it presents higher NO_x emissions [12, 13].

Brazil is a major world producer of biodiesel, ranking second in production volume in 2021, totaling approximately 6.9 billion liters, only behind the United States, which will produce more than 8.5 billion liters [14].

Another option to add value with deodorized distillate is direct esterification, that is, the synthesis of fatty acids to esters. This pathway was evaluated in a subcritical environment, as fatty acids are miscible in ethanol, unlike triglycerides, which are constituents of refined oils. These conditions, even though they demand high temperatures, dispense with the catalyst separation step, allowing for greater gains in the process. Some studies point to a 97% conversion of fatty acids into esters using subcritical environments at 100°C [15].

The economic potential of a complete process is available in the literature [16]. The work evaluated the potential of DSOD for the supercritical synthesis of esters and recovery of squalene. The squalene recovered was 31 g per kg DSOD with 98% purity by weight, the esters had a purity of 88% plus a mixture of tocopherols and sterols. The results of the economic analysis show, for the worst scenario, that is, with higher raw material values and lower product sales, a gross margin (MB) of 35% and the contribution margin index (BMI) of 29% for DSOD. These indicators are presented in Eq. 1 and 2.

$$MB = \frac{Product\ Revenue\,(PR) - Raw\ Material\ Cost\,(RMC)}{Product\ Revenue\,(PR)} \tag{1}$$

$$MB = \frac{Product\ Revenue\,(PR) - Operational\ Cost\,(OC)}{Product\ Revenue\,(PR)} \tag{2}$$

Other alternative routes are discussed in this chapter. The focus is on the synthesis of biodiesel via chemical reactions other than transesterification via heterogeneous, enzymatic and medium catalysis conventional supercritical, which promote the synthesis under different conditions.

As the formation of glycerol as a by-product is linked to the transesterification of triglycerides by alcohols, other supercritical routes propose to use other reagents for the same purpose – methyl acetate, acetic acid and dimethyl carbonate –, generating by-products of greater economic value than glycerol. These alternative routes, therefore, have all the advantages and disadvantages mentioned for the supercritical transesterification process, with the difference that they present more economically attractive by-products [13].

5. Interesterification with methyl acetate

This route consists of a single step, in which the interesterification between the methyl acetate and the triglyceride of the oil takes place, forming the fatty acid methyl esters and the triacetin as a by-product. The reaction, like the transesterification, occurs in three reversible steps: in the first, a triglyceride molecule reacts with a methyl acetate molecule to form a FAME (fatty acid methyl esters) and monoacetyl diglyceride molecule. Then, one molecule of monoacetyl diglyceride reacts with the second molecule of methyl acetate to form another of FAME and diacetyl monoglyceride; finally, diacetyl monoglyceride reacts with a methyl acetate molecule to form the third FAME molecule and triacetin, as can be seen in **Figure 2**.

If fatty acids are present in the oil, they are esterified by methyl acetate in FAME, generating acetic acid as a by-product. Other parallel reactions that can occur are the hydrolysis of triglycerides due to the presence of water, forming fatty acids and glycerol. This, in turn, reacts with acetic acid derived from esterification, forming triacetin and water [17, 18].

Triacetin formed as a by-product is the great advantage of this route. Although, its commercial value is greater than that of glycerol, the sale of biodiesel with this substance in its composition can be even more advantageous as it will increase the volume of biofuel generated, now a mixture of FAME and triacetin. Thus, there would be no need to purify triacetin in the process [18].

This final product alternative is only possible because triacetin, in addition to being miscible with fatty acid methyl esters, does not influence the fuel properties in a way that leaves them out of the pattern. Important properties of biodiesel as a fuel, such as kinematic viscosity, pour point, cloud point, cold filter plugging point,

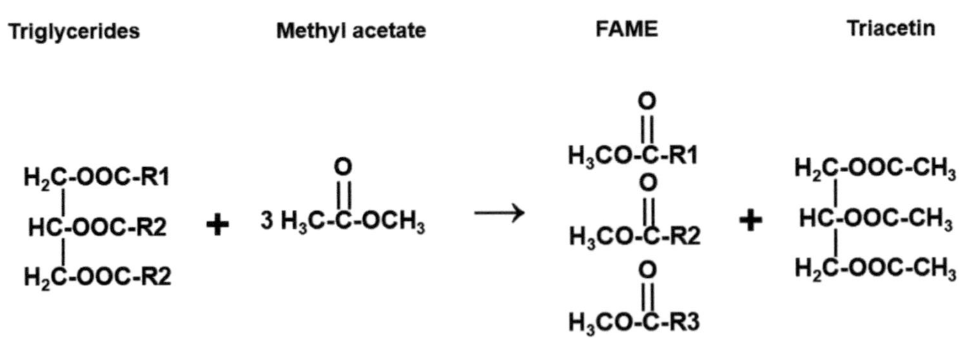

Figure 2.
Overall FAME synthesis reaction from triglycerides and methyl acetate.

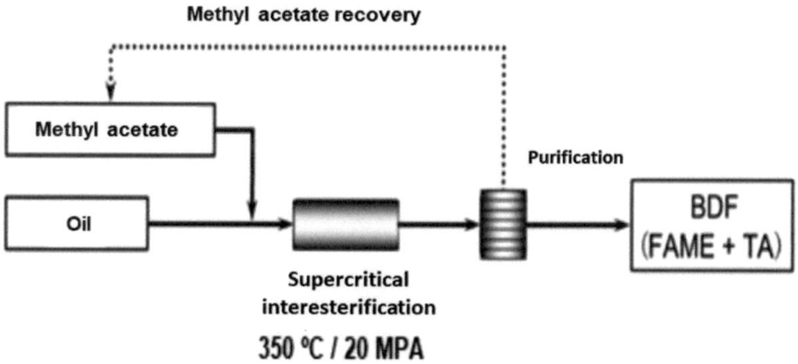

Figure 3.
Simplified diagram for the interesterification process with methyl acetate. Source: [18].

flash point and cetane number, are not changed with the addition of the by-product, remaining within European and North American standards [19]. However, due to the influence on the heat of combustion and density of the biofuel, the ideal is a mixture with a maximum of 10% by mass of triacetin - the European standard establishes 900 kg/m^3 as a density limit, as well as the Brazilian legislation on ANP [18, 20, 21].

Figure 3 shows a preliminary scheme for the process was proposed by Saka and Isayama [18].

6. Acetic acid hydrolysis and methanol esterification

This route consists of two steps: in the first, triglycerides, when reacting with acetic acid, are converted into fatty acids, which are esterified by methanol to FAME. The main justification for this route is the adoption of reaction temperatures of up to 300°C, avoiding thermal degradation of the esters.

6.1 Step 1: reaction with acetic acid

The route begins with the breakdown of triglycerides in the oil by acetic acid, a reaction in which one mole of triglyceride reacts with three moles of acid, forming one mole of triacetin and three moles of fatty acid, as shown in **Figure 4**. The reaction also takes place in three reversible steps: first, an acetic acid molecule reacts with a triglyceride molecule, forming a fatty acid molecule and a monoacetyl diglyceride molecule, which reacts with another acetic acid molecule to form the second fatty acid molecule and diacetyl monoglyceride; this reacts with the last acetic acid molecule to form again fatty acid and finally triacetin [22].

Regarding the cost of raw materials, acetic acid is a viable obtaining reagent because it is a commodity of the chemical industry, mainly used for the production of the acetate monometer of vinyl, for the production of polyvinyl acetate (PVA) [23].

6.2 Step 2: supercritical esterification with methanol

The second reaction of this route takes place in just one reversible step, in which a fatty acid molecule reacts with a methanol molecule, forming a FAME molecule and a water molecule, as shown in **Figure 5**.

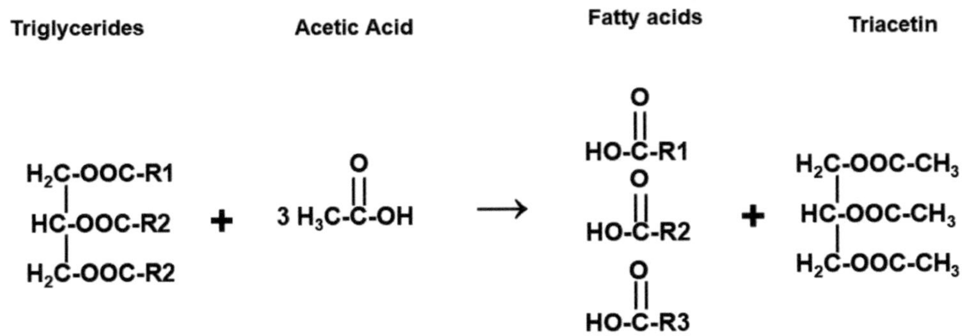

Figure 4.
Hydrolysis reaction between triglyceride and acetic acid. Source: [22].

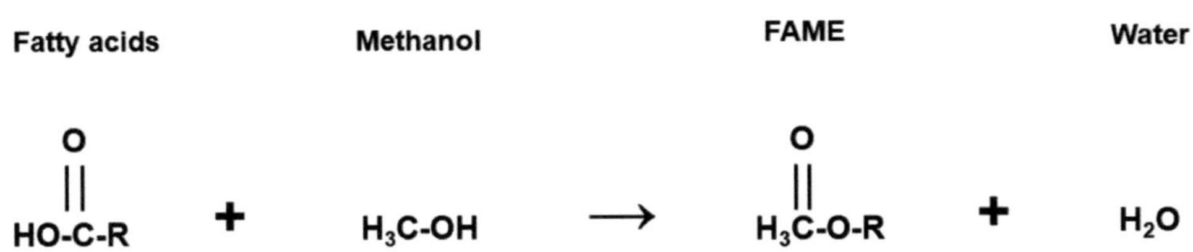

Figure 5.
Esterification reaction between fatty acids and methanol.

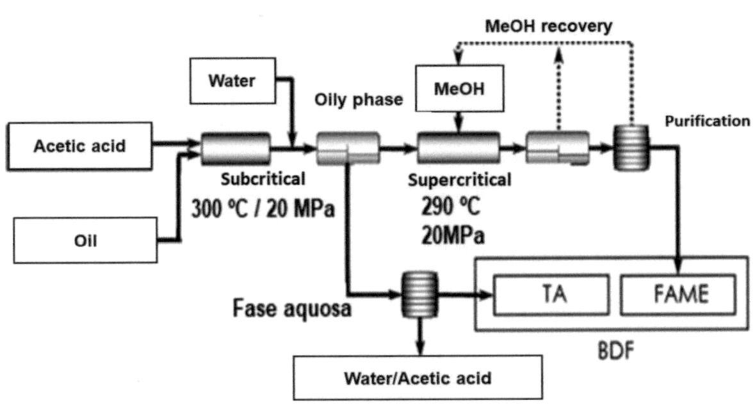

Figure 6.
Diagram of the process of hydrolysis with acetic acid and esterification with methanol. Source: [22].

After step 1, where the oil reacts with acetic acid through subcritical hydrolysis, the reaction products are separated by aqueous washing, from which triacetin and acetic acid are recovered in the step aqueous and fatty acids are taken to the second reactor with methanol under supercritical conditions for esterification. A phase separator is used to recover some of the excess methanol, followed by an unspecified scrubber that will separate the rest of the methanol and water from the FAME. The aqueous phase of the first stage is also subjected to unspecified purification to obtain triacetin, which will be mixed with FAME and, thus, composes the biodiesel in the final product. Performing an intermediate separation between the two steps can be advantageous to reduce the size of equipment in the second step. However, this imposes successive heating and cooling on the process, increasing energy expenditure, which can be one of the biggest disadvantages of this route [22].

Literature reports a yield of 95% in mass compared to the oil used in the first step, with only 2 minutes of reaction at 300°C and 15 minutes of reaction at 270°C. This approach to this process is interesting because, in addition to avoiding the saturation of the triacetin market, a phenomenon observed for glycerol, it is possible to increase the net production of FAME [24]. Saka et al. [22] suggested a diagram for this route, shown in **Figure 6**.

7. Reaction with dimethyl carbonate (DMC)

The third route is to obtain FAME from a reaction between triglycerides and dimethyl carbonate. Unlike all the other pathways described, this reaction takes place in two reversible steps: in the first, a DMC molecule reacts with a triglyceride molecule, releasing two FAME molecules and an intermediate, the fatty acid glycerol carbonate (FAGC - fatty). acid glycerol carbonate). In the second step, the FAGC molecule reacts with the second DMC molecule to form a FAME molecule and the by-product of the route, glycerol dicarbonate [25]. The reaction and its stoichiometry are shown in **Figure 7**.

DMC is an attractive reagent because it can be considered a green alternative in the organic synthesis industry and has several applications as a methylation and carbonylation agent. It is a non-toxic and biodegradable substance, which is obtained by environmentally interesting routes. The most used industrial route today is the oxidative carbonylation of methanol with carbon monoxide and oxygen, catalyzed by copper(I) chloride. As methanol and carbon monoxide can be obtained from biomass synthesis gas, obtaining biodiesel using DMC from this route would only use reagents of renewable origin [26–28].

Another important point is the fact that DMC does not decompose at temperatures lower than 390°C, so using it as a reagent in supercritical processes is viable [29].

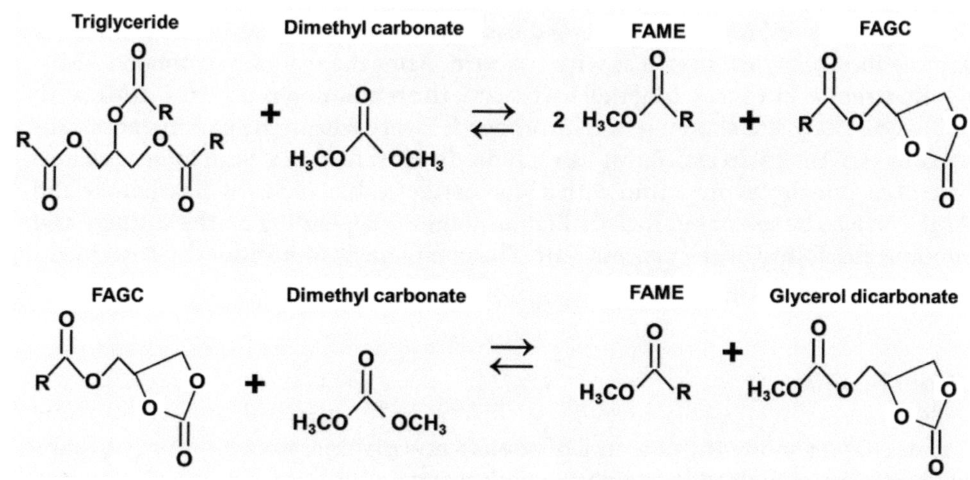

Figure 7.
Two-step reaction with dimethyl carbonate. Source: adapted from [25].

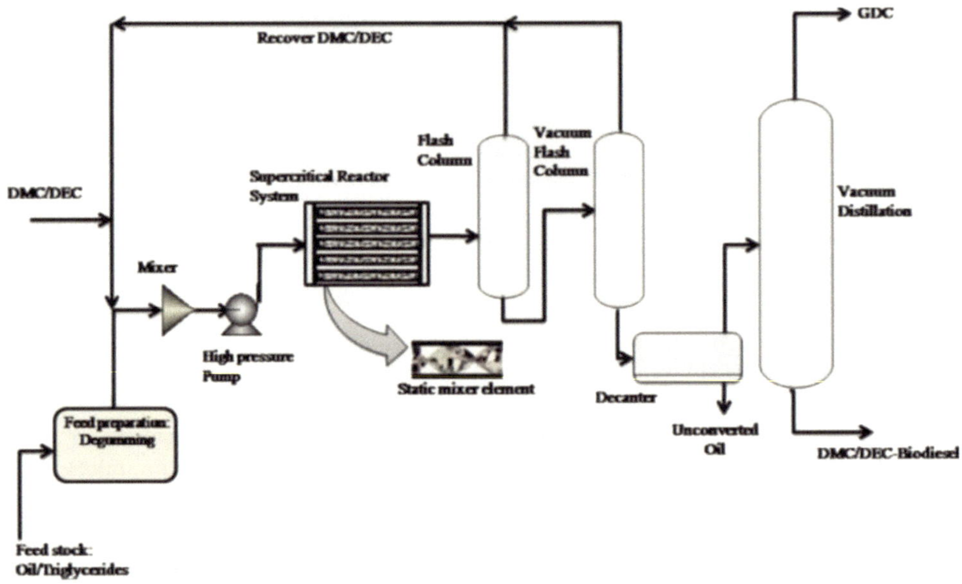

Figure 8.
Process using DMC as a reagent, producing FAME and glycerol dicarbonate (DCG) as products. Source: [30].

Biodiesel production by this route via supercritical process was studied using crude oils of *Jatropha curcas* and *pongamia pinnata* as raw material and DMC and diethyl carbonate as reagents, in a batch reactor, optimizing parameters such as molar ratio, temperature and reaction time. For all experiments, carried out at 15 MPa, the highest conversion obtained was for the system with *pongamia* oil and DMC, which shows that dimethyl carbonate is more reactive than diethyl carbonate [30].

The two-stage route, as well as for the methanol and acetic acid routes, has the advantage of being able to apply milder process conditions to avoid product degradation. However, this pathway generates glycerol as a by-product, although this can be converted to glycerol carbonate, a compound with greater added value, when reacting with DMC [29, 31].

The mixture of FAME and glycerol dicarbonate can be used as biodiesel, without harming its properties, as well as with triacetin. After the reaction to obtain FAME and the removal of excess dimethyl carbonate, the remaining product – FAME and DCG – was analyzed, showing that its properties were within the parameters established by the American and European standards. However, as it is an intermediate for glycerol carbonate, a compound with a high market value, glycerol dicarbonate and FAME can also be separated in a distillation column, depending on the strategy and economic feasibility of the process [30]. The complete flowsheet for the described process is shown in **Figure 8**.

8. Conclusions

This chapter shows the potential of soy as a raw material for a biorefinery, that is, obtaining several high added value products through different technological routes. The residues of this oilseed can be used to obtain these products with high added value without competing with the food sector. Routes that make the production of biodiesel viable are real and technically feasible, in addition to the generation of co-products of greater interest such as: tocopherols, sterols, squalene, triacetin, DCG. Another interesting point is the flexibility that some products allow when generated together with the esters, as they do not interfere with the quality of the biofuel. Soy has a wide range of exploration possibilities, and its production follows the pace of industrial production. Studies that make its development economically viable are the biggest challenge, as the technical feasibility is well consolidated.

Acknowledgements

The authors are grateful to SENAI CIMATEC for their financial support.

Conflict of interest

The authors have no conflict of interest.

Author details

Fernando Luiz Pellegrini Pessoa[1*], Ewerton Emmanuel da Silva Calixto[1], Hugo Gomes D'amatto Villardi[1], Ana Lucia Barbosa de Souza[1] and Bruna Aparecida Souza Machado[2]

1 Chemistry Department, National Service of Industrial Learning – SENAI, University Center SENAI CIMATEC, Salvador, Bahia, Brazil

2 Health Institute of Technologies (CIMATEC ITS), National Service of Industrial Learning – SENAI, University Center SENAI CIMATEC, Salvador, Bahia, Brazil

*Address all correspondence to: fernando.pessoa@fieb.org.br

References

[1] Hirakuri MH, Lazzarotto JJ. O agronegócio da soja nos contextos mundial e brasileiro. Embrapa Soja-Documentos (INFOTECA-E). Londrina: Embrapa Soja; 2014. p. 2014

[2] Sangoi L, de Almeida ML. Avaliação de genótipos de girassol em duas épocas de semeadura no Planalto Catarinense. Revista da Universidade do Estado de Santa Catarina. 1994;**2**:154-164

[3] Ma F, Hanna MA. Biodiesel production: A review. Bioresource Technology. 1999;**70**:1-15

[4] Dohrn R, Brunner G. Correlations of pure-component parameters of the Peng–Robinson equation of state. In: Proceedings of Second International Symposium on Supercritical Fluids, Boston; 1991. pp. 471-478

[5] Aranda M, Morlock G. New method for caffeine quantification by planar chromatography coupled with electropray ionization mass spectrometry using stable isotope dilution analysis. Rapid Communications in Mass Spectrometry. 2007;**21**:1297-1303

[6] FONTANA. Refino de Óleos Vegetais Desodorização DDOS – Destilados de Ácidos Graxos. AUSTENBIO. 2015;**1**:1-6

[7] ABIOVE. Balanço de Oferta/Demanda SOJA [Internet]. 2021 [cited January 4, 2022]. Available from: https://abiove.org.br/wp-content/uploads/2019/10/2021-10-01-Quadro-de-Oferta-e-Demanda.pdf

[8] Jafarian Asl P, Niazmand R, Jahani M. Theoretical and experimental assessment of supercritical CO2 in the extraction of phytosterols from rapeseed oil deodorizer distillates. Journal of Food Engineering. 2020;**269**:109748

[9] Maniet G, Jacquet N, Richel A. Recovery of sterols from vegetable oil distillate by enzymatic and non-enzymatic processes. Comptes Rendus Chimie. 2019;**22**:347-353

[10] de Mello M, Young A, Villardi H, Pessoa F, Salgado A. Biodiesel production by the methylic-alkaline and ethylic-enzymatic routes: Discussion of some environmental aspects. Journal of Cleaner Production. 2017;**144**:347-357

[11] Santos C, de Mello M, Gomes D'Amato Villardi H, Ferreira Young A, Luiz Pellegrini Pessoa F, Medeiros SA. Life cycle assessment of biodiesel produced by the methylic-alkaline and ethylic-enzymatic routes. Fuel. 2017;**208**:329-336

[12] Kumar A, Nerella VK, v. Experimental analysis of exhaust emissions from transit buses fuelled with biodiesel. The Open Environmental Engineering Journal. 2009;**2**:81-96

[13] Ang GT, Tan KT, Lee KT. Recent development and economic analysis of glycerol-free processes via supercritical fluid transesterification for biodiesel production. Renewable and Sustainable Energy Reviews. 2014;**31**:61-70

[14] USDA. BICO Reports [Internet]. Foreign Agriculture Service. 2021 [cited January 5, 2022]. Available from: https://apps.fas.usda.gov/Gats/BICOReport.aspx?type=bico-hs10

[15] D'Amato Villardi HG, Leal MF, Pellegrini Pessoa FL, Salgado AM. Synthesis of methyl esters through residual feedstock using acid and free catalyst – Proposal of new reactor. Renewable Energy. 2019;**131**:1146-1155

[16] Bouriakova A, Mendes PSF, Elst K, de Clercq J, Thybaut JW. Techno-economic

evaluation of squalene recovery from oil deodorizer distillates. Chemical Engineering Research and Design. 2020;**154**:122-134

[17] Niza NM, Tan KT, Lee KT, Ahmad Z. Influence of impurities on biodiesel production from Jatropha curcas L. by supercritical methyl acetate process. The Journal of Supercritical Fluids. 2013;**79**:73-75

[18] Saka S, Isayama Y. A new process for catalyst-free production of biodiesel using supercritical methyl acetate. Fuel. 2009;**88**:1307-1313

[19] Casas A, Ruiz JR, Ramos MJ, Pérez Á. Effects of Triacetin on biodiesel quality. Energy & Fuels. 2010;**24**:4481-4489

[20] Campanelli P, Banchero M, Manna L. Synthesis of biodiesel from edible, non-edible and waste cooking oils via supercritical methyl acetate transesterification. Fuel. 2010;**89**:3675-3682

[21] ANP. Resolução ANP N° 45 DE 25/08/2014 - Federal - LegisWeb [Internet]. [cited January 3, 2022]. Available from: https://www.legisweb.com.br/legislacao/?id=274064

[22] Saka S, Isayama Y, Ilham Z, Jiayu X. New process for catalyst-free biodiesel production using subcritical acetic acid and supercritical methanol. Fuel. 2010;**89**:1442-1446

[23] Yoneda N, Kusano S, Yasui M, Pujado P, Wilcher S. Recent advances in processes and catalysts for the production of acetic acid. Applied Catalysis A: General. 2001;**221**:253-265

[24] Ang GT, Tan KT, Lee KT. Recent development and economic analysis of glycerol-free processes via supercritical fluid transesterification for biodiesel production. Renewable and Sustainable Energy Reviews. 2014;**31**:61-70

[25] Farobie O, Matsumura Y. State of the art of biodiesel production under supercritical conditions. Progress in Energy and Combustion Science. 2017;**63**:173-203

[26] Delledonne D, Rivetti F, Romano U. Developments in the production and application of dimethylcarbonate. Applied Catalysis A: General. 2001;**221**:241-251

[27] Fabbri D, Bevoni V, Notari M, Rivetti F. Properties of a potential biofuel obtained from soybean oil by transmethylation with dimethyl carbonate. Fuel. 2007;**86**:690-697

[28] Tundo P, Selva M. The chemistry of dimethyl carbonate. Accounts of Chemical Research. 2002;**35**:706-716

[29] Ilham Z, Saka S. Dimethyl carbonate as potential reactant in non-catalytic biodiesel production by supercritical method. Bioresource Technology. 2009;**100**:1793-1796

[30] Rathore V, Tyagi S, Newalkar B, Badoni RP. Glycerin-free synthesis of Jatropha and Pongamia biodiesel in supercritical dimethyl and diethyl carbonate. Industrial & Engineering Chemistry Research. 2014;**53**:10525-10533

[31] Kim SC, Kim YH, Lee H, Yoon DY, Song BK. Lipase-catalyzed synthesis of glycerol carbonate from renewable glycerol and dimethyl carbonate through transesterification. Journal of Molecular Catalysis B: Enzymatic. 2007;**49**:75-78